# Plants Responses to Climate Change

# Plants Responses to Climate Change

Editors

**Isabel Marques**
**Ana I Ribeiro-Barros**
**José C. Ramalho**

MDPI

Basel • Beijing • Wuhan • Barcelona • Belgrade • Novi Sad • Cluj • Manchester

*Editors*
Isabel Marques
Forest Research Centre
Associate Laboratory TERRA
School of Agriculture
University of Lisbon
Lisbon
Portugal

Ana I Ribeiro-Barros
Forest Research Centre
Associate Laboratory TERRA
School of Agriculture
University of Lisbon
Lisbon
Portugal

José C. Ramalho
Forest Research Centre
Associate Laboratory TERRA
School of Agriculture
University of Lisbon
Lisbon
Portugal

*Editorial Office*
MDPI
St. Alban-Anlage 66
4052 Basel, Switzerland

This is a reprint of articles from the Special Issue published online in the open access journal *International Journal of Molecular Sciences* (ISSN 1422-0067) (available at: www.mdpi.com/journal/ijms/special_issues/CCCC_2).

For citation purposes, cite each article independently as indicated on the article page online and as indicated below:

Lastname, A.A.; Lastname, B.B. Article Title. *Journal Name* **Year**, *Volume Number*, Page Range.

**ISBN 978-3-0365-9938-0 (Hbk)**
**ISBN 978-3-0365-9937-3 (PDF)**
**doi.org/10.3390/books978-3-0365-9937-3**

# Contents

# About the Editors

**Isabel Marques**

Isabel Marques focuses on interdisciplinary research across plant biology (structure, function, diversity, genetics, evolution, systematics), all levels of organization (molecular to ecosystem), and all plant groups and allied organisms (cyanobacteria, algae, fungi, and lichens) to understand species diversification and extinction as well as how they can cope with the effects of environmental changes. ORCID: 0000-0001-9788-4831.

**Ana I Ribeiro-Barros**

Ana I. Ribeiro-Barros has a PhD in Plant Molecular Biology (1997, Wageningen University and Research). Her area of scientific activity includes Agrobiotechnology applied to the management and characterization of agro-forestry resources: Biodiversity; Conservation Genetics; Ethnobotany; Landscape genomics; Molecular Ecology; Plant–Environment Interactions (symbioses, pathogenesis and abiotic stresses); Soil diversity. ORCID: 0000-0002-6071-6460.

**José C. Ramalho**

José C. Ramalho holds a PhD in Plant Physiology and Biochemistry and has co-authored approximately 240 research papers, books, and book chapters, mostly addressing plant responses to abiotic stresses (cold, heat, drought, salinity, minerals) at physiological, biochemical, and molecular levels, as well agronomic biofortification issues. His work mainly focuses on the tropical Coffea species but also involves other species from cereals (rice, wheat) to perennials (carob, casuarina, oak, pear trees). ORCID: 0000-0002-7639-7214.

International Journal of
*Molecular Sciences*

MDPI

*Editorial*

# Plant Responses to Climate Change

**Isabel Marques [1,*], José C. Ramalho [1,2] and Ana I. Ribeiro-Barros [1,2]**

[1] Plant-Environment Interactions and Biodiversity Lab, Forest Research Centre & Associate Laboratory TERRA, Higher School of Agriculture, University of Lisbon, 1349-017 Lisboa, Portugal; cochichor@isa.ulisboa.pt (J.C.R.); anaifribeiro@campus.ul.pt (A.I.R.-B.)
[2] GeoBioSciences, GeoTechnologies and GeoEngineering, Faculdade de Ciências e Tecnologia, Universidade NOVA de Lisboa, 2829-516 Monte de Caparica, Portugal
* Correspondence: isabelmarques@isa.ulisboa.pt

## 1. Molecular Mechanisms of Plants to Climate Change

Ongoing climate change poses a great risk to the natural environment and the sustainability of agriculture [1,2]. Major abiotic stresses, such as extreme temperatures and drought, are already responsible for 51% to 82% of global annual losses in crop yield, a scenario expected to be aggravated in the future [3–5]. Thus, scientific advances have a special role in meeting the challenges of overcoming such impacts. In this context, this Special Issue covers basic and applied research aimed at understanding the molecular mechanisms associated with plant responses to abiotic stresses, including drought, cold, heat, high light, and salinity. A particular emphasis was placed on the influence of stresses at the whole genome or the transcriptome level, the characterization of tolerance across different accessions and genotypes, and the discovery of candidate genes that can improve the productivity and yield of crops under changing abiotic conditions. The role of soil microbiome in enhancing plant tolerance against stresses has also been revised. Taken together, the new information provided in these manuscripts not only increases our understanding of the molecular basis of plants' adaptive responses but also provides key fundamentals for the future successful selection and breeding of tolerant crops.

## 2. List of Contributions

Twenty-four manuscripts were submitted for consideration for this Special Issue, and after peer review, ten of them were finally accepted for publication (nine articles and one review). These studies were performed in multiple scientific institutions from Brazil, Croatia, China, Egypt, Portugal, Russia, the USA, and Vietnam. The contributions are listed below:

1. Zhao, W.; Song, J.; Wang, M.; Chen, X.; Du, B.; An, Y.; Zhang, L.; Wang, D.; Guo, C. Alfalfa MsATG13 Confers Cold Stress Tolerance to Plants by Promoting Autophagy. Int. J. Mol. Sci. 2023, 24, 12033. https://doi.org/10.3390/ijms241512033.
2. Lu, Z.; Liu, H.; Kong, Y.; Wen, L.; Zhao, Y.; Zhou, C.; Han, L. Late Elongated Hypocotyl Positively Regulates Salt Stress Tolerance in *Medicago truncatula*. Int. J. Mol. Sci. 2023, 24, 9948. https://doi.org/10.3390/ijms24129948.
3. Marques, I.; Fernandes, I.; Paulo, O.S.; Batista, D.; Lidon, F.C.; Partelli, F.; DaMatta, F.M.; Ribeiro-Barros, A.I.; Ramalho, J.C. Overexpression of Water-Responsive Genes Promoted by Elevated $CO_2$ Reduces ROS and Enhances Drought Tolerance in *Coffea* Species. Int. J. Mol. Sci. 2023, 24, 3210. https://doi.org/10.3390/ijms24043210.
4. Bauer, N.; Tkalec, M.; Major, N.; Talanga Vasari, A.; Tokić, M.; Vitko, S.; Ban, D.; Ban, S.G.; Salopek-Sondi, B. Mechanisms of Kale (*Brassica oleracea* var. *acephala*) Tolerance to Individual and Combined Stresses of Drought and Elevated Temperature. Int. J. Mol. Sci. 2022, 23, 11494. https://doi.org/10.3390/ijms231911494.

**Citation:** Marques, I.; Ramalho, J.C.; Ribeiro-Barros, A.I. Plant Responses to Climate Change. *Int. J. Mol. Sci.* **2023**, *24*, 15902. https://doi.org/10.3390/ijms242115902

Received: 21 October 2023
Accepted: 24 October 2023
Published: 2 November 2023

5. Yuan, P.; Poovaiah, B.W. Interplay between Ca2+/Calmodulin-Mediated Signaling and *AtSR1/CAMTA3* during Increased Temperature Resulting in Compromised Immune Response in Plants. Int. J. Mol. Sci. 2022, 23, 2175. https://doi.org/10.3390/ijms23042175.
6. Islam, M.M.; Qi, S.; Zhang, S.; Amin, B.; Yadav, V.; El-Sappah, A.H.; Zhang, F.; Liang, Y. Genome-Wide Identification and Functions against Tomato Spotted Wilt Tospovirus of PR-10 in *Solanum lycopersicum*. Int. J. Mol. Sci. 2022, 23, 1502. https://doi.org/10.3390/ijms23031502.
7. Wan, H.; Qian, J.; Zhang, H.; Lu, H.; Li, O.; Li, R.; Yu, Y.; Wen, J.; Zhao, L.; Yi, B.; et al. Combined Transcriptomics and Metabolomics Analysis Reveals the Molecular Mechanism of Salt Tolerance of Huayouza 62, an Elite Cultivar in Rapeseed (*Brassica napus* L.). Int. J. Mol. Sci. 2022, 23, 1279. https://doi.org/10.3390/ijms23031279.
8. Luo, Y.; Teng, S.; Yin, H.; Zhang, S.; Tuo, X.; Tran, L.-S.P. Transcriptome Analysis Reveals Roles of Anthocyanin- and Jasmonic Acid-Biosynthetic Pathways in Rapeseed in Response to High Light Stress. Int. J. Mol. Sci. 2021, 22, 13027. https://doi.org/10.3390/ijms222313027.
9. Bazhenov, M.S.; Chernook, A.G.; Bespalova, L.A.; Gritsay, T.I.; Polevikova, N.A.; Karlov, G.I.; Nazarova, L.A.; Divashuk, M.G. Alleles of the GRF3-2A Gene in Wheat and Their Agronomic Value. Int. J. Mol. Sci. 2021, 22, 12376. https://doi.org/10.3390/ijms222212376.
10. Chen, Y.; Yao, Z.; Sun, Y.; Wang, E.; Tian, C.; Sun, Y.; Liu, J.; Sun, C.; Tian, L. Current Studies of the Effects of Drought Stress on Root Exudates and Rhizosphere Microbiomes of Crop Plant Species. Int. J. Mol. Sci. 2022, 23, 2374. https://doi.org/10.3390/ijms23042374.


Contribution 1 explored the role that autophagy-related genes (ATGs) have in regulating and even improving plant tolerance to abiotic stresses. Autophagy is an evolutionarily conserved self-degradative process in all eukaryotic cells and plays many physiological roles in maintaining cellular homeostasis. However, the biological mechanisms and action of *ATGs* under abiotic stresses remain unclear. Specifically, the authors showed how the up-regulation of the gene *MsATG13* enhanced cold tolerance in alfalfa (*Medicago sativa* L.) through the modulation of autophagy and antioxidant levels. This gene seems to activate other key *ATGs*, increasing antioxidant levels and reducing the accumulation of ROS in response to cold. These results demonstrate that autophagy plays an important role in plant survival under abiotic stress.

Contribution 2 revealed how the core component of the circadian clock, *MtLHY* (the Late Elongated Hypocotyl orthologue) is responsible for salt stress tolerance in *Medicago truncatula*. The circadian clock is an internal time mechanism that synchronizes the physiological response of an organism to its surroundings, including changing environmental conditions. In this study, the authors showed how the expression of *MtLHY* increases salt stress tolerance in *M. truncatula* through flavonoid biosynthesis, while mediating $Na^+$ /$K^+$ homeostasis and inhibiting ROS production. These results show an existing link between the circadian clock and plant responses to stresses.

Contribution 3 studied the single and combined effects of drought and elevated air [$CO_2$] ($eCO_2$) on the transcriptomic machinery of two *Coffea* genotypes (from the two species that support global coffee trade): *Coffea canephora* Pierre ex *A. Froehner* cv. Conilon Clone 153 (CL153) and *C. arabica* L. cv. Icatu Vermelho (Icatu). The authors showed a predominance of protective and ROS-scavenging genes, directly or indirectly related to ABA signaling pathways involved in tolerance responses, although with different regulatory mechanisms in Icatu and CL153. Moderate drought had a minor impact on the number of transcripts differentially regulated in these plants, contrary to severe water deficit. Additionally, it was found that elevated $eCO_2$ attenuated the impacts of drought in the two genotypes but especially in Icatu, in agreement with the contrasting physiological tolerance responses previously reported in these genotypes.

Contribution 4 explored the effects of individual and combined stress (osmotic + heat stress) in the tolerance responses of 33 kale accessions (*Brassica oleracea* var. *acephala*). The authors showed that the more tolerant accessions had higher basal contents of proline, total soluble sugars, glucosinolates, and heat shock proteins and higher transcript levels of *NAC* and *DREB* transcription factors. On the contrary, sensitive accessions were characterized by a high basal content of fructans. These findings can be further used as markers for screening heat- and drought-tolerant kale cultivars.

Contribution 5 reported the effects that a moderate change in temperature had on plant immune response, through $Ca^{2+}$/calmodulin-mediated signaling. Plants use hormones to respond to environmental changes. $Ca^{2+}$ signaling plays an important role in the adaptation to abiotic stresses, but the underlying mechanisms involved in this process are still scarcely known. The authors showed how *AtSR1/CAMTA3*, a $Ca^{2+}$/CaM receptor, is involved in increased temperature-mediated stomatal defense and apoplastic immunity. Together with other studies, this suggests that $Ca^{2+}$ signaling acts as a general defense response to pathogen infection in the context of temperature.

Contribution 6 was focused on the action of pathogenesis-related (PR) proteins under pathogen attack or infection. Specifically, the authors aimed to understand if these genes act to defend tomato plants against infection by the spotted wilt virus (TSWV), one of the most important plant viruses in the world. This study identified and characterized forty-five candidate genes, with some, such as *PR-10* and *Sw-5b*, up-regulated upon infection. These findings lead to a better understanding of how systemic necrosis occurs upon infection by the tomato spotted wilt tospovirus, which could aid in the future development of new antiviral approaches.

Contribution 7 explored the molecular mechanism of salt tolerance in rapeseed (*Brassica napus* L.), comparing two varieties with significant differences in salt tolerance: Zhongshuang 11 (ZS11), a conventional rapeseed variety, which is sensitive to salt stress, and Huayouza 62 (H62), an elite rapeseed cultivar, widely planted in China. Several differentially expressed metabolites and key differentially expressed genes were identified through metabolomics and transcriptomics analysis, revealing vast differences in the two varieties and a high mechanism of tolerance to salt, especially in H62. Specifically, several candidate genes of non-specific lipid transporters (nsLTPs) were identified as important for salt tolerance since they were significantly up-regulated as the salt concentration increased, especially in the salt-tolerant H62.

Contribution 8 studied the molecular mechanisms underlying rapeseed's response to high light (HL) stress, namely the potential roles of anthocyanins and jasmonic acid (JA) in plant adaptations. Although light is an essential environmental factor for plant growth and development, HL is also a stress factor, although scarcely studied. This work showed that plants under HL, up-regulated genes that were involved in the regulation and biosynthesis of anthocyanins and JA. In accordance, the accumulation of anthocyanins, which act as photoprotectants, was significantly promoted under HL conditions. In agreement with the positive regulatory role of JA in anthocyanin biosynthesis, this study suggests that JA plays a key role in the responses of rapeseed seedlings to HL, contributing to the development of HL-tolerant rapeseed plants.

Contribution 9 studied how the allelic diversity of growth-regulating factors (GRF), a family of plant-specific transcription factors with roles in plant growth, development, and stress response, changed between bread wheat (*Triticum aestivum* L.) and ancestral wheat accessions. The authors identified a rare allele in a world wheat collection, associated with earlier heading and higher grain filling, which is suggested to be an ancestral adaptation of local landraces in the Black Sea region. Together with the finding of other unique mutations, this study reveals the importance of ex situ conservation for the preservation of unique genetic alleles.

Contribution 10 reviewed the impact of drought stress on root exudates and associated microbiomes, discussing how they help plants to successfully respond to drought. The authors highlight how root exudates play an important role in the recruitment of mycorrhizal fungi and plant growth-promoting rhizobacteria (PGPR), with implications for plant tolerance against environmental stresses. However, the crosstalk between root-associated microbiomes and root exudates, as well as the construction of beneficial microbial consortia in agriculture still presents great challenges. Therefore, a deeper understanding of root exudates and soil microbiomes is vital for disentangling their role in plant fitness and sustainable agriculture.

## 3. Future Perspectives

This Special Issue progresses our understanding of the molecular mechanisms underlying the responses of plants to abiotic stresses and highlights that multiple and complex processes are involved, including signaling pathways, transcription, translation, and post-translational modifications. Despite the remarkable research involved in this Special Issue combining genomics, transcriptomics, genetics, and other approaches, there are still much scope for plant research. Further studies should clarify the key relationships between abiotic stresses and how they interact individually and in combination [3]. Current research on plant tolerance is also, generally, focused on vegetative parts. Therefore, more studies are needed to understand the impacts of abiotic stressors on other key plant components such as the flowers, which are highly sensitive to environmental changes [6]. Together with emerging solutions such as genome editing and the advance in artificial intelligence, this would be key in aiding sustainable agriculture and the food requirements of the ever-increasing human population.

**Funding:** This work received funding from the European Union's Horizon 2020 research and innovation program under the grant agreement No. 727934 (project BreedCAFS) and from national funds through the FCT—Fundação para a Ciência e a Tecnologia, I.P., Portugal, under the project PTDC/ASP-AGR/31257/2017; the research units CEF (UIDB/00239/2020), GeoBioTec (UIDP/04035/2020); and the associate laboratory TERRA (LA/P/0092/2020). Isabel Marques received funding through the FCT—Fundação para a Ciência e a Tecnologia, I.P., Portugal, under the Scientific Employment Stimulus—Individual Call (CEEC Individual)—2021.01107.CEECIND/CP1689/CT0001.

**Acknowledgments:** As Guest Editors of this Special Issue, we would like to express our deep appreciation to all authors whose valuable work was published under this issue, as well as the time and efforts of all reviewers, which altogether, have contributed to the success of the edition.

**Conflicts of Interest:** The authors declare no conflict of interest. The funders had no role in the design of the study; in the collection, analyses, or interpretation of data; in the writing of the manuscript; or in the decision to publish the results.

## References

1. Pereira, A. Plant abiotic stress challenges from the changing environment. *Front. Plant Sci.* **2016**, *7*, 218865. [CrossRef] [PubMed]
2. Rivero, R.M.; Mittler, R.; Blumwald, E.; Zandalinas, S.I. Developing climate-resilient crops: Improving plant tolerance to stress combination. *Plant J.* **2022**, *109*, 373–389. [CrossRef] [PubMed]
3. Teshome, D.T.; Zharare, G.E.; Naidoo, S. The Threat of the Combined Effect of Biotic and Abiotic Stress Factors in Forestry Under a Changing Climate. *Front. Plant Sci.* **2020**, *11*, 601009. [CrossRef] [PubMed]
4. Dhankher, O.P.; Foyer, C.H. Climate resilient crops for improving global food security and safety. *Plant. Cell Environ.* **2018**, *41*, 877–884. [CrossRef] [PubMed]
5. DaMatta, F.M.; Ramalho, J.D.C. Impacts of drought and temperature stress on coffee physiology and production: A review. *Braz. J. Plant Physiol.* **2006**, *18*, 55–81. [CrossRef]
6. Borghi, M.; Perez de Souza, L.; Yoshida, T.; Fernie, A.R. Flowers and climate change: A metabolic perspective. *New Phytol.* **2019**, *224*, 1425–1441. [CrossRef] [PubMed]

International Journal of
*Molecular Sciences*

MDPI

*Article*

# Alfalfa *MsATG13* Confers Cold Stress Tolerance to Plants by Promoting Autophagy

**Weidi Zhao †, Jiayi Song †, Meijia Wang, Xiuxiu Chen, Binghao Du, Yimin An, Lishuang Zhang, Dan Wang and Changhong Guo ***

Key Laboratory of Molecular Cytogenetics and Genetic Breeding of Heilongjiang Province, College of Life Science and Technology, Harbin Normal University, No. 1 of Shida Road, Limin Development Zone, Harbin 150025, China
* Correspondence: kaku3008@hrbnu.edu.cn
† These authors contributed equally to this work.

**Abstract:** Autophagy is a conserved cellular process that functions in the maintenance of physiological and metabolic balance. It has previously been demonstrated to improve plant tolerance to abiotic stress. Numerous autophagy–related genes (ATGs) that regulate abiotic stress have been identified, but there have been few functional studies showing how ATGs confer cold stress tolerance. The cold transcriptome data of the crown buds that experienced overwintering of the alfalfa (*Medicago sativa* L.) showed that *MsATG13* is upregulated in response to cold stress. In the present study, we found that *MsATG13* transgenic tobacco enhanced cold tolerance compared to wild–type (WT) plants. Transmission electron microscopy demonstrated that transgenic tobacco overexpressing *MsATG13* formed more autophagosomes than WT plants in response to cold stress conditions. The transgenic tobacco increased autophagy levels due to upregulation of other ATGs that were necessary for autophagosome production under cold stress conditions. *MsATG13* transgenic tobacco also increased the proline contents and antioxidant enzyme activities, enhancing the antioxidant defense capabilities under cold stress conditions. Furthermore, *MsATG13* overexpression decreased levels of superoxide anion radicals and hydrogen peroxide under cold stress conditions. These findings demonstrate the role of *MsATG13* in enhancing plant cold tolerance through modulation of autophagy and antioxidant levels.

**Keywords:** *MsATG13*; *Medicago sativa*; autophagy; cold stress; ROS

**Citation:** Zhao, W.; Song, J.; Wang, M.; Chen, X.; Du, B.; An, Y.; Zhang, L.; Wang, D.; Guo, C. Alfalfa *MsATG13* Confers Cold Stress Tolerance to Plants by Promoting Autophagy. *Int. J. Mol. Sci.* **2023**, *24*, 12033. https://doi.org/10.3390/ijms241512033

Academic Editors: Isabel Marques, Ana I. Ribeiro-Barros and José C. Ramalho

Received: 26 June 2023
Revised: 25 July 2023
Accepted: 25 July 2023
Published: 27 July 2023



## 1. Introduction

Cold is a common abiotic stress condition that can seriously damage crops, inhibiting their growth and yield [1]. It is also a major environmental factor that limits the geographic range of key crops [2]. Cold stress typically causes adverse effects such as seedling stunting, chlorosis, reduced leaf expansion, wilting, and water loss [3]. Furthermore, it induces accumulation of reactive oxygen species (ROS) [4]. High levels of ROS can cause oxidative stress [5,6], which disrupts membrane systems and organelles, causes protein damage and aggregation, and ultimately leads to disordered plant physiology and metabolism [7]. To prevent such oxidative stress, plant cells must have the capacity to quickly remove excess ROS to maintain normal physiological and metabolic balance. Enzymatic and non–enzymatic systems are important methods for plants to remove excess ROS [8]. The primary antioxidant enzymes are catalase (CAT), peroxidase (POD), and superoxide dismutase (SOD) [9]. The non–enzymatic ROS protection system includes the osmotic protector proline [10]. Both systems are critical in preventing ROS–induced damage, and thus promote stress tolerance.

Oxidative stress can also induce autophagy [11,12], which contributes to the regulation of physiological and metabolic balance through autophagosomes in plant cells [13,14]. Previous studies have identified that autophagy–related genes (ATGs) play important roles in the key steps of autophagosome formation: initiation, nucleation, membrane extension, and maturation. The genes that control these steps therefore regulate an organism's capacity for autophagy, thereby regulating abiotic stress tolerance. For example, in an apple (*Malus domestica*), overexpression of *MdATG5* enhances autophagy levels; this may regulate antioxidant enzyme activity and promote antioxidant accumulation, improving plant defense capabilities [15]. Overexpression of *MdATG10* in an apple also enhances autophagy levels, which has been shown to improve salt tolerance, potentially by improving ion homeostasis [16]. *Arabidopsis thaliana* plants overexpressing *MaATG8* have higher endogenous abscisic acid (ABA) levels, autophagy levels, and drought tolerance than wild–type (WT) plants, which is likely due to the association between ABA biosynthesis and autophagy [17]. In pepper (*Capsicum annuum*) and wheat (*Triticum aestivum*), cold stress promotes autophagosome formation and increases expression levels of *CaATG13* and *TaATG8*, respectively [18,19]. Although several studies have demonstrated that ATGs are induced by cold stress, there have been no reports detailing the function of ATGs in terms of plant tolerance to cold stress.

Alfalfa (*Medicago sativa* L.) is a type of perennial forage legume that is cultivated throughout the world [20]. It has such excellent quality and palatability that it has been referred to as the "king of forage" [21]. However, overwintering is a key problem that restricts its sustainable growth and utilization. *M. sativa* cv. 'Zhaodong' is a native cultivar of alfalfa found in Heilongjiang Province, China, that has strong cold resistance; the survival rate of overwintering *M. sativa* 'Zhaodong' remains above 90%, even with temperatures below −30 °C [22]. Previous transcriptomic data from *M. sativa* 'Zhaodong' (accession number SRP060503) showed that a gene related to autophagy, *MsATG13*, was significantly upregulated under cold stress. *ATG13* plays a primary controlling role in the initiation process of induced autophagy. When plant cells sense stress stimuli, *ATG13* undergoes dephosphorylation and associates with *ATG1*. The formation of the *ATG13–ATG1* complex activates the kinase activity of *ATG1*, initiating the recruitment of other autophagy-related proteins to collectively form the pre-autophagosomal structures and participate in the initiation of autophagosome formation [23–25]. To determine the function of *MsATG13* in plant responses to cold stress, we here isolated *MsATG13* from *M. sativa* 'Zhaodong' and analyzed its expression patterns in response to cold stress. We also quantified parameters associated with stress responses, including autophagosome formation and antioxidant levels, among WT plants and transgenic tobacco overexpressing *MsATG13*. The results revealed the effects of *MsATG13*–mediated autophagy on the antioxidant system and cold tolerance in transgenic tobacco. This study not only demonstrates the functionality of a key *ATG* in cold tolerance but also provides a theoretical basis for further discovery and characterization of cold tolerance genes in alfalfa.

## 2. Results

### *2.1. Characterization and Expression Profiles of MsATG13*

*MsATG13* comprised a 1686–bp ORF encoding a predicted protein of 561 amino acids in length. Conserved protein domain prediction suggested that it contained an ATG13 domain, which is typical of ATG13 family members. The InterPro search results indicated an ATG13 domain (PF10033) where the amino acids located at positions 20–253 of MsATG13 belonged to the HORMA domain (Figure 1A). Members of theATG13 family are involved in cytoplasm to vacuole transport (Cvt), and more specifically in Cvt vesicle formation. They are probably involved in the switching machinery regulating the conversion between the Cvt pathway and autophagy [26–28]. Additionally, phylogenetic analysis showed that MsATG13 was closely related to ATG13 in several species of the family Leguminosae, including *Medicago truncatula*, *Trifolium pratense*, and *Trifolium medium*. MsATG13 shared the highest sequence identity (94.75%) with MtATG13 (Figure 1B).

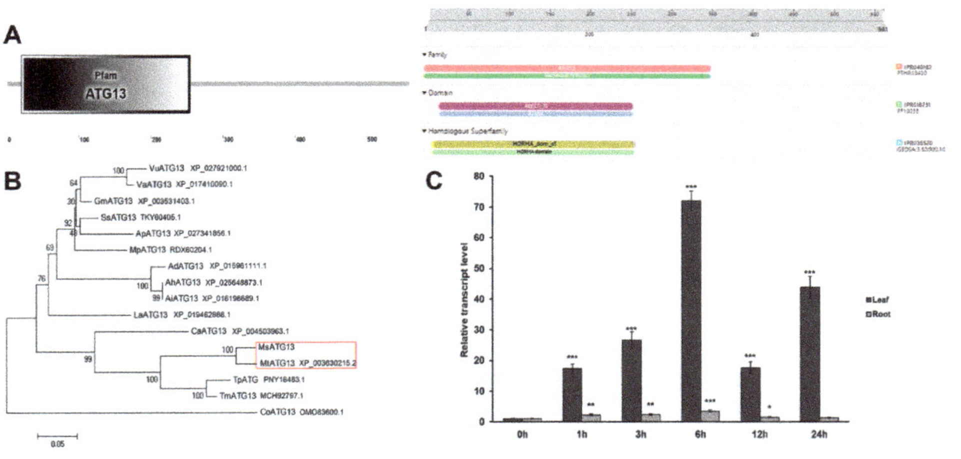


**Figure 1.** Characterization of *MsATG13*. (**A**) Prediction of the functional domain present in MsATG13. (**B**) Phylogenetic analysis of MsATG13 and other family members. MsATG13 (*Medicago sativa* L., UUH54286.1); AdATG13 (*Arachis duranensis*, XP_015961111.1); AhATG13 (*Arachis hypogaea*, XP_025648873.1); AiATG13 (*Arachis ipaensis*, XP_016198689.1); ApATG13 (*Abrus precatorius*, XP_027341856.1); CaATG13 (*Cicer arietinum*, XP_004503963.1); CoATG13 (*Corchorus olitorius*, OMO83600.1); GmATG13 (*Glycine max*, XP_003531403.1); LaATG13 (*Lupinus angustifolius*, XP_019462868.1); MpATG13 (*Mucuna pruriens*, RDX60204.1); MtATG13 (*Medicago truncatula*, XP_003630215.2); SsATG13 (*Spatholobus suberectus*, TKY60405.1); TmATG13 (*Trifolium medium*, MCH92797.1); TpATG (*Trifolium pratense*, PNY16483.1); VaATG13 (*Vigna angularis*, XP_017410090.1); and VuATG13 (*Vigna unguiculata*, XP_027921000.1). Red square represented the member with the highest sequence identity to the MsATG13. (**C**) Relative *MsATG13* expression levels in alfalfa leaves and roots in response to cold stress (4 °C). All data were normalized to the expression level in the control (0 h). Data are presented as the mean ± standard deviation from three biological replicates, the *MsActin2* (JQ028730.1) was used as the internal control gene. Asterisks indicate significant differences (* $p < 0.05$, ** $p < 0.01$, *** $p < 0.001$) compared to 0 h (Student's *t*-test).

Analysis of control and cold–stressed *M. sativa* plants demonstrated that *MsATG13* was upregulated by cold stress in both the leaves and the roots, although relative expression levels were significantly higher in the leaves than in the roots. *MsATG13* expression peaked twice in the leaves, first at 6 h (by 72.1–fold compared to 0 h), then at 24 h (by 43.9–fold). In the roots, *MsATG13* expression reached a maximum (3.4–fold higher than 0 h) at 6 h of cold stress (Figure 1C).

### *2.2. MsATG13 Overexpression Conferred Cold Stress Tolerance*

Tobacco plants were transformed with the recombinant overexpression vector pCaMV35S–*MsATG13* containing a screening gene with glufosinate–ammonium resistance (Figure 2A) to determine how *MsATG13* would affect cold tolerance. Out of 24 putative transformants, five transgenic lines were positively identified. Three lines (L6, L20, and L21) were selected for further analysis because they showed higher transcript levels of *MsATG13* expression levels (Figure 2B). We then compared the morphological and physiological differences of the WT and transgenic lines. There were no phenotypic differences between the WT and transgenic plants grown at 25 °C. After cold stress, WT plants were severely wilted, whereas there was no visible damage to transgenic plants (Figure 2C). This suggested that *MsATG13* conferred resistance to cold stress.

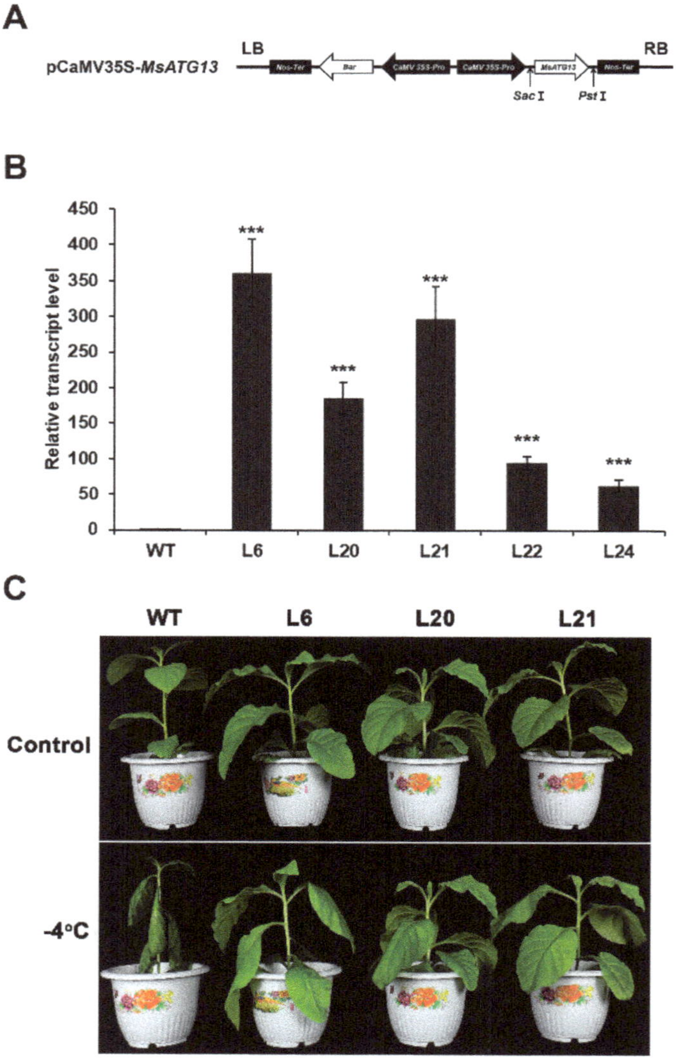


**Figure 2.** Evaluation of cold tolerance in transgenic tobacco. (**A**) Structure of the *MsATG13* overexpression vector pCaMV35S-*MsATG13*. (**B**) Relative *MsATG13* expression levels in transgenic tobacco lines. Expression levels were normalized to that in WT plant. Data are presented as the mean ± standard deviation from three biological replicates, the *NtGAPDH* (XM_016655379.1) was used as the internal control gene. Asterisks indicate significant differences (*** $p < 0.001$) compared to WT plants (Student's *t*–test). (**C**) Phenotype of WT and transgenic tobacco plants in the control (upper panel) and cold–treatment (lower panel) groups.

### 2.3. *MsATG13* Overexpression Increased Autophagosome Formation

To analyze potential differences in autophagosome formation between the WT and transgenic plants, we observed and quantified autophagosomes via TEM. Autophagosomes, which are double–membraned vesicles, sequester and eliminate damaged cellular components by fusing with the vacuole [29]. Degradation occurs after autophagosomes fuse with the vacuole, and the products are exported into the cytoplasm for recycling [30,31].

There were relatively few autophagosomes in any of the lines under control conditions. However, cold-treated transgenic plants had approximately 1.7–2.1 times more autophagosomes than cold-treated WT plants (Figure 3A,B). To determine whether the differences in autophagosome number corresponded to differences in autophagic function, we measured the expression levels of other key ATGs. Under control conditions, there were no significant differences (remained around 1–fold normalized to WT of the controls, respectively) in the expression of *NtATG1*, *NtATG6*, *NtATG8*, and *NtATG9* between transgenic and WT plants. Both the WT and the transgenic plants showed upregulation of all four ATGs after cold stress, but the expression of four NtATGs in transgenic tobacco were significantly higher than those in WT (Figure 3C–F). These results indicated that *MsATG13* promoted autophagosome formation and function in tobacco.

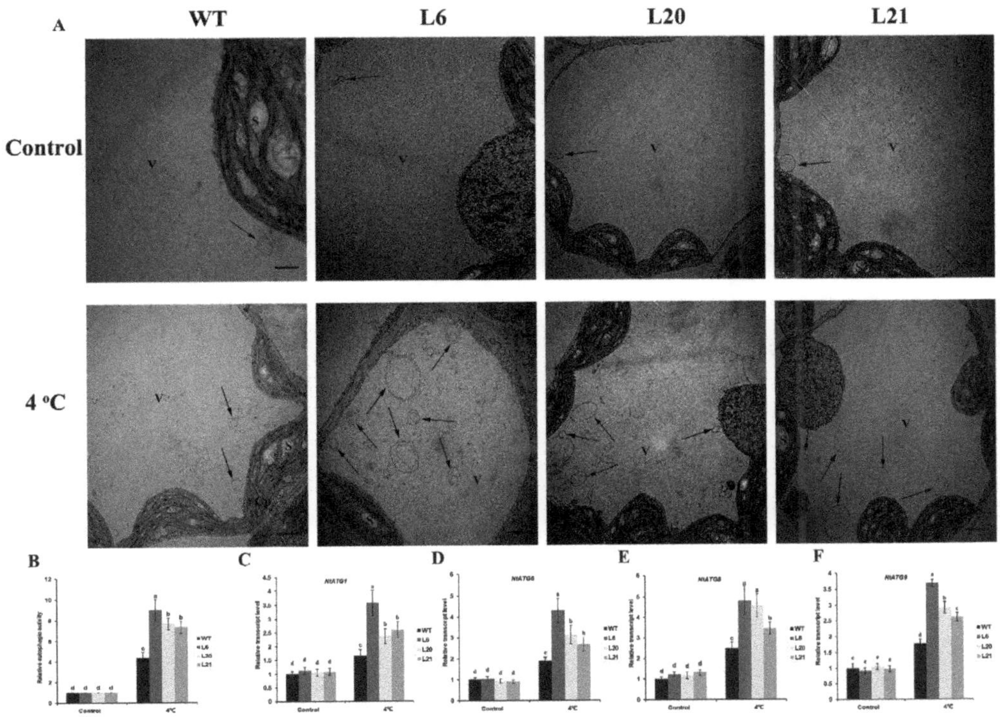


**Figure 3.** Analysis of autophagy in WT and *MsATG13* overexpressing tobacco plants. (**A**) Representative images of leaf cells in tobacco plants incubated at 4 °C. Cp, chloroplast; V, vacuole; S, starch. Autophagosomes are indicated with black arrows. Scale bar = 1 µm. (**B**) Relative autophagic activity normalized to the activity of transgenic tobacco lines or WT was shown in (**A**). More than 10 cells were used for statistics. (**C–F**) Expression levels of (**C**) *NtATG1*, (**D**) *NtATG6*, (**E**) *NtATG8*, and (**F**) *NtATG9* in WT and transgenic tobacco plants. All data were normalized to the expression level in WT of the controls, respectively. Data are presented as the mean ± standard deviation from three biological replicates, the *NtGAPDH* (XM_016655379.1) was used as the internal control gene. Letters above each bar indicate statistical significance groups at $p < 0.05$ (one–way analysis of variance).

### 2.4. *MsATG13 Reduced Cold–Induced Oxidative Damage*

Plant cold stress responses generally include ROS accumulation, which can result in oxidative damage to the cells. To determine whether ROS accumulation was affected by *MsATG13* expression, we performed histochemical staining for $O_2^-$ and $H_2O_2$. For both ROS compounds, leaves of all genotypes showed relatively light staining in the control samples. However, among the cold-treated group, leaves from WT plants were stained darker than those from transgenic plants (Figure 4A,B). These qualitative assessments were supported by quantitative measurements; $O_2^-$ levels were approximately 1.2–1.9 times higher in the WT than in the transgenic lines under cold stress conditions, and cold-stressed WT plants had approximately 1.2–2.1 times more $H_2O_2$ levels than the transgenic plants (Figure 4C,D). To assess the extent of membrane damage, which is associated with excess ROS, we measured MDA levels and electrolyte leakage. Both parameters showed similar values between the transgenic and WT plants under control conditions. In response to cold stress, MDA levels and electrolyte leakage were significantly reduced in transgenic plants compared to the WT (Figure 4E,F).

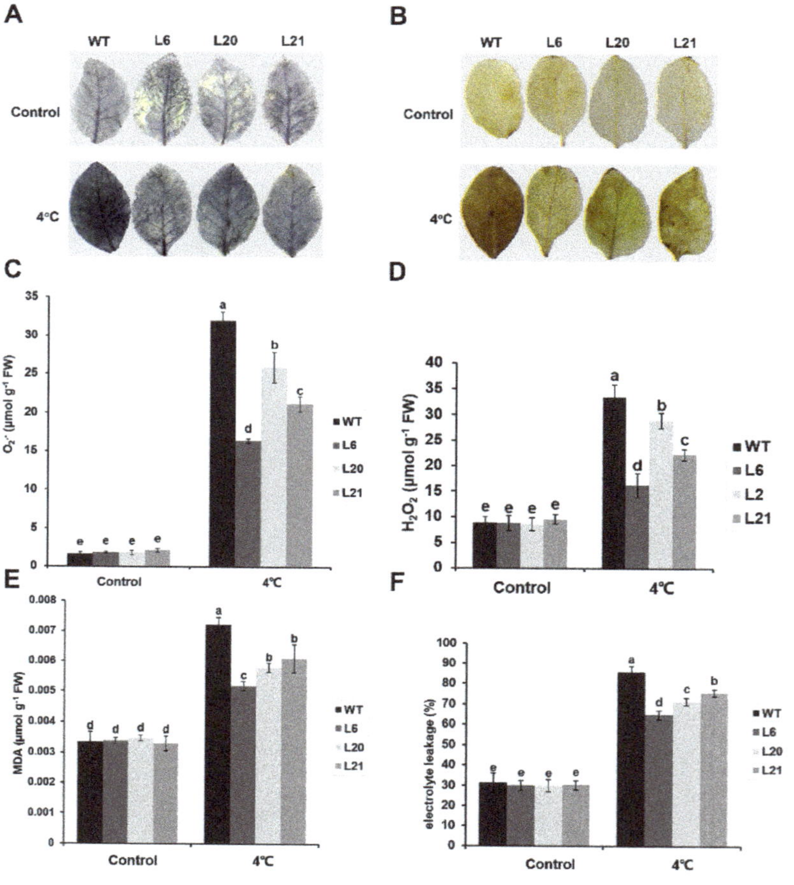


**Figure 4.** Evaluation of oxidative damage in transgenic tobacco. (**A**,**B**) The leaves of WT and transgenic tobacco plants overexpressing *MsATG13* incubated at 4 °C were stained with (**A**) p-Nitro blue tetrazolium chloride (NBT) and (**B**) diaminobenzidine (DAB). (**C**,**D**) Quantification of (**C**) $O_2^-$ and (**D**) $H_2O_2$ levels in the leaves of WT and transgenic tobacco plants incubated at 4 °C. (**E**,**F**) Quantification of (**E**) malondialdehyde (MDA) and (**F**) electrolyte leakage in the leaves of WT and transgenic tobacco plants incubated at 4 °C. Data are presented as the mean ± standard deviation from three biological replicates. Letters above each bar indicate statistical significance groups at $p < 0.05$ (oneway analysis of variance).

### 2.5. *MsATG13* Increased Antioxidant Enzymes Activity and Proline Content under Cold Stress

To identify a potential mechanism by which *MsATG13* expression may have reduced oxidative damage, we assessed the activities of three key antioxidant enzymes: SOD, POD, and CAT. Activities of all three were increased under cold stress conditions compared to the control plants. Furthermore, all three enzymes showed higher activity in the transgenic plants than in WT plants. Specifically, in the cold–stressed group, CAT and POD activities were approximately 1.3–2.0 times higher among transgenic plants than WT plants (Figure 5A,B). SOD activity was also significantly higher in the transgenic lines than that in WT plants (Figure 5C). Finally, we assessed the effects of *MsATG13* overexpression on the non–enzymatic antioxidant system, namely proline content. In control plants, proline levels were similar between the WT and the transgenic lines. However, among cold-treated plants, the transgenic lines had significantly higher proline content than WT plants (~1.7–1.9 times higher) (Figure 5D). These results suggested that *MsATG13* overexpression reduced oxidative damage in response to cold stress by influencing both the enzymatic and non–enzymatic antioxidant systems.

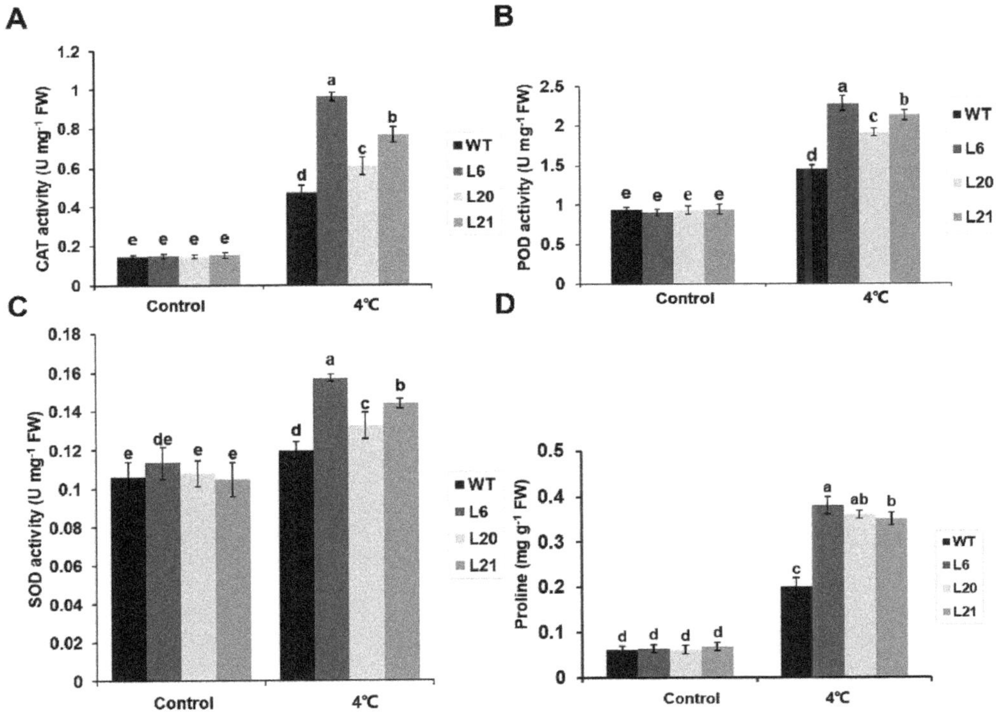


**Figure 5.** Antioxidant enzyme activity and proline content. (**A–C**) Activities of (**A**) catalase (CAT), (**B**) peroxidase (POD), and (**C**) superoxide dismutase (SOD) in the leaves of WT and transgenic tobacco plants incubated at 4 °C. (**D**) Proline content in the leaves of WT and transgenic tobacco plants incubated at 4 °C. Data are presented as the mean ± standard deviation from three biological replicates. Letters above each bar indicate statistical significance groups at $p < 0.05$ (one–way analysis of variance).

## 3. Discussion

Autophagy is a conserved intracellular degradation pathway that is crucial to plant survival in environmental stress conditions [32,33]. In the present study, we characterized the role of a *M. sativa* gene associated with autophagy, *MsATG13*, in the cold stress response; the survival of alfalfa under cold stress depends on the crown buds [34] and this gene was previously shown to be upregulated in the crown buds of a naturally cold-tolerant alfalfa cultivar during cold exposure. Initial expression profiling showed that *MsATG13* was upregulated in both the leaves and the roots in response to cold treatment and it was much more strongly induced in the leaves (Figure 1C). This is consistent with earlier studies demonstrating the important role of autophagy in maintaining organellar function and metabolite homeostasis in the leaves [35,36].

To further characterize this gene, we overexpressed it in tobacco. In response to cold stress, transgenic tobacco plants showed less damage than WT plants (Figure 2C). Furthermore, cold-stressed transgenic plants contained significantly more autophagosomes than WT plants (Figure 3A), indicating that *MsATG13* overexpression increased levels of autophagy in cold-stressed tobacco plants. Similar phenomena have been observed in previous studies; for example, transgenic apples overexpressing *MdATG8i* have more autophagosomes and increased salt stress tolerance compared to WT apples [37].

The key steps of autophagosome production are initiation, nucleation, membrane extension, and maturation. *ATG13* reportedly mediates the initiation step, and overexpression of this gene may also promote expression of other ATGs. Indeed, we here found that exogenous overexpression of *MsATG13* in cold-stressed tobacco plants triggered upregulation of the endogenous ATGs. *ATG1* serves as a crucial constituent of the upstream *ATG13–ATG1* complex, which also plays a critical role in the initiation of autophagosomes formation [38]. In this study, it was found that the expression level of *NtATG1* in transgenic lines was significantly higher than that of WT under cold stress conditions (Figure 3C). *ATG9* is an important membrane source for inducing the membrane elongation of autophagosomes [39], *ATG6* is responsible for the nucleation of autophagosomes [40], and we found that *MsATG13* overexpression significantly increased the expression levels of *NtATG6* and *NtATG9* in transgenic lines compared with WT under cold stress (Figure 3D,F). *ATG8* is an important member of the ubiquitin-like conjugation systems which are responsible for the maturation of autophagosomes [41]. Additionally, *NtATG8* was upregulated in both the transgenic lines and WT, showing higher expression level in the transgenic lines under cold stress (Figure 3E). There was no significant difference in the expression levels of the four NtATGs genes of all plants under control conditions, and their expression was induced by cold stress. After cold stress, their expression levels were significantly higher in transgenic tobacco than in WT, which is consistent with the changes in the number of autophagosomes. The *MsATG13* overexpression promoted the expression level of other key NtATGs which are necessary for the production of autophagosomes, which therefore promoted the formation of more autophagosomes in transgenic lines, enhancing the level of autophagy and thus participating in the regulation of cold stress tolerance.

Autophagy appears to be a necessary process by which cells clear excess ROS [42]. The primary forms of ROS accumulated by cold-stressed plants are $O_2^-$ and $H_2O_2$ [43]. Accumulation of compounds such as these can lead to membrane damage via lipid peroxidation [44]. MDA and electrolyte leakage are two commonly-used proxies used to evaluate abiotic stress tolerance based on the degree of membrane damage [45,46]. We here found that *MsATG13*-overexpressing tobacco plants, which had higher autophagy levels, accumulated lower levels of ROS compared with WT plants (Figure 4A–D). Furthermore, MDA contents and electrolyte leakage levels of *MsATG13* transgenic tobacco were significantly lower than those of WT (Figure 4E,F), indicating decreased oxidative damage among transgenic plants.

Autophagy may also improve the antioxidant system by regulating the activity of antioxidant enzymes [15]. *MsATG13* transgenic tobacco, which had higher autophagy levels than WT plants, also showed higher CAT, POD, and SOD activities in response to cold stress (Figure 5A–C). The non–enzymatic antioxidant system includes proline, accumulation of which improves cold stress tolerance in plant cells by stabilizing osmotic resistance, increasing cell turgor, and controlling water consumption [47]. In addition to acting as an osmotic protector, proline also plays an important role in ROS scavenging and the promotion of antioxidant enzyme activity [48]. Here, proline levels were shown to be significantly higher in transgenic tobacco than WT under cold stress conditions (Figure 5D). This indicated that *MsATG13* overexpression improved the antioxidant system (perhaps via the enhanced autophagy levels mediated by *MsATG13*), promoting ROS degradation in response to cold stress.

In summary, we here studied the roles of *MsATG13* in mediating autophagy and regulating plant tolerance to cold stress. Our results showed that *MsATG13* was shown to promote autophagy by upregulating other key ATGs necessary for autophagosome production, which may have also enhanced the antioxidant system. Enhanced antioxidant levels served to reduce ROS accumulation in response to cold treatment, preventing cellular damage due to excess ROS. Overall, these results demonstrated that *MsATG13* plays a significant role in enhancing plant cold tolerance through its regulation of autophagy, which is inseparable from its association with other autophagy related genes. *MsATG13* regulates autophagy by influencing other autophagy–related genes, thereby enhancing the cold tolerance of plants. However, the molecular mechanisms involved in this process require further investigation. *MsATG13* regulates autophagy by influencing other autophagy–related genes, thereby enhancing the cold tolerance of plants. However, the molecular mechanisms involved in this process require further investigation. This study increases our understanding of the role that autophagy plays in promoting plant cold tolerance. Furthermore, identification of *MsATG13* as a critical regulator of resistance to cold stress and offers future opportunities for breeding plants with cold tolerance.

## 4. Materials and Methods

### *4.1. Plant Materials*

*M. sativa* 'Zhaodong' seeds were provided by the Institute of Animal Husbandry, Heilongjiang Province, China. The seeds were placed in culture dishes containing distilled water and incubated at 18 °C in the dark for 48 h to sprout. The sprouted seeds were transferred to jars containing vermiculite and $\frac{1}{2}\times$ Hoagland solution and were incubated under a 14/10 h light/dark photoperiod (24/18 °C) with 60–80% relative humidity.

Tobacco seeds (*Nicotiana tabacum*) were disinfected with alcohol and sodium hypochlorite and placed on culture dishes containing Murashige and Skoog [49] medium (pH 5.8). The plates were incubated under the conditions described above for the sprouted *M. sativa* seedlings. After 1 week, germinated seedlings were transplanted into culture flasks containing Murashige and Skoog medium, then grown under the same conditions. After another 3 weeks, the seedlings were transferred to jars containing vermiculite and $\frac{1}{2}\times$ Hoagland solution, then returned to the growth chamber until they were collected for cold tolerance assessments.

### *4.2. Gene Cloning and Sequence Analysis*

Total RNA was isolated from *M. sativa* crown buds using the RNAprep Pure Plant Kit (TianGen Biotech, DP432, Beijing, China). cDNA was synthesized using One–Step gDNA Removal and cDNA Synthesis SuperMix (TransGen Biotech, AE311, Beijing, China). The coding sequence (CDS) of *MsATG13* (MW774897.1) was amplified from alfalfa cDNA using primer pair P1 (Supplementary Table S1) and Ex Taq DNA Polymerase (TaKaRa Biotech, RR001B, Beijing, China). The PCR amplification procedure was as follows: initial denaturation for 5 min at 95 °C; 35 cycles of 30 s at 95 °C, 30 s at 50 °C, and 2 min at 72 °C; and final extension for 10 min at 72 °C. The *MsATG13* CDS was inserted into the pMD18–T vector for subsequent analysis with sequencing. The open reading frame (ORF) and amino acid sequence of MsATG13 (UUH54286.1) were downloaded from NCBI: https://www.ncbi.nlm.nih.gov/orffinder (accessed on 20 April 2023). Amino acid domains of MsATG13 were identified with SMART: http://smart.embl-heidelberg.de (accessed on 20 April 2023) and the specific type and function description of the domains were searched with InterPro: http://www.ebi.ac.uk/interpro (accessed on 17 July 2023). Relevant amino acid sequences were downloaded from NCBI: https://blast.ncbi.nlm.nih.gov/Blast.cgi (accessed on 20 April 2023) and these sequences were aligned with ClustalW and a phylogenetic tree was constructed in MEGA 5.0 [50] using the neighbor–joining (NJ) method with 1000 bootstrap replicates.

### *4.3. Gene Expression Analysis*

Four–weeks–old alfalfa seedlings were incubated at 4 °C and the leaves and roots were collected at 0, 1, 3, 6, 12, and 24 h. RNA was extracted and cDNA was synthesized as described above. Expression levels of *MsATG13* were then analyzed with quantitative reverse transcription (qRT)–PCR using the TransStart Tip Green qPCR SuperMix (TransGen Biotech, AQ141, Beijing, China) and the primer pair P2 (Supplementary Table S1). Reactions were carried out on a 7300 Real–Time PCR System (Applied Biosystems, Waltham, MA, USA) with the following program: initial denaturation for 30 s at 94 °C, 40 cycles of 5 s at 94 °C, 31 s at 54 °C, and 31 s at 72 °C. Average gene expression levels were calculated from three technical replicates of each sample. The relative expressing levels were determined on the basis of the $2^{-\Delta\Delta Ct}$ method [51], and the *MsActin2* (JQ028730.1) was used as the internal control gene. All data were normalized to the expression level in the control (0 h). There were three independent replicates of this experiment.

### *4.4. Production of Transgenic Tobacco Lines and Cold Tolerance Assessments*

An expression vector containing the cauliflower mosaic virus (CaMV) 35S promoter [52] was digested with *Sac* I and *Pst* I, then the *MsATG13* CDS was inserted to produce the pCaMV35S–*MsATG13* construct; this included the bar gene, encoding phosphinothricin acetyltransferase, to allow transformant screening. The *Agrobacterium tumefaciens* strain EHA105 was transformed with the resulting construct using the freeze–thaw method. Transgenic tobacco plants were generated with the *Agrobacterium*–mediated leaf disc transformation method [53]. Transformants were verified through PCR with primer pair P3 (Supplementary Table S1) and EasyTaq DNA Polymerase (TransGen Biotech, AP111, Beijing, China). The PCR amplification procedure was as follows: initial denaturation for 5 min at 95 °C; 30 cycles of 30 s at 95 °C, 30 s at 52 °C, and 40 s at 72 °C; and final extension for 10 min at 72 °C. *MsATG13* expression was quantified in the transgenic and WT plants via qRT–PCR with the primer pair P2 and the following program: initial denaturation for 30 s at 94 °C, 40 cycles of 5 s at 94 °C, 31 s at 54 °C, and 31 s at 72 °C. Average gene expression levels were calculated from three technical replicates of each sample. The relative expressing levels were determined on the basis of the $2^{-\Delta\Delta Ct}$ method, and *NtGAPDH* (XM_016655379.1) was used as the internal control gene. All data were normalized to that in the WT plant. There were three independent replicates of this experiment.

To assess the effects of *MsATG13* overexpression on cold tolerance, transgenic and WT tobacco plants were raised to the six–weeks stage, then transferred to an artificial climate incubator. Plants were incubated at 4 °C for 4 h, followed by −4 °C for 3 h [54]. Control plants were maintained at 25 °C for all 7 h. A visual assessment of plant growth was used to determine cold tolerance.

### *4.5. Analyses of Oxidative Damage in Cold–Treated Plants*

#### 4.5.1. Histochemical Staining

WT and transgenic tobacco plants were raised to the six–weeks stage, then incubated at 4 °C (cold–stress group) or 25 °C (control group) for 24 h, then leaves were collected. Levels of superoxide anion radicals ($O_2^-$) and hydrogen peroxide ($H_2O_2$) were measured via histochemical staining with p-Nitro blue tetrazolium chloride (NBT) and diaminobenzidine (DAB), respectively, as described by [55] (with slight modifications). Leaves were fully immersed in NBT dye solution, vacuum infiltrated for 10 min, then incubated for 24 h in the dark at room temperature. After the dye solution was discarded, leaves were decolorized for 15 min in a rinse solution (3:1:1 ethanol:glycerol:lactic acid) in a boiling water bath. After the samples cooled, the rinse solution was replaced, and samples were incubated for another 15 min in the boiling water bath. After samples once again cooled, the rinse solution was replaced and samples were incubated in the dark at room temperature for 1–2 h. After chlorophyll decolorization was complete, the leaves were imaged. DAB staining was performed using similar methods, using DAB dye solution and a rinse solution (3:1:1 ethanol:acetic acid:glycerol).

4.5.2. ROS Quantification

$O_2^-$ and $H_2O_2$ contents were next measured to quantify ROS levels. The $O_2^-$ content was measured as described by [56]. Briefly, leaves were ground in pre–cooled phosphate buffer (pH 7.8). Samples were centrifuged, then the supernatant was removed and mixed with hydroxylamine hydrochloride and incubated for 1 h at 25 °C. P–amino benzene sulfonic acid and a–naphthylamine were added and samples were incubated at 30 °C for 30 min. Finally, absorbance was measured at 530 nm to calculate the $O_2^-$ content.

$H_2O_2$ content was determined using the titanium sulfate method [57] with slight modifications. Leaves were ground in pre–cooled acetone and centrifuged. The resulting supernatant was mixed well with titanium sulfate and concentrated ammonia, then centrifuged. The supernatant was discarded, and the pellet was fully dissolved in sulfuric acid. Absorbance was then measured at 405 nm and the $H_2O_2$ content was calculated.

4.5.3. Electrolyte Leakage Measurement

Electrolyte leakage was measured [58] with slight modifications. Tobacco leaf discs were placed into 20 mL deionized water, vacuum infiltrated for 15 min, then incubated at room temperature for 20 min. Initial conductivity (C1) was measured using a conductivity meter. The leaf discs were then boiled for 20 min and cooled to 25 °C before conductivity was measured again (C2). C1 and C2 were used to calculate electrolyte leakage.

4.5.4. Malondialdehyde (MDA) Measurement

MDA was measured [59] with some modifications. Briefly, leaves were ground in trichloroacetic acid, then centrifuged. The supernatant was mixed with thiobarbituric acid and incubated in a boiling water bath for 15 min. After cooling to room temperature, the absorbance at wavelengths of 532 and 450 nm were measured to calculate MDA content.

### *4.6. Proline Content and Antioxidant Enzyme Activity Assays*

Leaves were collected from the tobaccos of cold–treated and control groups. Proline was measured [60] with slight modifications. Leaves were ground in sulfosalicylic acid and incubated in a boiling water bath for 10 min, then centrifuged. The supernatant was mixed with glacial acetic acid and acidic ninhydrin, then incubated in a boiling water bath for 15 min. Samples were cooled to room temperature, then mixed with toluene. After standing and layering, the absorbance of the upper red extraction solution was measured at 520 nm to calculate proline contents.

The activities of SOD, CAT, and POD were measured as described by [61–63], respectively, with slight modifications. Leaves were ground in pre–cooled phosphate buffer (pH 7.8) and centrifuged to obtain the supernatant. To measure SOD activity, the supernatant was mixed with a SOD reaction solution (composed of methionine, EDTA–Na2, NBT, riboflavin, and phosphate buffer at pH 7.8). The reaction was allowed to proceed under illumination for 25 min, and the absorbance at the wavelength of 560 nm was measured to calculate SOD activity. To measure CAT activity, the supernatant described above was mixed with $H_2O_2$ and the absorbance at 240 nm was measured to calculate CAT activity. For the detection of the POD activity, leaves were ground in pre–cooled phosphate buffer (pH 6.0) and centrifuged to obtain the supernatant. The supernatant was mixed with POD reaction solution (composed of guaiacol, $H_2O_2$, and phosphate buffer at pH 6.0) and the change in absorbance at the wavelength of 470 nm was measured to calculate POD activity.

### 4.7. Autophagosome Quantification

Autophagosomes were observed as previously described [64]. Briefly, leaves were collected from the cold–treated and control plants, immediately cut into small pieces, and fixed with 2.5% glutaraldehyde in the dark for 12 h at 4 °C. After washing with phosphate buffer (pH 7.4), leaf fragments were fixed for 2.5 h in 1% (*v/v*) osmium tetroxide, then dehydrated in a graded ethanol series and embedded in Epon 812. Ultrathin sections (50–80 nm) were prepared with an EM UC7 ultramicrotome (Leica, Wetzlar, Germany) and collected on grids. The sections were visualized via transmission electron microscopy (TEM) on an H–7650 microscope (Hitachi, Tokyo, Japan) to observe and count the autophagosomes.

### 4.8. Autophagy-Related Genes Expression Analysis

Leaves were collected from tobaccos of cold–treated and control groups. Total RNA was isolated, and cDNA was synthesized as described above. Expression levels of autophagy–related genes *NtATG1* (KR336558.1), *NtATG6* (KP316403.1), *NtATG8* (KR336564.1), and *NtATG9* (KR336569.1) were measured with qRT–PCR. The thermal cycling program was as follows: initial denaturation for 30 s at 94 °C, 40 cycles of 5 s at 94 °C, 31 s at 54 °C, and 31 s at 72 °C. Average gene expression levels were calculated from three technical replicates of each sample. The relative expressing levels were determined on the basis of the $2^{-\Delta\Delta Ct}$ method, *NtGAPDH* (XM_016655379.1) was used as the internal control gene. All data were normalized to the expression level in WT of the controls, respectively. There were three independent replicates of this experiment. Primers for each gene are listed in Supplementary Table S1.

### 4.9. Statistical Analysis

There were three independent replicates of each experiment, from which the mean values and standard deviations were calculated. Significant differences between pairs of samples were analyzed with Student's *t*-test (* $p < 0.05$, ** $p < 0.01$, *** $p < 0.001$). Differences between three or more samples were assessed with one–way analysis of variance (ANOVA, $p < 0.05$).

**Supplementary Materials:** The following supporting information can be downloaded at: https://www.mdpi.com/article/10.3390/ijms241512033/s1.

**Author Contributions:** Conceptualization, D.W. and C.G.; methodology, B.D. and Y.A.; formal analysis, M.W. and X.C.; investigation, W.Z. and J.S.; resources, C.G.; writing—original draft preparation, W.Z. and J.S.; writing—review and editing, M.W., X.C. and D.W.; visualization, L.Z.; supervision, C.G.; funding acquisition, D.W. and C.G. All authors have read and agreed to the published version of the manuscript.

**Funding:** The project was sponsored by National Natural Science Foundation of China (U21A20182, 31972507); Science and Technology Major Project of Heilongjiang Province (GA19B103); Excellent Youth Project of Natural Science Foundation of Heilongjiang Province (YQ2020C033); Doctor Starting Fund Project of Harbin Normal University (XKB202001).

**Institutional Review Board Statement:** Not applicable.

**Informed Consent Statement:** Not applicable.

**Data Availability Statement:** The data presented in this study are available on request from the corresponding author.

**Conflicts of Interest:** The authors declare that they have no known competing financial interest or personal relationships that could have appeared to influence the work reported in this paper.

## References


1. Atayee, A.R.; Noori, M.S. Alleviation of cold stress in vegetable crops. *J. Sci. Agric.* **2020**, *4*, 38–44. [CrossRef]
2. Repo, T.; Mononen, K.; Alvila, L.; Pakkanen, T.T.; Hänninen, H. Cold acclimation of pedunculate oak (*Quercus robur* L.) at its northernmost distribution range. *Environ. Exp. Bot.* **2008**, *63*, 59–70. [CrossRef]
3. Yadav, S.K. Cold stress tolerance mechanisms in plants. A review. *Agron. Sustain. Dev.* **2010**, *30*, 515–527. [CrossRef]
4. Chan, Z.; Yokawa, K.; Kim, W.-Y.; Song, C.-P. ROS regulation during plant abiotic stress responses. *Front. Plant Sci.* **2016**, *7*, 1536. [CrossRef]
5. Avin-Wittenberg, T. Autophagy and its role in plant abiotic stress management. *Plant Cell Environ.* **2019**, *42*, 1045–1053. [CrossRef]
6. Waszczak, C.; Carmody, M.; Kangasjärvi, J. Reactive oxygen species in plant signaling. *Annu. Rev. Plant Biol.* **2018**, *69*, 209–236. [CrossRef]
7. Han, S.; Yu, B.; Wang, Y.; Liu, Y. Role of plant autophagy in stress response. *Protein Cell* **2011**, *2*, 784–791. [CrossRef]
8. Ahmad, P.; Jaleel, C.A.; Salem, M.A.; Nabi, G.; Sharma, S. Roles of enzymatic and nonenzymatic antioxidants in plants during abiotic stress. *Crit. Rev. Biotechnol.* **2010**, *30*, 161–175. [CrossRef]
9. Alici, E.H.; Arabaci, G. Determination of SOD, POD, PPO and cat enzyme activities in *Rumex obtusifolius* L. *Annu. Res. Rev. Biol.* **2016**, *11*, 1–7. [CrossRef]
10. Rejeb, K.B.; Abdelly, C.; Savouré, A. How reactive oxygen species and proline face stress together. *Plant Physiol. Biochem.* **2014**, *80*, 278–284. [CrossRef]
11. Malaviya, R.; Laskin, J.D.; Laskin, D.L. Oxidative stress-induced autophagy: Role in pulmonary toxicity. *Toxicol. Appl. Pharmacol.* **2014**, *275*, 145–151. [CrossRef]
12. Filomeni, G.; De Zio, D.; Cecconi, F. Oxidative stress and autophagy: The clash between damage and metabolic needs. *Cell Death Differ.* **2015**, *22*, 377–388. [CrossRef]
13. Marshall, R.S.; Vierstra, R.D. Autophagy: The master of bulk and selective recycling. *Annu. Rev. Plant Biol.* **2018**, *69*, 173–208. [CrossRef]
14. Li, F.; Vierstra, R.D. Autophagy: A multifaceted intracellular system for bulk and selective recycling. *Trends Plant Sci.* **2012**, *17*, 526–537. [CrossRef]
15. Jia, X.; Jia, X.; Li, T.; Wang, Y.; Sun, X.; Huo, L.; Wang, P.; Che, R.; Gong, X.; Ma, F. MdATG5a induces drought tolerance by improving the antioxidant defenses and promoting starch degradation in apple. *Plant Sci.* **2021**, *312*, 111052. [CrossRef]
16. Huo, L.; Guo, Z.; Jia, X.; Sun, X.; Wang, P.; Gong, X.; Ma, F. Increased autophagic activity in roots caused by overexpression of the autophagy-related gene MdATG10 in apple enhances salt tolerance. *Plant Sci.* **2020**, *294*, 110444. [CrossRef]
17. Li, B.; Liu, G.; Wang, Y.; Wei, Y.; Shi, H. Overexpression of banana ATG8f modulates drought stress resistance in *Arabidopsis*. *Biomolecules* **2019**, *9*, 814. [CrossRef]
18. Zhai, Y.; Guo, M.; Wang, H.; Lu, J.; Liu, J.; Zhang, C.; Gong, Z.; Lu, M. Autophagy, a conserved mechanism for protein degradation, responds to heat, and other abiotic stresses in *Capsicum annuum* L. *Front. Plant Sci.* **2016**, *7*, 131. [CrossRef]
19. Valitova, J.; Renkova, A.; Mukhitova, F.; Dmitrieva, S.; Beckett, R.P.; Minibayeva, F.V. Membrane sterols and genes of sterol biosynthesis are involved in the response of *Triticum aestivum* seedlings to cold stress. *Plant Physiol. Biochem.* **2019**, *142*, 452–459. [CrossRef]
20. Bai, Y.; Yang, Q.; Kang, J.; Sun, Y.; Gruber, M.; Chao, Y. Isolation and functional characterization of a *Medicago sativa* L. gene, MsLEA3-1. *Mol. Biol. Rep.* **2012**, *39*, 2883–2892. [CrossRef]
21. Kumar, S. Biotechnological advancements in alfalfa improvement. *J. Appl. Genet.* **2011**, *52*, 111–124. [CrossRef] [PubMed]
22. Hui, Y.; Yao, J.; Liu, R.; Cui, G.; Hou, S. Comprehensive evaluation on forage yield, nutrition quality and winter surviving rate of different alfalfa varieties. *Chin. J. Grassl.* **2010**, *32*, 108–111. [CrossRef]
23. Qi, H.; Lei, X.; Wang, Y.; Yu, S.; Liu, T.; Zhou, S.-K.; Chen, J.-Y.; Chen, Q.-F.; Qiu, R.-L.; Jiang, L. 14-3-3 proteins contribute to autophagy by modulating SINAT-mediated degradation of ATG13. *Plant Cell* **2022**, *34*, 4857–4876. [CrossRef] [PubMed]
24. Suttangkakul, A.; Li, F.; Chung, T.; Vierstra, R.D. The ATG1/ATG13 protein kinase complex is both a regulator and a target of autophagic recycling in Arabidopsis. *Plant Cell* **2011**, *23*, 3761–3779. [CrossRef] [PubMed]
25. Mizushima, N. The role of the Atg1/ULK1 complex in autophagy regulation. *Curr. Opin. Cell Biol.* **2010**, *22*, 132–139. [CrossRef]
26. Funakoshi, T.; Matsuura, A.; Noda, T.; Ohsumi, Y. Analyses of APG13 gene involved in autophagy in yeast, Saccharomyces cerevisiae. *Gene* **1997**, *192*, 207–213. [CrossRef]
27. Tsukada, M.; Ohsumi, Y. Isolation and characterization of autophagy-defective mutants of Saccharomyces cerevisiae. *FEBS Lett.* **1993**, *333*, 169–174. [CrossRef]
28. Scott, S.V.; Nice, D.C.; Nau, J.J.; Weisman, L.S.; Kamada, Y.; Keizer-Gunnink, I.; Funakoshi, T.; Veenhuis, M.; Ohsumi, Y.; Klionsky, D.J. Apg13p and Vac8p are part of a complex of phosphoproteins that are required for cytoplasm to vacuole targeting. *J. Biol. Chem.* **2000**, *275*, 25840–25849. [CrossRef]
29. Han, S.; Wang, Y.; Zheng, X.; Jia, Q.; Zhao, J.; Bai, F.; Hong, Y.; Liu, Y. Cytoplastic glyceraldehyde-3-phosphate dehydrogenases interact with ATG3 to negatively regulate autophagy and immunity in Nicotiana benthamiana. *Plant Cell* **2015**, *27*, 1316–1331. [CrossRef]
30. Rabinowitz, J.D.; White, E. Autophagy and Metabolism. *Science* **2010**, *330*, 1344–1348. [CrossRef]
31. Li, F.; Chung, T.; Vierstra, R.D. AUTOPHAGY-RELATED11 plays a critical role in general autophagy-and senescence-induced mitophagy in Arabidopsis. *Plant Cell* **2014**, *26*, 788–807. [CrossRef]

32. Michaeli, S.; Galili, G.; Genschik, P.; Fernie, A.R.; Avin-Wittenberg, T. Autophagy in plants—What's new on the menu? *Trends Plant Sci.* **2016**, *21*, 134–144. [CrossRef]
33. Zhai, Y.; Wang, H.; Liang, M.; Lu, M. Both silencing-and over-expression of pepper CaATG8c gene compromise plant tolerance to heat and salt stress. *Environ. Exp. Bot.* **2017**, *141*, 10–18. [CrossRef]
34. Pembleton, K.G.; Volenec, J.J.; Rawnsley, R.P.; Donaghy, D.J. Partitioning of taproot constituents and crown bud development are affected by water deficit in regrowing alfalfa (*Medicago sativa* L.). *Crop. Sci.* **2010**, *50*, 989–999. [CrossRef]
35. Wang, Y.; Yu, B.; Zhao, J.; Guo, J.; Li, Y.; Han, S.; Huang, L.; Du, Y.; Hong, Y.; Tang, D. Autophagy contributes to leaf starch degradation. *Plant Cell* **2013**, *25*, 1383–1399. [CrossRef]
36. Nakamura, S.; Hagihara, S.; Otomo, K.; Ishida, H.; Hidema, J.; Nemoto, T.; Izumi, M. Autophagy contributes to the quality control of leaf mitochondria. *Plant Cell Physiol.* **2021**, *62*, 229–247. [CrossRef]
37. Huo, L.; Guo, Z.; Wang, P.; Zhang, Z.; Jia, X.; Sun, Y.; Sun, X.; Gong, X.; Ma, F. MdATG8i functions positively in apple salt tolerance by maintaining photosynthetic ability and increasing the accumulation of arginine and polyamines. *Environ. Exp. Bot.* **2020**, *172*, 103989. [CrossRef]
38. Noda, N.N.; Fujioka, Y. Atg1 family kinases in autophagy initiation. *Cell. Mol. Life Sci.* **2015**, *72*, 3083–3096. [CrossRef]
39. Matoba, K.; Kotani, T.; Tsutsumi, A.; Tsuji, T.; Mori, T.; Noshiro, D.; Sugita, Y.; Nomura, N.; Iwata, S.; Ohsumi, Y. Atg9 is a lipid scramblase that mediates autophagosomal membrane expansion. *Nat. Struct. Mol. Biol.* **2020**, *27*, 1185–1193. [CrossRef]
40. Baskaran, S.; Carlson, L.-A.; Stjepanovic, G.; Young, L.N.; Grob, P.; Stanley, R.E.; Nogales, E.; Hurley, J.H. Architecture and dynamics of the autophagic phosphatidylinositol 3-kinase complex. *eLife* **2014**, *3*, e05115. [CrossRef]
41. Lane, J.D.; Nakatogawa, H. Two ubiquitin-like conjugation systems that mediate membrane formation during autophagy. *Essays Biochem.* **2013**, *55*, 39–50. [CrossRef] [PubMed]
42. Qi, H.; Xia, F.N.; Xiao, S. Autophagy in plants: Physiological roles and post-translational regulation. *J. Integr. Plant Biol.* **2021**, *63*, 161–179. [CrossRef] [PubMed]
43. Jaleel, C.A.; Riadh, K.; Gopi, R.; Manivannan, P.; Ines, J.; Al-Juburi, H.J.; Chang-Xing, Z.; Hong-Bo, S.; Panneerselvam, R. Antioxidant defense responses: Physiological plasticity in higher plants under abiotic constraints. *Acta Physiol. Plant.* **2009**, *31*, 427–436. [CrossRef]
44. Blokhina, O.; Virolainen, E.; Fagerstedt, K.V. Antioxidants, oxidative damage and oxygen deprivation stress: A review. *Ann. Bot.* **2003**, *91*, 179–194. [CrossRef]
45. Liao, X.; Guo, X.; Wang, Q.; Wang, Y.; Zhao, D.; Yao, L.; Wang, S.; Liu, G.; Li, T. Overexpression of Ms DREB 6.2 results in cytokinin-deficient developmental phenotypes and enhances drought tolerance in transgenic apple plants. *Plant J.* **2017**, *89*, 510–526. [CrossRef]
46. Tang, L.; Cai, H.; Ji, W.; Luo, X.; Wang, Z.; Wu, J.; Wang, X.; Cui, L.; Wang, Y.; Zhu, Y. Overexpression of GsZFP1 enhances salt and drought tolerance in transgenic alfalfa (*Medicago sativa* L.). *Plant Physiol. Biochem.* **2013**, *71*, 22–30. [CrossRef]
47. Gao, C.; Mumtaz, M.A.; Zhou, Y.; Yang, Z.; Shu, H.; Zhu, J.; Bao, W.; Cheng, S.; Yin, L.; Huang, J. Integrated transcriptomic and metabolomic analyses of cold-tolerant and cold-sensitive pepper species reveal key genes and essential metabolic pathways involved in response to cold stress. *Int. J. Mol. Sci.* **2022**, *23*, 6683. [CrossRef]
48. Ghosh, U.; Islam, M.; Siddiqui, M.; Cao, X.; Khan, M. Proline, a multifaceted signalling molecule in plant responses to abiotic stress: Understanding the physiological mechanisms. *Plant Biol.* **2022**, *24*, 227–239. [CrossRef]
49. Murashige, T.; Skoog, F. A revised medium for rapid growth and bio assays with tobacco tissue cultures. *Physiol. Plant.* **1962**, *15*, 473–497. [CrossRef]
50. Tamura, K.; Peterson, D.; Peterson, N.; Stecher, G.; Nei, M.; Kumar, S. MEGA5: Molecular evolutionary genetics analysis using maximum likelihood, evolutionary distance, and maximum parsimony methods. *Mol. Biol. Evol.* **2011**, *28*, 2731–2739. [CrossRef]
51. Livak, K.J.; Schmittgen, T.D. Analysis of relative gene expression data using real-time quantitative PCR and the $2^{-\Delta\Delta CT}$ method. *Methods* **2001**, *25*, 402–408. [CrossRef]
52. Sham, A.; Aly, M.A. Bioinformatics based comparative analysis of omega-3 fatty acids in desert plants and their role in stress resistance and tolerance. *Int. J. Plant Res.* **2012**, *2*, 80–89. [CrossRef]
53. Hannah, M.A.; Wiese, D.; Freund, S.; Fiehn, O.; Heyer, A.G.; Hincha, D.K. Natural genetic variation of freezing tolerance in Arabidopsis. *Plant Physiol.* **2006**, *142*, 98–112. [CrossRef]
54. Hara, M.; Terashima, S.; Fukaya, T.; Kuboi, T. Enhancement of cold tolerance and inhibition of lipid peroxidation by citrus dehydrin in transgenic tobacco. *Planta* **2003**, *217*, 290–298. [CrossRef]
55. Romero-Puertas, M.; Rodríguez-Serrano, M.; Corpas, F.; Gomez, M.; Del Rio, L.; Sandalio, L. Cadmium-induced subcellular accumulation of $O_2^{\cdot-}$ and $H_2O_2$ in pea leaves. *Plant Cell Environ.* **2004**, *27*, 1122–1134. [CrossRef]
56. Lei, Y.; Yin, C.; Li, C. Differences in some morphological, physiological, and biochemical responses to drought stress in two contrasting populations of Populus przewalskii. *Physiol. Plant.* **2006**, *127*, 182–191. [CrossRef]
57. Bravo, M.; Wang, D.I. Enzymatic oxidation of methanol to produce formaldehyde and hydrogen peroxide. *Fermentation Products.* **1981**, 329–334. [CrossRef]
58. Liu, J.-H.; Nada, K.; Honda, C.; Kitashiba, H.; Wen, X.-P.; Pang, X.-M.; Moriguchi, T. Polyamine biosynthesis of apple callus under salt stress: Importance of the arginine decarboxylase pathway in stress response. *J. Exp. Bot.* **2006**, *57*, 2589–2599. [CrossRef]
59. Hodges, D.M.; DeLong, J.M.; Forney, C.F.; Prange, R.K. Improving the thiobarbituric acid-reactive-substances assay for estimating lipid peroxidation in plant tissues containing anthocyanin and other interfering compounds. *Planta* **1999**, *207*, 604–611. [CrossRef]

60. Bates, L.; Waldren, R.; Teare, I. Rapid determination of free proline for water-stress studies. *Plant Soil* **1973**, *39*, 205–207. [CrossRef]
61. Beauchamp, C.; Fridovich, I. Superoxide dismutase: Improved assays and an assay applicable to acrylamide gels. *Anal. Biochem.* **1971**, *44*, 276–287. [CrossRef] [PubMed]
62. Patra, H.; Kar, M.; Mishra, D. Catalase activity in leaves and cotyledons during plant development and senescence. *Biochem. Und Physiol. Der Pflanz.* **1978**, *172*, 385–390. [CrossRef]
63. Zaharieva, T.; Yamashita, K.; Matsumoto, H. Iron deficiency induced changes in ascorbate content and enzyme activities related to ascorbate metabolism in cucumber roots. *Plant Cell Physiol.* **1999**, *40*, 273–280. [CrossRef]
64. Sun, X.; Wang, P.; Jia, X.; Huo, L.; Che, R.; Ma, F. Improvement of drought tolerance by overexpressing MdATG18a is mediated by modified antioxidant system and activated autophagy in transgenic apple. *Plant Biotechnol. J.* **2018**, *16*, 545–557. [CrossRef] [PubMed]

International Journal of
*Molecular Sciences*

MDPI

*Article*

# *Late Elongated Hypocotyl* Positively Regulates Salt Stress Tolerance in *Medicago truncatula*

**Zhichao Lu [1], Haiyang Liu [1], Yiming Kong [2], Lizhu Wen [1], Yang Zhao [1], Chuanen Zhou [1] and Lu Han [1,*]**

1 The Key Laboratory of Plant Development and Environmental Adaptation Biology, Ministry of Education, School of Life Sciences, Shandong University, Qingdao 266237, China; lzcsdhz@163.com (Z.L.); 15949537052@163.com (H.L.); wlizhu@163.com (L.W.); aimee.yangzhao@sdu.edu.cn (Y.Z.); czhou@sdu.edu.cn (C.Z.)
2 College of Life Sciences, Shandong Normal University, Jinan 250014, China; lck7love@163.com
* Correspondence: lhan@sdu.edu.cn

**Abstract:** Abiotic stress, such as drought, osmotic, and salinity stresses, seriously affects plant growth and crop production. Studying stress-resistant genes that enhance plant stress tolerance is an efficient way to facilitate the breeding of crop species with high stress tolerance. In this study, we reported that the core circadian clock component, the *LATE ELONGATED HYPOCOTYL* (*LHY*) orthologue *MtLHY*, plays a positive role in salt stress response in *Medicago truncatula*. The expression of *MtLHY* was induced by salt stress, and loss-of-function mutants of *MtLHY* were shown to be hypersensitive to salt treatment. However, overexpression of *MtLHY* improved salt stress tolerance through a higher accumulation of flavonoids. Consistently, exogenous flavonol application improved the salt stress tolerance in *M. truncatula*. Additionally, *MtLHY* was identified as a transcriptional activator of the flavonol synthase gene, *MtFLS*. Our findings revealed that *MtLHY* confers plant salt stress tolerance, at least by modulating the flavonoid biosynthesis pathway, which provides insight into salt stress tolerance that links the circadian clock with flavonoid biosynthesis.

**Keywords:** circadian clock; *MtLHY*; salt tolerance; *MtFLS*; flavonoids

**Citation:** Lu, Z.; Liu, H.; Kong, Y.; Wen, L.; Zhao, Y.; Zhou, C.; Han, L. *Late Elongated Hypocotyl* Positively Regulates Salt Stress Tolerance in *Medicago truncatula*. *Int. J. Mol. Sci.* **2023**, *24*, 9948. https://doi.org/10.3390/ijms24129948

Academic Editors: Ana I Ribeiro-Barros, Isabel Marques and José C. Ramalho

Received: 28 April 2023
Revised: 26 May 2023
Accepted: 27 May 2023
Published: 9 June 2023



## 1. Introduction

Plants have evolved intricate molecular systems and physiological strategies to cope with unfavorable environmental conditions, such as drought, osmotic stress, and salinity. In recent years, it has been discovered that the circadian clock system plays a crucial role in responding to abiotic stress [1,2]. As an endogenous time-keeping mechanism, the circadian clock components synchronize the developmental and physiological behaviors with external environmental cycles, providing plants with adaptive strategies to cope with environmental oscillations [1,2]. The circadian clock is composed of multiple transcriptional feedback regulatory loops that induce the transcription of numerous abiotic stress-associated genes, enabling plant adaptation and survival in unfavorable conditions [3,4]. For example, in Arabidopsis, the evening-phased clock component *GIGANTEA* (*GI*) regulates salt stress tolerance through the salt overly sensitive (SOS) pathway [5]. Another core circadian clock component, *EARLY FLOWERING3* (*ELF3*), enhances salt tolerance by post-transcriptionally promoting GI degradation and inhibiting the transcription of *PHYTOCHROME-INTERACTING FACTOR4* (*PIF4*) [6]. Additionally, two core components of the circadian clock, *CIRCADIAN CLOCK-ASSOCIATED1* (*CCA1*) and *LATE ELONGATED HYPOCOTYL* (*LHY*), directly bind to the promoter of *C-REPEAT BINDING FACTOR* (*CBF1*, *CBF2*, and *CBF3*) and activate their expression in response to low temperature [7]. In soybean, two homologous *LHY* pairs negatively control drought tolerance by repressing the abscisic acid responses [8]. In rice, the *LHY/CCA1* ortholog, *OsCCA1*, synchronously confers multiple abiotic stress tolerance by transcriptionally regulating ABA signaling [9]. However, the involvement of *MtLHY*, the *LHY/CCA1* ortholog in *Medicago truncatula*, in abiotic stress responses remains to be determined, despite the extensive research on nodulation and leaf movement mechanisms in *M. truncatula*. [10].

Soil salinity is a major limiting factor to plant growth and crop production in agriculture [11–13]. To cope with salt stress, plants have developed various mechanisms to improve salt tolerance, including osmoregulation, ionic balance, and antioxidant defense systems [14–19]. In response to salt stress, plants accumulate osmoprotectants, such as glycine, betaine, and proline to adjust osmotic pressure [14]. As excessive $Na^+$ accumulation in plants leads to membrane damage due to oxidative stress, cells must maintain ionic balance, particularly $Na^+/K^+$ homeostasis, by regulating ion transporters to achieve a low cytoplasmic $Na^+/K^+$ ratio [14]. Reactive oxygen species (ROS), including hydroxyl radical ($OH^-$), superoxide anion ($O_2^-$), and hydrogen peroxide ($H_2O_2$) are highly reactive molecules that are generally induced in response to abiotic stress [20]. Excessive ROS accumulation leads to oxidative stress, causing membrane damage and even cell death [20–22]. Therefore, plants activate antioxidant protection systems to remove excessive ROS and protect the membrane structure. ROS detoxification is regulated by cytosolic enzymatic antioxidants, including peroxidase (POD), catalase (CAT), and superoxide dismutase (SOD) [23]. Additionally, excessive ROS can be eliminated by nonenzymatic antioxidants, such as glutathione (GSH), alkaloids, carotenoids, and flavonoids [24].

Flavonoids are a major group of secondary metabolites found in most plants and their organs. They are known for their ability to scavenge ROS efficiently and are divided into various subgroups, including chalcones, flavones, flavonols, and isoflavones [25–27]. Recently, there has been increasing interest in the role of flavonoids in protecting plants against environmental stresses such as UV radiation, cold, drought, and salinity [26,28–30]. For example, in the pigeon pea, the CcCIPK14-CcCBL1 complex regulates flavonoid biosynthesis and plays a crucial role in drought stress tolerance [31]. In soybean, silencing of the flavone synthase gene, *GmFNSII*, resulted in reduced flavone content and hypersensitivity to salt treatment in hairy roots [32]. Heat shock factor *HSFB2b* inhibited the expression of *GmNACs* to promote flavonoid biosynthesis, which conferred soybean salt stress tolerance [33]. Additionally, exogenous flavonoid application reduced MDA content and increased salt stress tolerance [33]. Therefore, flavonoid accumulation is strongly associated with salt stress responses.

In this study, we discovered the core component of the circadian clock, *MtLHY*, is responsible for salt stress tolerance in *M. truncatula*. Loss of function in *MtLHY* results in a hypersensitivity phenotype to salt treatment, while the ectopic expression of *MtLHY* increases the ability of salt stress tolerance in *M. truncatula*. *MtLHY* plays a crucial role in inhibiting ROS production and mediating $Na^+/K^+$ homeostasis under salt stress. Additionally, we found that *MtLHY* plays a key role in modulating flavonoid biosynthesis. *MtLHY* directly binds to the promoters of *MtFLS* to activate its expression, and a high content of flavonols accumulates in *MtLHY*-overexpressing plants to resist salt stress. Consistently, exogenous flavonols application also enhances the salt stress tolerance in *M. truncatula*. Thus, our results support the role of *MtLHY* as a positive regulator in salt stress tolerance, linking the circadian clock and flavonoid biosynthesis to coordinate plant growth and environmental adaptation.

## 2. Results

### 2.1. Loss-of-Function Mutants of MtLHY Is Hypersensitive to Salt Stress

Previous research has demonstrated that *LHY* orthologues are a crucial component of the circadian clock and play a vital role in plant abiotic stress responses, such as drought and cold stress [7,8]. However, there is limited information regarding the involvement of *LHY* in response to salt stress. To investigate the role of *LHY* in the response to salinity stress in *M. truncatula*, we initially examined the expression patterns of *MtLHY* in response to salt treatment. The results show that the transcript of *MtLHY* could be rapidly induced by the salt stress treatment. Its transcript level was increased by only 3 h (hours) after treatment with 150 mM NaCl and remained at higher levels for 9 h more than the control without salt treatment (Figure 1A), suggesting it may be involved in response to salt stress.

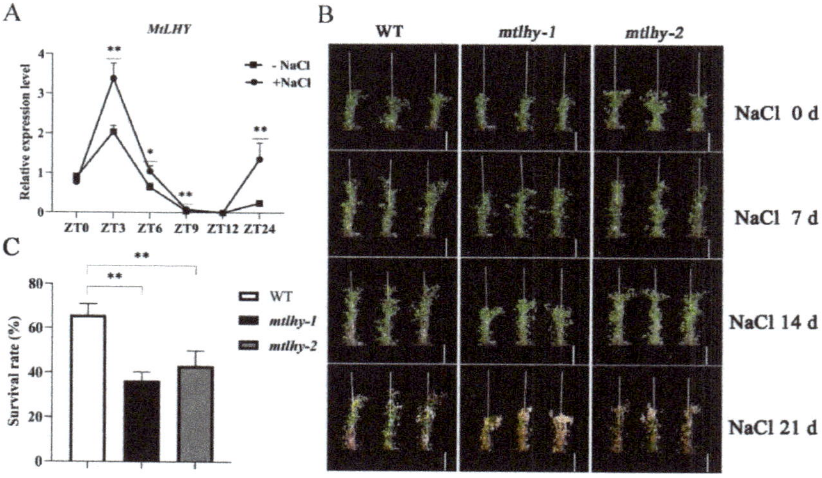


**Figure 1.** Loss-of-function mutants of *MtLHY* were shown to be hypersensitive to salt treatment: The expression of *MtLHY* was induced by salt stress (**A**). Loss-of-function mutants of *MtLHY* were shown to be hypersensitive to salt treatment with 150 mM NaCl for 3 weeks (**B**). Statistical analysis of the survival rate of WT and *mtlhy* mutants at treatment with 150 mM NaCl for 3 weeks (**C**). Data are shown as means ± SD ($n = 15$; * $p < 0.05$, ** $p < 0.01$, based on Dunnett test). Scale bar, 10 cm.

To further investigate whether *MtLHY* plays a key role in salt tolerance, we screened and obtain two independent *Tnt1* insertion mutants of *MtLHY* from the *Tnt1*-tagged mutant population [10]. We then subjected 4-week-old *mtlhy* mutants and wild-type (WT) plants to salt stress treatments and analyzed their survival rates after treatment with 150 mM NaCl for 3 weeks (Figure 1B,C). Our statistical analysis showed that approximately 65.9% of WT plants survived, while only 36.5–43.1% of the two *mtlhy* alleles recovered to live (Figure 1C), suggesting the loss of function of *MtLHY* resulted in a hypersensitivity phenotype to salt stress.

### 2.2. *MtLHY Mutation Impaired Seed Germination and Seedling Growth under NaCl Stress*

Salt is a crucial factor that impacts seed germination and seedling growth [34]. To confirm the involvement of *MtLHY* in these processes under salt stress, we studied the phenotype of *mtlhy* mutants and WT with or without a 150 mM NaCl treatment. Firstly, we tested the seed germination phenotype of mutants and WT (Figure 2A). The WT seeds sprouted completely on the sixth day without treatment, but the *mtlhy* mutants required 9 d (days) (Figure 2A,B), indicating that the loss of function of *MtLHY* delayed seed germination. After treatment with 150 mM NaCl, salt stress significantly inhibited seed germination rates in both *mtlhy* mutants and WT (Figure 2A–C). In particular, the mutants exhibited significantly lower seed germination rates than WT seeds after NaCl treatment for 9 d (Figure 2A–C). For example, the germination rate of WT seeds was approximately 66.9% at 150 mM NaCl, whereas the mutants were 12.8–32.3% at that point (Figure 2C).

Root growth was more affected by salt stress than shoot growth [35]. To further identify the function of *MtLHY* on the postgermination seedling under salt stress, we performed experiments to measure root length in both WT and *mtlhy* mutant seedlings. When grown on 1/2MS medium, both WT and *mtlhy* mutant seedlings exhibited similar root lengths (Figure 2D,E). However, the root length of the mutant seeding was significantly shorter than that of WT at 150 mM NaCl for a week (Figure 2D,E). These findings collectively suggest that *MtLHY* plays a crucial role in seed germination and seedling growth under salt stress conditions.

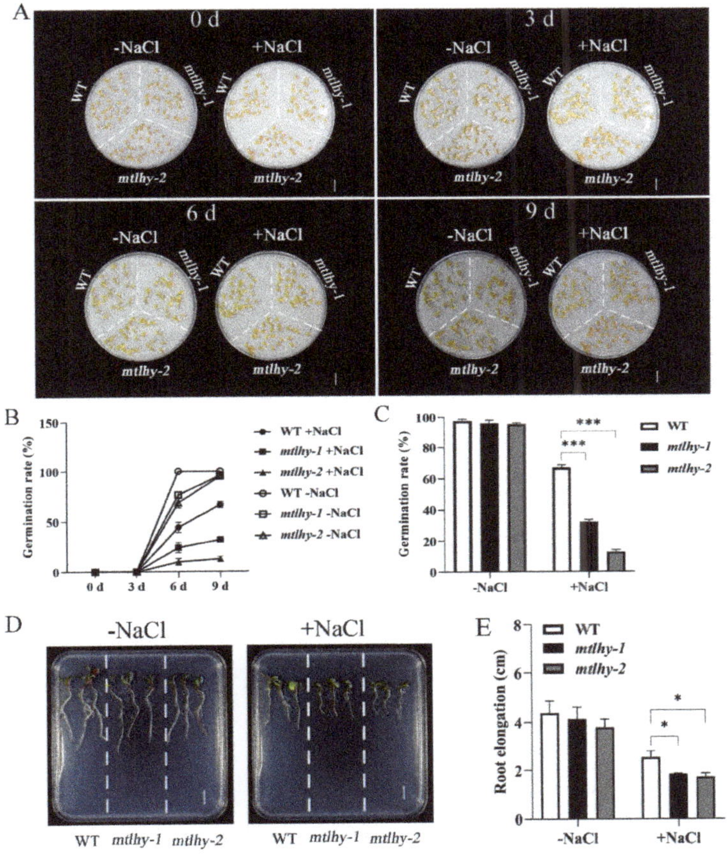


**Figure 2.** *MtLHY* mutation impaired seed germination and seedling growth under NaCl stress: Seed germination phenotype of *mtlhy* mutants and WT with or without a 150 mM NaCl treatment (**A**). Statistical analysis of seed germination rate of *mtlhy* mutants and WT with or without a 150 mM NaCl treatment (**B**,**C**). The seedling phenotype of *mtlhy* mutants and WT with or without a 150 mM NaCl treatment (**D**). Statistical analysis of root length of *mtlhy* mutants and WT with or without a 150 mM NaCl treatment (**E**). Data are shown as means ± SD ($n = 3$; * $p < 0.05$, *** $p < 0.001$, based on Dunnett test). Scale bar, 1 cm.

### 2.3. Physiological Responses of MtLHY Mutation to Salt Stress

To assess the physiological response of *mtlhy* mutants and WT to salt stress, we initially compared the levels of malondialdehyde (MDA), which is an indicator of lipid peroxidation [36]. Our findings showed that *mtlhy* mutant leaves had a higher accumulation of MDA than WT at 150 mM NaCl (Figure 3A) for a week, suggesting that the loss of function of *MtLHY* resulted in increased cell membrane damage in response to NaCl stress. Subsequently, we observed that the production of $H_2O_2$ in *mtlhy* leaves was significantly higher than in WT leaves under NaCl stress (Figure 3B), indicating high levels of ROS in the mutants. This was further confirmed by both 3,3′-diaminobenzidine (DAB) and nitroblue tetrazolium (NBT) staining in roots (Figure 3C,D). While the rapid accumulation of ROS in plant cells is an effective strategy in response to stress, elevated ROS levels can be harmful to the plant [37]. Therefore, the balance between ROS production and detoxification is crucial for plant tolerance to stress. Notably, the reduced magnitude of SOD activity in the *mtlhy* mutants indicates that the terms of ROS detoxification were impaired in the *mtlhy* mutants (Figure 3E). As $Na^+/K^+$ homeostasis in the cytoplasm is critical for salt tolerance, and a low $Na^+/K^+$ ratio is an indicator of salt tolerance [38], we analyzed the $Na^+/K^+$ ratio in the WT leaves and *mtlhy* mutants subjected to salt treatment. The $Na^+/K^+$ ratio of *mtlhy* leaves was significantly higher than that of WT (Figure 3F). Conversely, the chlorophyll content of *mtlhy* leaves was lower than that of WT under salt stress (Figure 3G). These data collectively indicated that *mtlhy* mutants were hypersensitive to salt stress.

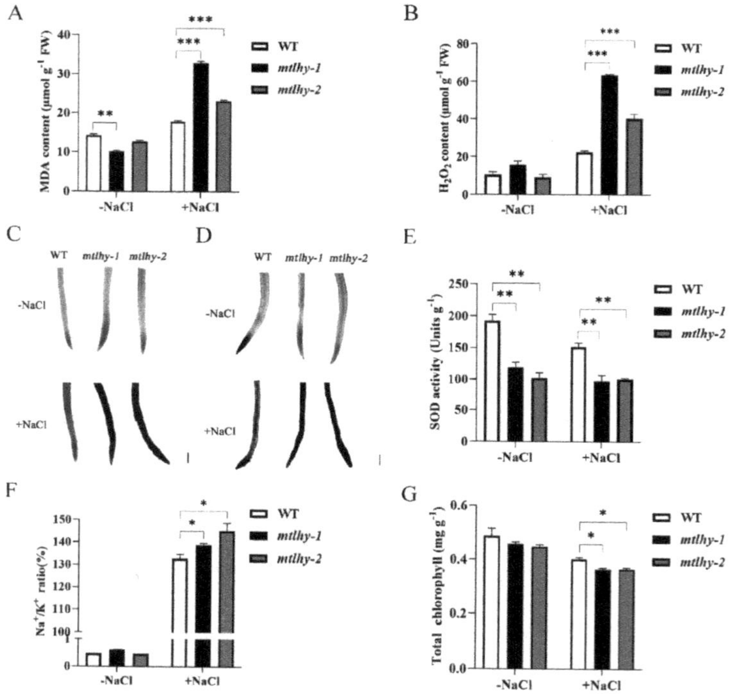


**Figure 3.** Physiological responses of *mtlhy* mutant to salt stress: The concentrations of MDA were determined in the leaves of *mtlhy* mutants and WT with or without a 150 mM NaCl treatment for 7 d (**A**). The content of $H_2O_2$ was determined in the leaves of *mtlhy* mutants and WT with or without a 150 mM NaCl treatment for 7 d (**B**). The accumulation of hydrogen peroxide ($H_2O_2$) in the root of *mtlhy* mutants and WT was visualized by staining with 3,3′-diaminobenzidine (DAB) with or without a 150 mM NaCl treatment for 7 d (**C**). The accumulation of superoxide anion ($O_2^-$) in the root of *mtlhy* mutants and WT was visualized by staining with nitroblue tetrazolium (NBT) with or without a 150 mM NaCl treatment for 7 d (**D**). SOD activity was examined in the leaves of *mtlhy* mutants and WT with or without a 150 mM NaCl treatment for 7 d (**E**). $Na^+/K^+$ ratio and total chlorophyll content were examined in the leaves of *mtlhy* mutants and WT with or without a 150 mM NaCl treatment for 7 d (**F,G**). Data are shown as means ± SD ($n$ = 3; * $p < 0.05$, ** $p < 0.01$ *** $p < 0.001$, based on Dunnett test). Scale bar, 1 mm.

### 2.4. MtLHY Overexpression Improves Salt Stress Tolerance in M. truncatula

To further investigate the role of *MtLHY* in regulating salt stress tolerance in *M. truncatula*, we generated transgenic plants overexpressing *MtLHY*, resulting in significantly elevated transcript levels of *MtLHY* in leaves (Figure 4A). These plants were selected for further study to explore the salt stress tolerance conferred by *MtLHY* expression. *MtLHY*-overexpressing transgenic plants and WT plants were subjected to salt treatments (Figure 4B), and statistical data showed that the survival rate of WT plants was approximately 65.6% at 150 mM NaCl for three weeks (Figure 4C), while that of *MtLHY*-overexpressing lines was over 92.3% (Figure 4C), indicating that elevated *MtLHY* transcript levels enhanced salt stress tolerance in *M. truncatula*. Additionally, we measured the root lengths of seedlings from both the overexpression plants and WT grown on 1/2MS medium with or without 150 mM NaCl treatment for a week (Figure 4D). Under normal conditions, the root lengths of *MtLHY*-overexpressing seedlings were comparable to those of WT (Figure 4D,E). However, in the presence of salt stress, the root lengths of *MtLHY*-overexpressing transgenic lines were significantly longer than those of WT (Figure 4D,E), indicating that *MtLHY* overexpression promotes efficient seedling growth under salt stress conditions. Overall, our results suggest that *MtLHY* overexpression improves salt stress tolerance in *M. truncatula*.

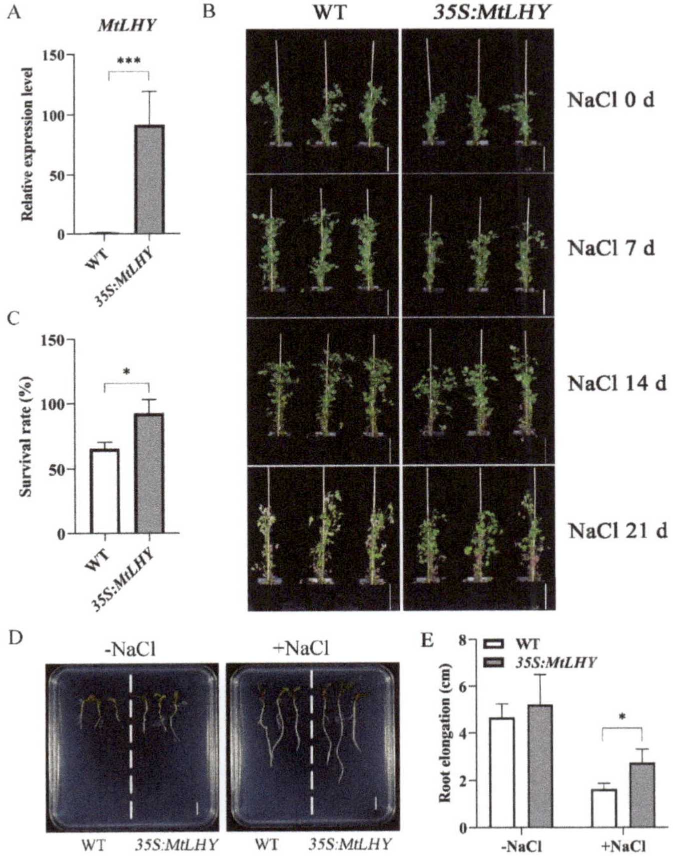


**Figure 4.** *MtLHY* overexpression improves salt stress tolerance in *M. truncatula*: The expression level of *MtLHY* in *MtLHY*-overexpressing plants (**A**). *MtLHY*-overexpressing plants were shown to be resistant to salt treatment with 150 mM NaCl for 3 weeks (**B**). Statistical analysis of the survival rate of WT and *MtLHY*-overexpressing plants at treatment with 150 mM NaCl for 3 weeks (**C**). The seedling phenotype of *MtLHY*-overexpressing plants and WT with or without a 150 mM NaCl treatment (**D**). Statistical analysis of root length of *MtLHY*-overexpressing plants and WT with or without a 150 mM NaCl treatment (**E**). Data are shown as means ± SD ($n = 3$; * $p < 0.05$, *** $p < 0.001$, based on two-tailed *t*-tests). Scale bars, 10 cm in (**B**) and 1 cm in (**D**).

To further validate our findings, we examined the physiological responses of *MtLHY*-overexpressing plants to NaCl stress. Compared with the WT plants, *MtLHY*-overexpressing plants exhibited lower levels of MDA accumulation (Figure 5A), regardless of whether they were treated with 150 mM NaCl or not. Additionally, the $H_2O_2$ content in *MtLHY*-overexpressing plants was significantly lower than that in WT (Figure 5B), indicating that high levels of *MtLHY* expression enhance the capacity for ROS scavenging. This was further confirmed by DAB and NBT staining in roots (Figure 5C,D). Moreover, *MtLHY* overexpression increased SOD activity under salt conditions (Figure 5E), suggesting that *MtLHY* plays a key role in regulating SOD activity. Notably, the $Na^+/K^+$ ratio in the leaves of *MtLHY*-overexpressing lines was lower than that in WT (Figure 5F), while the chlorophyll content in the leaves of *MtLHY*-overexpressing plants was higher than that in WT (Figure 5G), suggesting that *MtLHY* mediates $Na^+/K^+$ homeostasis in response to salt stress. In summary, *MtLHY* plays a positive role in salt tolerance.

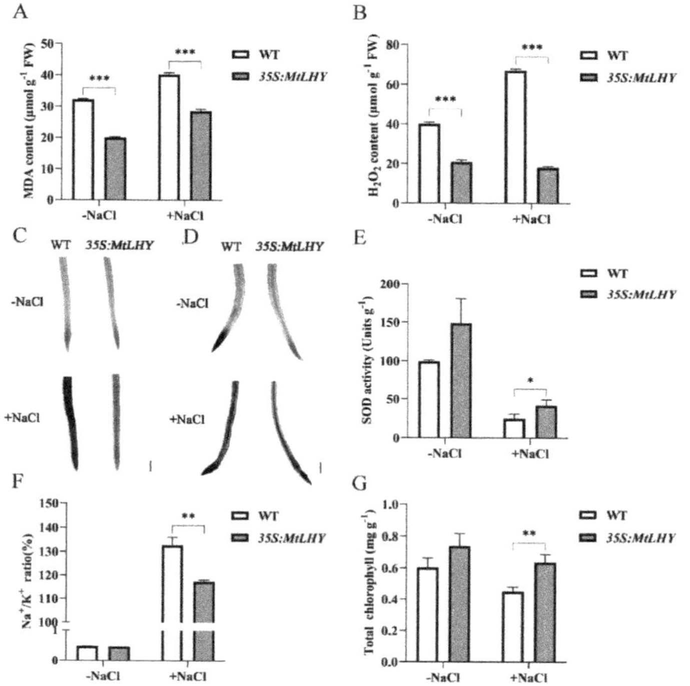


**Figure 5.** Physiological responses of *MtLHY*-overexpressing plants to salt stress: The concentrations of MDA were determined in the leaves of *MtLHY*-overexpressing plants and WT with or without a 150 mM NaCl treatment for 7 d (**A**). The content of $H_2O_2$ was determined in the leaves of *MtLHY*-overexpressing plants and WT with or without a 150 mM NaCl treatment for 7 d (**B**). The accumulation of hydrogen peroxide ($H_2O_2$) in the root of *MtLHY*-overexpressing plants and WT was visualized by staining with 3,3′-diaminobenzidine (DAB) with or without a 150 mM NaCl treatment for 7 d (**C**). The accumulation of superoxide anion ($O_2^-$) in the root of *MtLHY*-overexpressing plants and WT was visualized by staining with nitroblue tetrazolium (NBT) with or without a 150 mM NaCl treatment for 7 d (**D**). SOD activity was examined in the leaves of *MtLHY*-overexpressing plants and WT with or without a 150 mM NaCl treatment for 7 d (**E**). $Na^+/K^+$ ratio and total chlorophyll content were examined in the leaves of *MtLHY*-overexpressing plants and WT with or without a 150 mM NaCl treatment for 7 d (**F**,**G**). Data are shown as means ± SD ($n = 3$; * $p < 0.05$, ** $p < 0.01$ *** $p < 0.001$, based on two-tailed *t*-tests). Scale bar, 1 mm.

## 2.5. *MtLHY* Is Involved in Flavonoid Biosynthesis

To further identify the direct targets of *MtLHY* in transcriptional regulation, we compared the differentially expressed genes (DEGs) in the leaves of WT and *mtlhy* mutants from our previous RNA-sequencing transcriptome data [10]. Kyoto Encyclopedia of Genes and Genomes (KEGG) analysis revealed that 63 DEGs were enriched in the flavonoid biosynthesis pathway, which is one of the top biological processes. Flavonoid compounds have been shown to possess antioxidant activity that can prevent damage caused by free radicals by scavenging ROS, activating antioxidant enzymes, and inhibiting NADPH oxidases. To investigate whether *MtLHY* is involved in flavonoid biosynthesis, we selected key enzyme genes involved in flavonoid biosynthesis from the DEGs and verified their expression levels using RT-qPCR in both WT and *mtlhy* mutant plants, as well as in *MtLHY*-overexpressing plants. The transcript levels of known flavonoid biosynthesis genes, such as *MtCHS, MtCHI, MtFLS, MtF3, MtF3'H,* and *MtIFS*, were significantly altered in both *mtlhy* mutants and *MtLHY*-overexpressing plants compared with those in WT (Figure 6). Notably, *MtFLS*, the gene responsible for flavonol synthase, was significantly downregulated in *mtlhy* mutants but significantly upregulated in *MtLHY*-overexpressing plants (Figure 6), suggesting that the expression of *MtFLS* is likely induced by MtLHY.

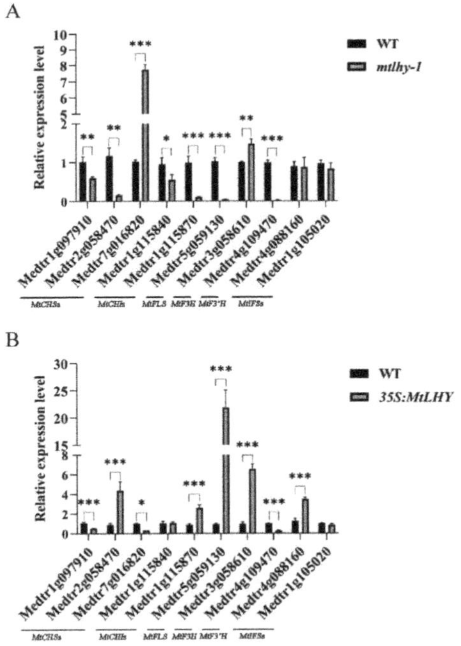


**Figure 6.** *MtLHY* is involved in flavonoid biosynthesis: The transcript levels of known flavonoid biosynthesis genes, such as *MtCHS, MtCHI, MtFLS, MtF3, MtF3'H,* and *MtIFS*, were examined in both *mtlhy* mutants and *MtLHY*-overexpressing plants compared with those in WT (**A**,**B**). Data are shown as means ± SD ($n = 3$; * $p < 0.05$, ** $p < 0.01$ *** $p < 0.001$, based on two-tailed $t$-tests).

## 2.6. *MtLHY* Overexpression Enhances Flavonoid Accumulation

As we know, *FLS* encodes flavonol synthase that catalyzes the formation of flavonols from dihydroflavonols. Based on the fact that *MtFLS* is likely induced by *MtLHY*, we hypothesize that *MtLHY* plays a crucial role in flavonol biosynthesis. To test this hypothesis, we investigated the flavonol content in WT, *MtLHY*, and *MtLHY*-overexpressing plants. We used diphenylboric acid 2-aminoethyl ester (DPBA) to image flavonol accumulation in leaves, as DPBA is a fluorescent dye that binds specifically to two flavonols, kaempferol and quercetin [39]. Significant DPBA fluorescence signals were detected in the guard cells of *MtLHY*-overexpressing plants (Figure 7A). However, weak DPBA fluorescence signals were observed in both WT and *mtlhy* mutants, with no significant difference (Figure 7A). These data suggest that overexpression of *MtLHY* leads to flavonol accumulation in guard cells. To demonstrate changes in flavonol and flavonoid abundance under salt stress, we performed high-pressure liquid chromatography–mass spectroscopy (LC-MS) to quantify flavonol and total flavonoids concentrations in WT, *mtlhy* mutants, and *MtLHY*-overexpressing plant leaves. We found no difference in the content of total flavonoids between WT and *mtlhy* mutants (Figure 7D). The two flavonol components, kaempferol and quercetin, also showed no difference between WT and *mtlhy* mutants (Figure 7B,C), indicating that the content of both kaempferol and quercetin in *mtlhy* mutants is not the primary factor causing the hypersensitivity phenotype to salt stress. However, the content of total flavonoids, including kaempferol and quercetin, was significantly higher in the *MtLHY*-overexpressing plant compared with WT (Figure 7E–G). Taken together, overexpression of *MtLHY* enhances the accumulation of flavonoids, including kaempferol and quercetin.

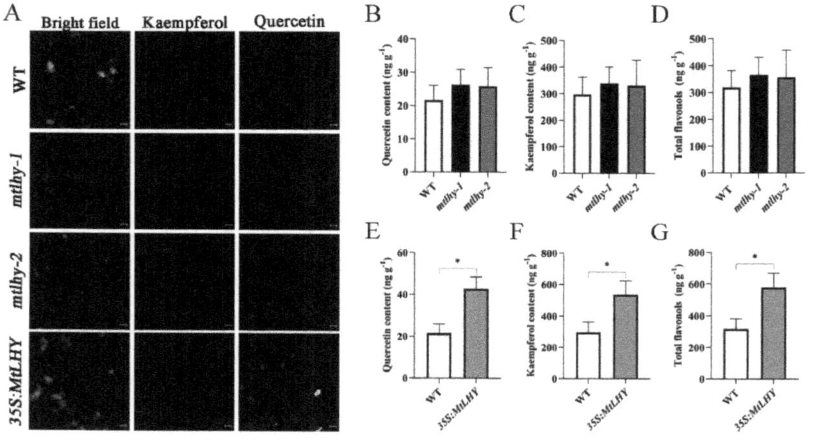


**Figure 7.** *MtLHY* overexpression enhances flavonoid accumulation in leaves: Significant DPBA fluorescence signals were detected in the guard cells of *MtLHY*-overexpressing plants (**A**). The content of quercetin (**B**), kaempferol (**C**), and total flavonoids (**D**) was examined in the leaves of *mtlhy* mutants and WT plants under a 150 mM NaCl treatment for 7 d (**B–D**); data are shown as means ± SD ($n$ = 3; based on Dunnett test). The content of quercetin (**E**), kaempferol (**F**), and total flavonoids (**G**) was examined in the leaves of *MtLHY*-overexpressing plants and WT plants under a 150 mM NaCl treatment for 7 d (**E–G**); data are shown as means ± SD ($n$ = 3; * $p < 0.05$, based on two-tailed $t$-tests). Scale bar, 20 μm.

### 2.7. Application of Exogenous Flavonols Enhances Salt Stress Tolerance in *M. truncatula*

Based on the known ability of flavonols to prevent cellular damage caused by salt stress and the fact that overexpression of *MtLHY* can increase flavonol accumulation, we hypothesize that high levels of flavonols can enhance salt stress tolerance in *M. truncatula*. Previous studies have shown that exogenous flavonoids can confer salt stress tolerance in soybean [33]. To investigate whether flavonols play a similar role in salt stress tolerance in *M. truncatula*, we treated WT plants with exogenous kaempferol and quercetin under salt stress conditions (Figure 8A). After 40 d, we assessed the survival rate of plants with and without the application of exogenous flavonols. Statistical analysis revealed that approximately 56.3% of WT plants survived with the application of exogenous flavonols at 150 mM NaCl, while all plants without the application died (Figure 8B). Therefore, the application of exogenous flavonols enhances salt tolerance in *M. truncatula*.

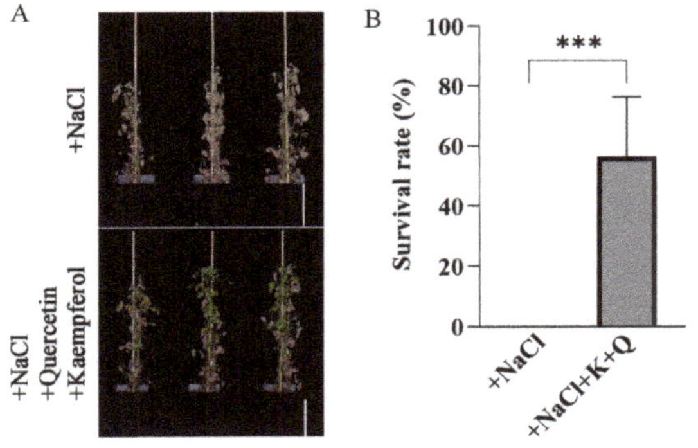


**Figure 8.** Application of exogenous flavonols enhances salt stress tolerance in *M. truncatula*: Application treatment of WT plants with exogenous 15 μm kaempferol and 15 μm quercetin in the Hoagland nutrient solution under 150 mM salt stress conditions. Kaempferol and quercetin were dissolved in ethanol separately, and the control sample was treated with an equal volume of ethanol (**A**). Under salt stress conditions, statistical analysis of the survival rate of the WT plants with or without the exogenous application of kaempferol (K) and quercetin (Q) (**B**). Data are shown as means ± SD ($n$ = 3; *** $p < 0.001$, based on two-tailed $t$-tests). Scale bar, 10 cm.

## *2.8. MtLHY Transcriptionally Activates MtFLS*

As the expression of *MtFLS* was found to be regulated by MtLHY, we hypothesize that *MtFLS* is a potential direct target of MtLHY. To test this hypothesis, we performed a yeast one-hybrid (Y1H) experiment, which demonstrated that the MtLHY protein binds to the promoter sequences of *MtFLS* (Figure 9A). Subsequently, we performed a dual luciferase assay to investigate the effects of MtLHY on the promoter of *MtFLS*. The results indicate that MtLHY increased the expression of LUC driven by the promoter of *MtFLS* (Figure 9B,C). Therefore, MtLHY transcriptionally activates the expression of *MtFLS*.

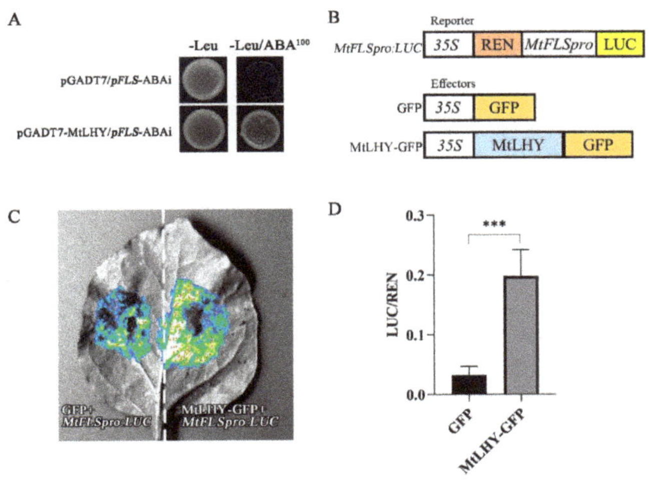


**Figure 9.** MtLHY transcriptionally activates *MtFLS*: A yeast one-hybrid (Y1H) assay showed that the *MtLHY* protein binds to the promoter sequences of *MtFLS* (**A**). Schematic representations of the reporter and effector constructs were used in the transient expression assay (**B**). A dual luciferase assay showed MtLHY increased the expression of LUC driven by the promoter of *MtFLS* (**C**,**D**). Data are shown as means ± SD ($n$ = 4; *** $p < 0.001$, based on two-tailed $t$-tests).

## 3. Discussion

Soil salinization is a global issue that is becoming increasingly severe. Improper fertilization, insufficient irrigation, and the intrusion of seawater all contribute to the accumulation of salt in the soil, which impairs plant growth [40]. Thus, the most effective and ultimate solution to cope with salt stress is to utilize salt-resistant genes to cultivate high-resistance species. In this study, we proposed a working model in which the core component of the circadian clock, *MtLHY*, plays a positive role in salt stress tolerance by regulating ROS homeostasis, $Na^+/K^+$ homeostasis, and flavonoid accumulation (Figure 10). Our findings suggested that the circadian clock may serve as a potential target for improving salt stress tolerance in crops.

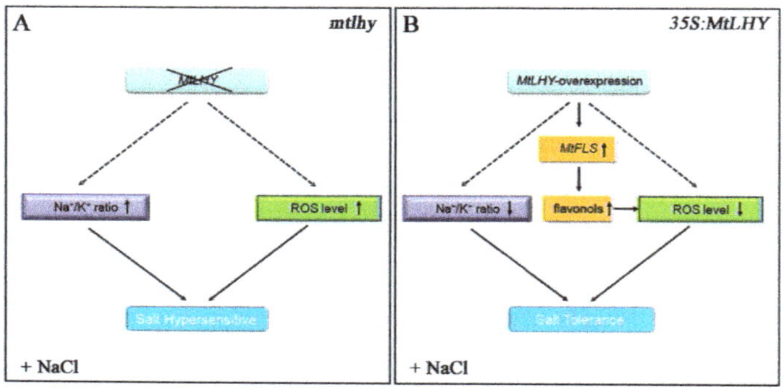


**Figure 10.** A proposed working model for *mtlhy* mutants and *MtLHY*-overexpressing plants under salt stress: A proposed working model for *mtlhy* mutants under salt stress (**A**). A proposed working model for *MtLHY*-overexpressing plants under salt stress (**B**). The dotted arrow means the indirect ation and the solid arrow means the direct ation.

ROS are induced by abiotic stress, and their excessive accumulation leads to membrane oxidation, DNA damage, and even cell death [20–22]. In recent years, it has been shown that *LHY/CCA1* orthologs inhibit the production of ROS [9,41]. In Arabidopsis, overexpression of *CCA1* has been found to enhance tolerance to oxidative stress, while a mutation in *CCA1* results in a hypersensitive phenotype [41]. In rice, loss-of-function mutants of *OsCCA1/OsLHY* accumulate high levels of ROS when subjected to salt treatment [9]. Our study found that *MtLHY*-overexpressing plants have a lower ROS level than WTs under salt stress, while *mtlhy* mutants have a higher ROS level, indicating a conserved role of *LHY/CCA1* orthologs in maintaining ROS homeostasis to cope with abiotic stress. However, the function of *LHY/CCA1* orthologs in abiotic stress has been subdivided and neofunctionalized during seed plant evolution. For example, mutations in two homologous pairs of *LHY*, *GmLHY1a* and *GmLHY1b*, have been found to confer drought tolerance in soybeans [8]. On the other hand, mutations in the other two homologous pairs, *GmLHY2a* and *GmLHY2b*, are still sensitive to drought, indicating the different roles of the involvement of *GmLHY1* and *GmLHY2* in drought tolerance. Interestingly, overexpression of *GmLHY1a* and *GmLHY1b* enhances abscisic acid resistance in both Arabidopsis and soybean, suggesting the conserved role of *GmLHY1a* and *GmLHY1b* in regulating the ABA signaling pathway [8]. Conversely, loss-of-function mutants of *OsCCA1/OsLHY* are hypersensitive to drought stress, indicating that the role of *LHY* orthologs involved in drought stress depends on the specific species. However, the biological functions of *LHY* orthologs involved in salt stress are conserved in rice and *M. truncatula*. Overall, the biological function of *LHY/CCA1* orthologs involved in abiotic stress has been subdivided and neofunctionalized, depending on the specific species, types of abiotic stresses, or their respective downstream targets.

*MtLHY*, a member of the MYB transcription factor family, is postulated to exert its biological functions by modulating the expression of target genes. Our RNA-seq data revealed that *MtLHY* plays a role in regulating genes involved in flavonoid biosynthesis, specifically by directly activating the expression of *MtFLS*, a flavonol synthase gene. Subsequently, overexpression of *MtLHY* resulted in an increased accumulation of flavonoids in *M. truncatula*. Flavonoids, known for their efficient ROS scavenging properties, have been extensively studied in recent years. In plants, flavonoids have been demonstrated to play a protective role against damage caused by biotic and abiotic stresses. We observed a decrease in ROS levels in the *MtLHY*-overexpressing plant under salt treatment, indicating that *MtLHY* overexpression may enhance salt tolerance by increasing the accumulation of flavonoids, particularly kaempferol and quercetin, in leaves. This prediction is supported by the fact that the application of kaempferol and quercetin has been shown to improve plant survival under salt stress. Therefore, it is likely that *MtLHY* overexpression improves salt tolerance, at least in part, through the antioxidant activity of flavonols. In addition, the application of quercetin has a similar effect on soybean under salt stress [33], indicating the critical role of flavonols in enhancing leguminous salt tolerance. Unfortunately, the levels of both kaempferol and quercetin in *mtlhy* mutants have not changed, suggesting that their contents are not responsible for the hypersensitive phenotype of the mutants. Currently, there are two hypotheses proposed to explain this observation. One hypothesis posits that other MtFLS orthologous or homologous proteins have redundant roles in flavonol biosynthesis. In *M. truncatula*, there are four orthologous of MtFLS, with unknown functions. MtFLS orthologous proteins likely play a redundant role in controlling the biosynthesis of kaempferol and quercetin to maintain their levels. The alternative hypothesis suggests that other flavonol compounds may be responsible for the hypersensitive phenotype of mutants. Flavanols, a subgroup of flavonoids, include compounds such as kaempferol, quercetin, fisetin, and myricetin. Other types of flavonols may be responsible for the hypersensitive phenotype of mutants. Furthermore, based on our data, we hypothesize that *MtLHY* contributes to plant salt tolerance by regulating various pathways, including stress-related signaling pathways, scavenging of ROS, and osmotic adjustment. Performing RNA-seq, DNA affinity purification sequencing (DAP-seq), and chromatin immunoprecipitation sequencing (ChIP-seq) experiments under salt stress conditions will enable us to investigate this hypothesis and identify the direct targets of *MtLHY* involved in controlling salt stress response on a genome-wide scale. Additionally, *MtFLS* is directly activated by the expression of *MtLHY*, and the application of exogenous flavonols has been shown to enhance the survival of *Medicago* plants under salt stress, indicating the crucial involvement of flavonol synthase genes in salt stress response. In *Arabidopsis thaliana*, ectopic expression of *DoFLS1*, a flavonol synthase gene from *Dendrobium officinale*, leads to an increase in flavonol production and improved tolerance to abiotic stress [42]. Ectopic expression of *EkFLS* from *Euphorbia kansui Liou* in Arabidopsis promotes flavonoid accumulation, significantly enhancing the activities of POD and SOD, which in turn improves abiotic stress tolerance in plants [42]. Furthermore, overexpression of an *Apocynum venetum* flavonol synthase gene has been found to enhance salinity stress tolerance in transgenic tobacco plants [43]. Therefore, these findings suggest that overexpression of flavonol synthase genes may be a promising strategy for improving plant tolerance to salt stress. Characterization of the loss-of-function *mtfls* mutants and *MtFLS*-overexpression plants under salt stress in *M. truncatula* could provide insights into the functions of *MtFLS* in the response to salt stress.

## 4. Materials and Methods

### 4.1. Plant Materials and Growth Environments

*Medicago truncatula* ecotype R108 was used as the WT plant in this study. The mutant lines of *MtLHY* and *MtLHY*-overexpressing plants were obtained from our previous work [10]. *M. truncatula* seedlings were initially cultured in artificial climate incubators at a 22 °C 16 h (daytime)/8 h (night) photoperiod cycle and 100 $\mu mol \cdot m^{-2} \cdot s^{-1}$ light, under 70 to 80% relative humidity. After two weeks of growth, similar seedlings were transplanted into pots (10 × 10 × 8 cm) filled with soil and Hoagland nutrient solution. The plants were then grown in a greenhouse under environmental conditions of 22 °C 16 h (daytime)/8 h (night) photoperiod cycle, 150 μmol·$m^{-2}$·$s^{-1}$ light, and 70 to 80% relative humidity.

### 4.2. Salt Treatments and Sampling

After being cultivated in the greenhouse for 2 weeks, 4-week-old plants were subjected to a salt stress test by irrigating them with 20 mL of 150 mM NaCl solution every 3 days for 4 weeks. To conduct transcriptional analysis, leaves were collected at 0 h, 3 h, 9 h, 12 h, and 24 h on the first day with or without 150 mM NaCl treatment. For physiological measurements, leaves and roots were sampled at 7th days.

### 4.3. Measurements of Survival Rate and Physiological Index

After 150 mM NaCl treatment, the survival rate statistics of plants were determined according to the performance of leaves. Plants with at least five green trifoliate leaves were considered to have survived. The experiments were performed on at least 3 independent biological replicates, and each replicate included 15 plants. The MDA content was determined as previously reported [44]. $H_2O_2$ accumulation was measured using the Hydrogen Peroxide Assay Kit (S0038, Beyotime, Haimen, China) to determine the $H_2O_2$ content. For $Na^+$ and $K^+$ concentrations, samples were heated at 200 °C for 8 h, and their contents were measured using an inductively coupled plasma optical emission spectrometer (ICAP6300). SOD activity was determined using the Total Superoxide Dismutase Assay Kit with WST-8 (S0101S, Beyotime, China). To measure chlorophyll content, fresh leaves were ground and transferred to an extract solution (80% acetone with 1 μM KOH), and the total chlorophyll content was measured using a UV–Vis spectrophotometer.

### 4.4. Histochemical Detection of ROS

The root samples were subjected to DAB staining by immersing them in a 1 mg/mL DAB solution for 8 h at 22 °C in the dark. After that, the samples were treated with 95% ethanol and boiled to remove excess stains. Similarly, for NBT staining, the root samples were immersed in NBT stain solution for 3 h at 22 °C in the dark, followed by treatment with 95% ethanol and boiling to decolorize. The samples were then cooled and transferred to 75% ethanol before being observed under a fluorescence microscope (Olympus, Tokyo, Japan).

### 4.5. RNA Extraction and RT-qPCR Analysis

Total RNA was extracted from leaves using RNAiso Plus (TaKaRa, Tokyo, Japan). The plant materials were ground into a fine powder using the Tissuelyser-48 (Shanghai Jingxin, Shanghai, China). Three biological samples were collected, and RT-qPCR analysis was conducted as described previously [45]. The UBQ gene was utilized as the internal reference, and all primers used are listed in Supplemental Table S1.

### 4.6. DPBA Staining

The samples were stained with DPBA and gently rotated for 5 min, followed by careful washing with $H_2O$ to remove the DPBA. The fluorescence signal was immediately detected using the LSM880 confocal laser scanning microscope (Zeiss, Oberkochen, Germany). The emission spectrum for Kaempferol-DPBA was set to 475–500 nm, while for Quercetin-DPBA, it was set to 585–619 nm.

### 4.7. Flavonoid Content

The flavonoid content was evaluated through the use of liquid chromatography–tandem mass spectrometry (LC-MS), following previously established methods [46]. The relative content of flavonol was quantified with daidzin as an external standard, and the unit of measurement for the relative content is presented as ng/g dry weight. Three biological replicates were obtained for LC-MS analysis.

### 4.8. Luciferase Imaging Assay

The full-length coding sequence (CDS) of *MtLHY* was amplified by PCR and fused to pEarleyGate 103 plasmids as the effector. To create the reporter construct, a promoter fragment of approximately 2.5 kb upstream of *MtFLS* was PCR-amplified and cloned into the pGreenII-0800-LUC vector. These constructs were then introduced into A. tumefaciens GV3101, along with the pSoup19 helper plasmid, and coexpressed in 4-week-old N. benthamiana leaves. After 48 h, the infiltrated tobacco leaves were harvested and luciferase fluorescence signals were examined using a plant living imaging system (Tanon 5200, Shanghai, China) after being sprayed with Luciferin. The remaining leaves were then punched and powdered in liquid nitrogen to measure LUC and REN activity using a dual-luciferase reporter (DLR) assay system (Promega, Madison, WI, USA) on the Centro XS LB960 (Berthold, Schwarzwald, Germany). The ratio of LUC to REN (LUC/REN) was used to demonstrate the activity of transactivation. The primer sequences are listed in Supplemental Table S1.

### 4.9. Yeast One-Hybrid (Y1H) Assay

The yeast one-hybrid assays were conducted using the Matchmaker Gold Yeast One-Hybrid System (Clontech, Mountain View, CA, USA). The full-length CDS of *MtLHY* was amplified by PCR and fused to the pGADT7 vector to create the pGADT7-*MtLHY* prey. The DNA fragment from the *MtFLS* promoter was amplified and inserted into the pAbAi vector to generate the pFLS-pAbAi bait. The pFLS-pAbAi vector was linearized and cotransformed with pGADT7-*MtLHY* into the Y1H Gold yeast strain. Transformants were selected on SD/-Leu/-Ura/AbA media using aureobasidin A (AbA). The primer sequences can be found in Supplemental Table S1.

### 4.10. Statistical Analysis

Most of the pairwise comparisons between the means were performed using a two-sided Student's *t*-test method (* $p < 0.05$, ** $p < 0.01$, *** $p < 0.001$) with GraphPad Prism version 9.0 software. The statistical comparison of various experimental groups and the control was conducted by a one-way ANOVA tool based on the Dunnett test (* $p < 0.05$, ** $p < 0.01$, *** $p < 0.001$).

## 5. Conclusions

In this study, our data indicate that the core circadian clock component, *MtLHY*, plays a positive role in the response to salt stress in *M. truncatula*. Our results demonstrate that loss-of-function mutants of *MtLHY* were more sensitive to salt stress treatment, while *MtLHY*-overexpressing lines exhibited increased salt stress tolerance. Based on the physiological indices obtained under salt treatment, it is likely that *MtLHY* plays a crucial role in the response to salt stress by regulating ROS and $Na^+/K^+$ homeostasis. Additionally, overexpression of *MtLHY* enhances salt stress tolerance by increasing flavonol accumulation through the regulation of flavonol synthase gene expression. Furthermore, the application of exogenous flavonols improved salt stress tolerance in *M. truncatula*. Therefore, our study has identified the important roles of *MtLHY* in the response to salt stress, which links the circadian clock with flavonoid biosynthesis in *M. truncatula*.

**Supplementary Materials:** The supporting information can be downloaded at: https://www.mdpi.com/article/10.3390/ijms24129948/s1.

**Author Contributions:** Z.L., H.L., Y.Z. and H.L. designed the research; Z.L., H.L., Y.K. and L.W. performed the research and analyzed data; Z.L., C.Z. and L.H. wrote the paper. All authors have read and agreed to the published version of the manuscript.

**Funding:** This work was supported by grants from the National Natural Science Foundation of China (U1906201 and 32100284) and Shandong Province (ZR2020KC018, ZR2020QC035 and ZR202103010073).

**Institutional Review Board Statement:** Not applicable.

**Informed Consent Statement:** Not applicable.

**Data Availability Statement:** The data presented in this study are available in supplementary material here.

**Acknowledgments:** We thank Haiyan Yu, Yuyu Guo, and Xiaomin Zhao from the Analysis and Testing Center of SKLMT (State Key Laboratory of Microbial Technology, Shandong University) for assistance with the laser scanning confocal microscopy.

**Conflicts of Interest:** The authors declare no conflict of interest.

## References

1. Imaizumi, T. Arabidopsis circadian clock and photoperiodism: Time to think about location. *Curr. Opin. Plant Biol.* **2010**, *13*, 83–89. [CrossRef] [PubMed]
2. Martin, L.B.B.; Sherwood, R.W.; Nicklay, J.J.; Yang, Y.; Muratore-Schroeder, T.L.; Anderson, E.T.; Thannhauser, T.W.; Rose, J.K.C.; Zhang, S. Application of wide selected-ion monitoring data-independent acquisition to identify tomato fruit proteins regulated by the CUTIN DEFICIENT2 transcription factor. *Proteomics* **2016**, *16*, 2081–2094. [CrossRef] [PubMed]
3. Pokhilko, A.; Fernandez, A.P.; Edwards, K.D.; Southern, M.M.; Halliday, K.J.; Millar, A.J. The clock gene circuit in Arabidopsis includes a repressilator with additional feedback loops. *Mol. Syst. Biol.* **2012**, *8*, 574. [CrossRef] [PubMed]
4. Sanchez, S.E.; Kay, S.A. The Plant Circadian Clock: From a Simple Timekeeper to a Complex Developmental Manager. *Csh. Perspect. Biol.* **2016**, *8*, a027748. [CrossRef]
5. Kim, W.Y.; Ali, Z.; Park, H.J.; Park, S.J.; Cha, J.Y.; Perez-Hormaeche, J.; Quintero, F.J.; Shin, G.; Kim, M.R.; Qiang, Z.; et al. Release of SOS2 kinase from sequestration with GIGANTEA determines salt tolerance in Arabidopsis (vol 4, 1352, 2013). *Nat. Commun.* **2013**, *4*, 1352. [CrossRef]
6. Sakuraba, Y.; Bulbul, S.; Piao, W.L.; Choi, G.; Paek, N.C. Arabidopsis EARLY FLOWERING3 increases salt tolerance by suppressing salt stress response pathways. *Plant J.* **2017**, *92*, 1106–1120. [CrossRef]
7. Dong, M.A.; Farre, E.M.; Thomashow, M.F. CIRCADIAN CLOCK-ASSOCIATED 1 and LATE ELONGATED HYPOCOTYL regulate expression of the C-REPEAT BINDING FACTOR (CBF) pathway in Arabidopsis. *Proc. Natl. Acad. Sci. USA* **2011**, *108*, 7241–7246. [CrossRef]
8. Wang, K.; Bu, T.T.; Cheng, Q.; Dong, L.D.; Su, T.; Chen, Z.M.; Kong, F.J.; Gong, Z.Z.; Liu, B.H.; Li, M.N. Two homologous LHY pairs negatively control soybean drought tolerance by repressing the abscisic acid responses. *New Phytol.* **2021**, *229*, 2660–2675. [CrossRef]
9. Wei, H.; Xu, H.; Su, C.; Wang, X.L.; Wang, L. Rice CIRCADIAN CLOCK ASSOCIATED 1 transcriptionally regulates ABA signaling to confer multiple abiotic stress tolerance. *Plant Physiol.* **2022**, *190*, 1057–1073. [CrossRef]
10. Kong, Y.; Han, L.; Liu, X.; Wang, H.; Wen, L.; Yu, X.; Xu, X.; Kong, F.; Fu, C.; Mysore, K.S. The nodulation and nyctinastic leaf movement is orchestrated by clock gene LHY in Medicago truncatula. *J. Integr. Plant Biol.* **2020**, *62*, 1880–1895. [CrossRef]
11. Godfray, H.C.J.; Beddington, J.R.; Crute, I.R.; Haddad, L.; Lawrence, D.; Muir, J.F.; Pretty, J.; Robinson, S.; Thomas, S.M.; Toulmin, C. Food Security: The Challenge of Feeding 9 Billion People. *Science* **2010**, *327*, 812–818. [CrossRef] [PubMed]
12. Tester, M.; Langridge, P. Breeding technologies to increase crop production in a changing world. *Science* **2010**, *327*, 818–822. [CrossRef] [PubMed]
13. Agarwal, P.K.; Shukla, P.S.; Gupta, K.; Jha, B. Bioengineering for salinity tolerance in plants: State of the art. *Mol. Biotechnol.* **2013**, *54*, 102–123. [CrossRef] [PubMed]
14. Munns, R.; Tester, M. Mechanisms of salinity tolerance. *Annu. Rev. Plant Biol.* **2008**, *59*, 651–681. [CrossRef] [PubMed]
15. Jaleel, C.A.; Riadh, K.; Gopi, R.; Manivannan, P.; Ines, J.; Al-Juburi, H.; Zhao, C.X.; Shao, H.B.; Panneerselvam, R. Antioxidant defense responses: Physiological plasticity in higher plants under abiotic constraints. *Acta Physiol. Plant* **2009**, *31*, 427–436. [CrossRef]
16. Arbona, V.; Manzi, M.; de Ollas, C.; Gomez-Cadenas, A. Metabolomics as a Tool to Investigate Abiotic Stress Tolerance in Plants. *Int. J. Mol. Sci.* **2013**, *14*, 4885–4911. [CrossRef]
17. Zhang, B.G.; Liu, K.D.; Zheng, Y.; Wang, Y.X.; Wang, J.X.; Liao, H. Disruption of AtWNK8 Enhances Tolerance of Arabidopsis to Salt and Osmotic Stresses via Modulating Proline Content and Activities of Catalase and Peroxidase. *Int. J. Mol. Sci.* **2013**, *14*, 7032–7047. [CrossRef]
18. Tang, X.L.; Mu, X.M.; Shao, H.B.; Wang, H.Y.; Brestic, M. Global plant-responding mechanisms to salt stress: Physiological and molecular levels and implications in biotechnology. *Crit. Rev. Biotechnol.* **2015**, *35*, 425–437. [CrossRef]
19. Fu, L.B.; Shen, Q.F.; Kuang, L.H.; Yu, J.H.; Wu, D.Z.; Zhang, G.P. Metabolite profiling and gene expression of Na/K transporter analyses reveal mechanisms of the difference in salt tolerance between barley and rice. *Plant Physiol. Bioch.* **2018**, *130*, 248–257. [CrossRef]
20. Gill, S.S.; Tuteja, N. Reactive oxygen species and antioxidant machinery in abiotic stress tolerance in crop plants. *Plant Physiol. Bioch.* **2010**, *48*, 909–930. [CrossRef]
21. Mittler, R. Oxidative stress, antioxidants and stress tolerance. *Trends Plant Sci.* **2002**, *7*, 405–410. [CrossRef] [PubMed]

22. Vranova, E.; Inze, D.; Van Breusegem, F. Signal transduction during oxidative stress. *J. Exp. Bot.* **2002**, *53*, 1227–1236. [CrossRef] [PubMed]
23. Das, K.; Roychoudhury, A. Reactive oxygen species (ROS) and response of antioxidants as ROS-scavengers during environmental stress in plants. *Front. Environ. Sci.* **2014**, *2*, 53. [CrossRef]
24. Noctor, G.; Mhamdi, A.; Foyer, C.H. The roles of reactive oxygen metabolism in drought: Not so cut and dried. *Plant Physiol.* **2014**, *164*, 1636–1648. [CrossRef]
25. Winkel-Shirley, B. Flavonoid biosynthesis. A colorful model for genetics, biochemistry, cell biology, and biotechnology. *Plant Physiol.* **2001**, *126*, 485–493. [CrossRef]
26. Winkel-Shirley, B. Biosynthesis of flavonoids and effects of stress. *Curr. Opin. Plant Biol.* **2002**, *5*, 218–223. [CrossRef]
27. Panche, A.N.; Diwan, A.D.; Chandra, S.R. Flavonoids: An overview. *J. Nutr. Sci.* **2016**, *5*, e47. [CrossRef]
28. Ahmed, N.U.; Park, J.I.; Jung, H.J.; Hur, Y.; Nou, I.S. Anthocyanin biosynthesis for cold and freezing stress tolerance and desirable color in Brassica rapa. *Funct. Integr. Genom.* **2015**, *15*, 383–394. [CrossRef]
29. Schulz, E.; Tohge, T.; Zuther, E.; Fernie, A.R.; Hincha, D.K. Flavonoids are determinants of freezing tolerance and cold acclimation in Arabidopsis thaliana. *Sci. Rep.* **2016**, *6*, 34027. [CrossRef]
30. Shomali, A.; Das, S.; Arif, N.; Sarraf, M.; Zahra, N.; Yadav, V.; Aliniaeifard, S.; Chauhan, D.K.; Hasanuzzaman, M. Diverse Physiological Roles of Flavonoids in Plant Environmental Stress Responses and Tolerance. *Plants* **2022**, *11*, 3158. [CrossRef]
31. Meng, D.; Dong, B.Y.; Niu, L.L.; Song, Z.H.; Wang, L.T.; Amin, R.; Cao, H.Y.; Li, H.H.; Yang, Q.; Fu, Y.J. The pigeon pea CcCIPK14-CcCBL1 pair positively modulates drought tolerance by enhancing flavonoid biosynthesis. *Plant J.* **2021**, *106*, 1278–1297. [CrossRef] [PubMed]
32. Yan, J.H.; Wang, B.A.; Jiang, Y.N.; Cheng, L.J.; Wu, T.L. GmFNSII-Controlled Soybean Flavone Metabolism Responds to Abiotic Stresses and Regulates Plant Salt Tolerance. *Plant Cell Physiol.* **2014**, *55*, 74–86. [CrossRef] [PubMed]
33. Bian, X.H.; Li, W.; Niu, C.F.; Wei, W.; Hu, Y.; Han, J.Q.; Lu, X.; Tao, J.J.; Jin, M.; Qin, H.; et al. A class B heat shock factor selected for during soybean domestication contributes to salt tolerance by promoting flavonoid biosynthesis. *New Phytol.* **2020**, *225*, 268–283. [CrossRef] [PubMed]
34. Katembe, W.J.; Ungar, I.A.; Mitchell, J.P. Effect of salinity on germination and seedling growth of twoAtriplexspecies (Chenopodiaceae). *Ann. Bot.* **1998**, *82*, 167–175. [CrossRef]
35. Bernstein, N.; Kafkafi, U. Root growth under salinity stress. In *Plant Roots*; CRC Press: Boca Raton, FL, USA, 2002; pp. 1222–1250.
36. Hodges, D.M.; DeLong, J.M.; Forney, C.F.; Prange, R.K. Improving the thiobarbituric acid-reactive-substances assay for estimating lipid peroxidation in plant tissues containing anthocyanin and other interfering compounds. *Planta* **1999**, *207*, 604–611. [CrossRef]
37. Sunkar, R.; Bartels, D.; Kirch, H.H. Overexpression of a stress-inducible aldehyde dehydrogenase gene from Arabidopsis thaliana in transgenic plants improves stress tolerance. *Plant J.* **2003**, *35*, 452–464. [CrossRef]
38. Zheng, M.; Lin, J.; Liu, X.; Chu, W.; Li, J.; Gao, Y.; An, K.; Song, W.; Xin, M.; Yao, Y. Histone acetyltransferase TaHAG1 acts as a crucial regulator to strengthen salt tolerance of hexaploid wheat. *Plant Physiol.* **2021**, *186*, 1951–1969. [CrossRef]
39. Watkins, J.M.; Chapman, J.M.; Muday, G.K. Abscisic Acid-Induced Reactive Oxygen Species Are Modulated by Flavonols to Control Stomata Aperture. *Plant Physiol.* **2017**, *175*, 1807–1825. [CrossRef]
40. Wang, X.; He, Y.; Wei, H.; Wang, L. A clock regulatory module is required for salt tolerance and control of heading date in rice. *Plant Cell Environ.* **2021**, *44*, 3283–3301. [CrossRef]
41. Lai, A.G.; Doherty, C.J.; Mueller-Roeber, B.; Kay, S.A.; Schippers, J.H.M.; Dijkwel, P.P. CIRCADIAN CLOCK-ASSOCIATED 1 regulates ROS homeostasis and oxidative stress responses. *Proc. Natl. Acad. Sci. USA* **2012**, *109*, 17129–17134. [CrossRef]
42. Wang, M.; Zhang, Y.; Zhu, C.Y.; Yao, X.Y.; Zheng, Z.; Tian, Z.N.; Cai, X. EkFLS overexpression promotes flavonoid accumulation and abiotic stress tolerance in plant. *Physiol. Plant.* **2021**, *172*, 1966–1982. [CrossRef] [PubMed]
43. Wang, M.; Ren, T.T.; Huang, R.H.; Li, Y.Q.; Zhang, C.S.; Xu, Z.C. Overexpression of an Apocynum venetum flavonols synthetase gene confers salinity stress tolerance to transgenic tobacco plants. *Plant Physiol. Bioch.* **2021**, *162*, 667–676. [CrossRef] [PubMed]
44. Zhang, Z.; Huang, R. Analysis of malondialdehyde, chlorophyll proline, soluble sugar, and glutathione content in Arabidopsis seedling. *Bio-Protocol* **2013**, *3*, e817. [CrossRef]
45. Wang, H.; Lu, Z.; Xu, Y.; Zhang, J.; Han, L.; Chai, M.; Wang, Z.Y.; Yang, X.; Lu, S.; Tong, J.; et al. Roles of very long-chain fatty acids in compound leaf patterning in Medicago truncatula. *Plant Physiol.* **2023**, *191*, 1751–1770. [CrossRef] [PubMed]
46. Chang, Y.W.; Zhao, C.X.; Wu, Z.M.; Zhou, J.; Zhao, S.M.; Lu, X.; Xu, G.W. Chip-based nanoflow high performance liquid chromatography coupled to mass spectrometry for profiling of soybean flavonoids. *Electrophoresis* **2012**, *33*, 2399–2406. [CrossRef] [PubMed]

International Journal of
*Molecular Sciences*

MDPI

*Article*

# Overexpression of Water-Responsive Genes Promoted by Elevated $CO_2$ Reduces ROS and Enhances Drought Tolerance in *Coffea* Species

**Isabel Marques [1,2,†], Isabel Fernandes [3,†], Octávio S. Paulo [3], Dora Batista [2,4], Fernando C. Lidon [5], Fábio Partelli [6], Fábio M. DaMatta [7], Ana I. Ribeiro-Barros [1,2,5,*] and José C. Ramalho [1,2,5,*]**

1 Plant-Environment Interactions and Biodiversity Lab (PlantStress & Biodiversity), Forest Research Centre (CEF), Instituto Superior de Agronomia (ISA), Universidade de Lisboa, 1349-017 Lisboa, Portugal
2 Associate Laboratory TERRA, Instituto Superior de Agronomia (ISA), Universidade de Lisboa, 1349-017 Lisboa, Portugal
3 cE3c—Center for Ecology, Evolution and Environmental Changes and CHANGE—Global Change and Sustainability Institute, Faculdade de Ciências, Universidade de Lisboa, 1749-016 Lisboa, Portugal
4 Linking Landscape, Environment, Agriculture and Food (LEAF), Instituto Superior de Agronomia (ISA), Universidade de Lisboa, 1349-017 Lisboa, Portugal
5 Unidade de Geobiociências, Geoengenharias e Geotecnologias (GeoBioTec), Faculdade de Ciências e Tecnologia (FCT), Universidade NOVA de Lisboa (UNL), 2829-516 Caparica, Portugal
6 Centro Universitário do Norte do Espírito Santo (CEUNES), Departmento Ciências Agrárias e Biológicas (DCAB), Universidade Federal Espírito Santo (UFES), São Mateus 29932-540, ES, Brazil
7 Departamento de Biologia Vegetal, Universidade Federal Viçosa (UFV), Viçosa 36570-900, MG, Brazil
* Correspondence: aribeiro@isa.ulisboa.pt (A.I.R.-B.); cochichor@isa.ulisboa.pt or cochichor@mail.telepac.pt (J.C.R.)
† These authors contributed equally to this work.

**Abstract:** Drought is a major constraint to plant growth and productivity worldwide and will aggravate as water availability becomes scarcer. Although elevated air [$CO_2$] might mitigate some of these effects in plants, the mechanisms underlying the involved responses are poorly understood in woody economically important crops such as *Coffea*. This study analyzed transcriptome changes in *Coffea canephora* cv. CL153 and *C. arabica* cv. Icatu exposed to moderate (MWD) or severe water deficits (SWD) and grown under ambient (a$CO_2$) or elevated (e$CO_2$) air [$CO_2$]. We found that changes in expression levels and regulatory pathways were barely affected by MWD, while the SWD condition led to a down-regulation of most differentially expressed genes (DEGs). e$CO_2$ attenuated the impacts of drought in the transcripts of both genotypes but mostly in Icatu, in agreement with physiological and metabolic studies. A predominance of protective and reactive oxygen species (ROS)-scavenging-related genes, directly or indirectly associated with ABA signaling pathways, was found in *Coffea* responses, including genes involved in water deprivation and desiccation, such as protein phosphatases in Icatu, and aspartic proteases and dehydrins in CL153, whose expression was validated by qRT-PCR. The existence of a complex post-transcriptional regulatory mechanism appears to occur in *Coffea* explaining some apparent discrepancies between transcriptomic, proteomic, and physiological data in these genotypes.

**Keywords:** ABA signaling; coffee; functional analysis; ROS; stress; tolerance

**Citation:** Marques, I.; Fernandes, I.; Paulo, O.S.; Batista, D.; Lidon, F.C.; Partelli, F.; DaMatta, F.M.; Ribeiro-Barros, A.I.; Ramalho, J.C. Overexpression of Water-Responsive Genes Promoted by Elevated $CO_2$ Reduces ROS and Enhances Drought Tolerance in *Coffea* Species. *Int. J. Mol. Sci.* **2023**, *24*, 3210. https://doi.org/10.3390/ijms24043210

Academic Editor: Abir U. Igamberdiev

Received: 24 November 2022
Revised: 23 January 2023
Accepted: 29 January 2023
Published: 6 February 2023



## 1. Introduction

Drought events have become more frequent, severe, and erratic nowadays, affecting the quality and yield of most crops [1,2]. Under the initial stages of drought, stomatal closure usually occurs to reduce the loss of water through transpiration, but at the same time, decreasing the entrance of $CO_2$ into leaves, limiting photosynthesis, and ultimately plant growth [3]. With an increase in the severity of drought, the functioning of photosynthesis can be further impaired by photochemical and biochemical dysfunctions occurring in pigments, photosystems performance, enzyme activities (namely RuBisCO), and cell membrane integrity [4–8]. The reduction in the photochemical energy use will also impose a secondary stress associated with the generation of reactive oxygen species (ROS) that oxidize, impair, and damage multiple cellular components, ultimately causing cell death [9].

In some C3 plants, the harsh effects of drought are sometimes counteracted by elevated [$CO_2$] ($eCO_2$), associated with a direct stimulation of photosynthesis, a reduction in stomatal conductance [10,11], and the strengthening of defense mechanisms and photosynthetic components, which altogether contribute to the preservation of the photosynthetic performance [12]. However, the positive effects of $eCO_2$ in attenuating the impact of drought depend on the stress severity and its duration, and also on the species/genotypes involved [13]. For instance, $eCO_2$ reduced the negative effects that drought imposed on the quality of sorghum grains by delaying physiological and metabolic responses to this stress [14]. In contrast, in soybean, $eCO_2$ did not counteract the negative impacts of drought on photosynthesis and yield, and the minor benefits that were initially observed were progressively lost with increasing drought severity [15]. This highlights the potential for complex interactions among the abiotic factors of global change, which have been poorly investigated in woody plants, despite the urgency to develop adaptive strategies considering the future climate conditions.

Coffee is one of the most important agricultural commodities worldwide, generating about USD 200.0000 million [16], and constituting a crucial source of income for 20–25 million smallholder farmers, which are mostly based in the tropical region [17,18]. Despite the recognized resilience and metabolic flexibility of some coffee genotypes to environmentally stressful conditions, adverse temperatures and limited water availability are the major causes of crop failure, affecting the yield and quality of coffee beans, promoting livelihood insecurity, and constraining the value chain of coffee [19,20]. *Coffea arabica* is native to Ethiopian tropical forests at altitudes of 1600–2800 m, with an annual average of about 20 °C, whereas *C. canephora* is native to the lowland forests of the Congo River basin, growing from sea level up to 1200 m and under average temperatures between 24 and 26 °C, although without large oscillations. Currently, the optimum mean annual temperature range for arabica coffee is considered to be between 18 and 21 °C, although elite cultivars under intensive management allow the spread of arabica coffee to regions with average temperatures as high as 24–25 °C. *Coffea canephora* can grow under higher temperatures with optimum annual mean temperatures ranging from 22 to 30 °C, depending on authors [18]. Despite drought is a concern for the crop, some genotypes can maintain high photosynthetic rates, especially under $eCO_2$ conditions [12,21], reducing physiological constraints imposed by drought (e.g., overcoming diffusional $CO_2$ limitations due to stomatal closure), and reinforcing some defense mechanisms, contributing to maintaining photosynthetic performance and, likely, crop yield, at least under moderate levels of drought [12,17,22–24].

Here, we explore the underlying transcriptomic mechanisms by which coffee genotypes adjust to increasing drought severity and how $eCO_2$ can modify such adjustments. Based on the fact that $eCO_2$ improves resilience to drought stress at the physiological and biochemical levels [12], we hypothesized that $eCO_2$ interacts at the transcriptomic level to promote a greater metabolic performance, and acclimation mechanisms, namely at the photosynthetic level. To test these hypotheses, we assessed the impacts of drought on two genotypes from the most important coffee-traded species, *Coffea canephora* Pierre ex A. Froehner cv. Conilon Clone 153 (CL153) and *C. arabica* L. cv. Icatu Vermelho (Icatu), grown under ambient ($aCO_2$) or elevated ($eCO_2$) air [$CO_2$], and gradually subjected to moderate water deficit (MWD) or severe water deficit (SWD) conditions in comparison with well-watered (WW) plants. CL153 is a late maturation/ripening diploid clonal variety created from Emcapa 8131 (also known as Vitória 13) by Incaper (Vitória, ES, Brazil) that already showed some relevant drought tolerance while Icatu is an introgressed tetraploid variety originated from a cross between *C. canephora* and *C. arabica* cv. Bourbon Vermelho that was further crossed with *C. arabica* cv. Mundo Novo by IAC. The two genotypes display a relevant response ability to drought, although having different degrees of resilience as CL153 suffers a higher negative impact than Icatu in the photochemical and biochemical components of C-assimilation under severe drought [12,25]. Here, we explore the molecular mechanisms beyond such striking physiological and biochemical differences. Understanding the molecular mechanisms that ultimately determine the response of *Coffea* to climatic changes is crucial to mitigate their harmful effects and establish better scenarios to maintain the sustainability of the value chain of coffee.

## 2. Results

### 2.1. Overview of the RNA-Seq Data from the Two Coffee Genotypes

RNA sequencing resulted in an average of 21.8 million paired-end reads per sample, generating an average of 16.9 and 19.3 million high-quality unique reads in Icatu and CL153, respectively (Table S2). A high proportion of reads was mapped to the corresponding reference genome since only an average of 16% and 14% of reads from Icatu and CL153, respectively, were not mapped. Statistical details for each replicate are depicted in Table S2.

The number of expressed genes varied in Icatu from 29199 (SWD-$aCO_2$) to 31041 (WW-$eCO_2$), while much lower values were found in CL153, ranging from 19558 (SWD-$eCO_2$) to 20463 (WW-$aCO_2$) (Figure 1). A principal component analysis (PCA) based on gene expression generally clustered the samples according to the different water treatments (Figure S1). In addition, MWD plants were usually closer to WW plants under $eCO_2$ but with SWD plants under $aCO_2$ conditions, especially when considering Icatu plants (Figure S1).

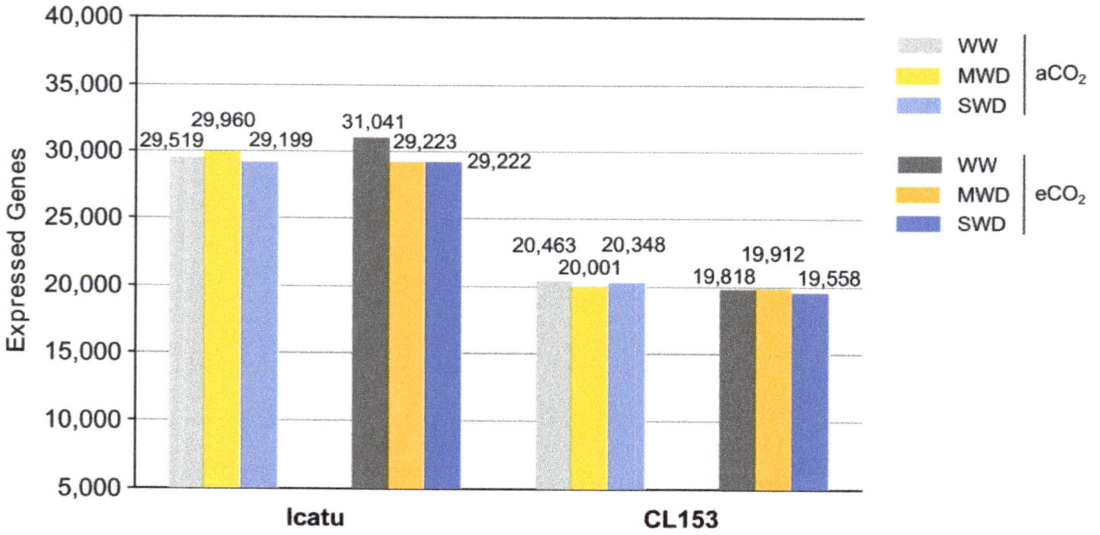


**Figure 1.** Total number of expressed genes in Icatu and CL153 plants grown under different watering conditions (well-watered, WW; moderate water deficit, MWD; and severe water deficit, SWD), and under ambient air 380 μL $L^{-1}$ [$CO_2$] ($aCO_2$) or elevated 700 μL $L^{-1}$ [$CO_2$] ($eCO_2$), at 25/20 °C.

### 2.2. Response of Differentially Expressed Genes (DEGs) to Drought and $eCO_2$

The impact of MWD on the total number of DEGs was minimal under $eCO_2$ in comparison with plants under $aCO_2$ (Figure 2A,B). The number of DEGs commonly triggered by both water deficits was much lower under $eCO_2$ than under $aCO_2$. In fact, under MWD, a high number of DEGs were up-regulated under $eCO_2$ plants when compared to their $aCO_2$ counterparts, especially in Icatu.

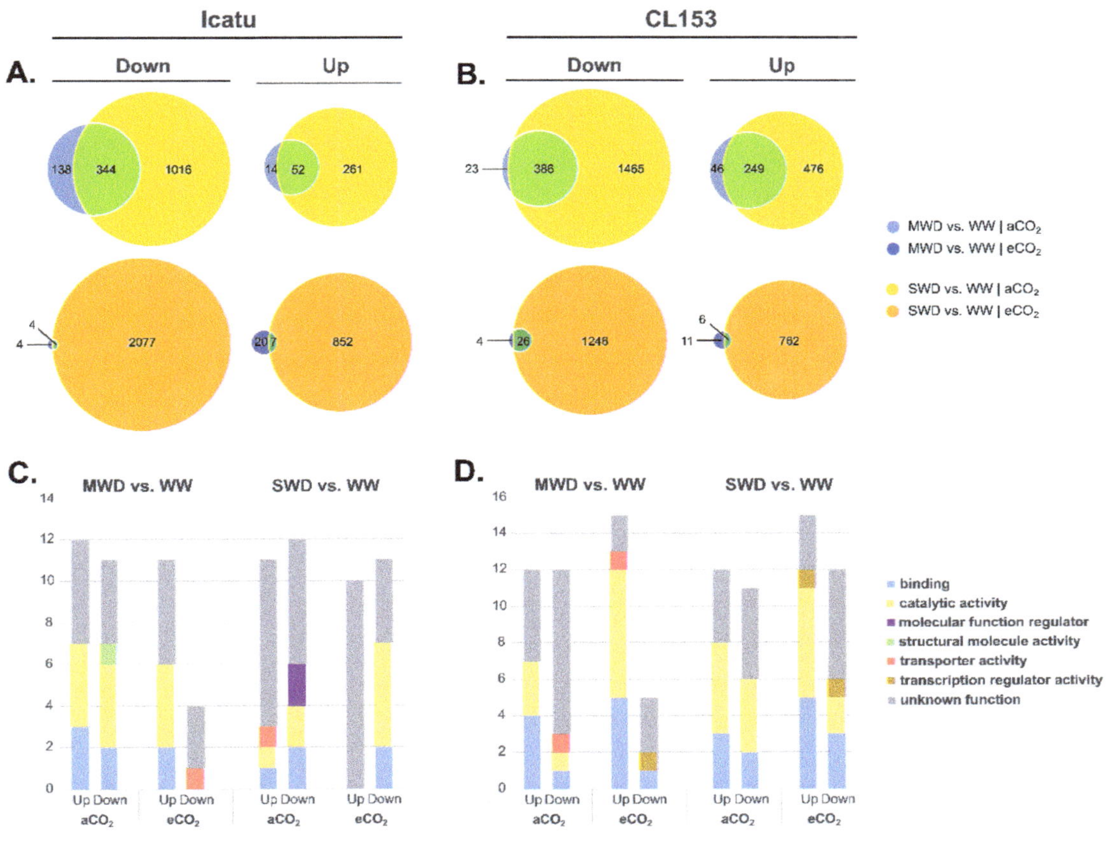


**Figure 2.** Patterns of differentially expressed genes (DEGs) at MWD or SWD in comparison with WW in (**A**) Icatu and (**B**) CL153 plants grown under either $aCO_2$ (light colors) or $eCO_2$ (dark colors), at 25/20 °C (day/night). Blue: DEGs specifically found under MWD. Yellow: DEGs specifically found under SWD. Green: DEGs expressed by both water conditions. Gene ontology (GO) terms found among the top up- and down-regulated differentially expressed genes (DEGs) in (**C**) Icatu and (**D**) CL153 plants. GO terms were selected according to UniProtKB and QuickGO databases.

The harshest drought level triggered a high down-regulation of DEGs, in the two genotypes, independently of [$CO_2$] levels (Figure 2A,B; Table S3). The strong impact of the SWD was also depicted on heatmaps visualizing the expression profile of all significant DEGs, where a high degree of variation was found under this harsher drought level in contrast with MWD, which promoted only minor effects (Figure S2). The full list of DEGs can be found in Tables S4 and S5.

The majority of the up- and down-regulated DEGs found had no annotated functions (Figure 2C,D). The remaining ones were mostly involved in 'catalytic activities', followed by 'binding' in the two genotypes regardless of [$CO_2$] levels (Figure 2C,D). Under $aCO_2$, MWD triggered the down-regulation of DEGs involved in 'structural molecule activity' in Icatu. Instead, in CL153 'transporter activity' DEGs were down-regulated under $aCO_2$ but up-regulated under $eCO_2$. Still, under MWD and $eCO_2$ there was a down-regulation of DEGs involved in 'transporter activity' in Icatu and 'transcription regulator activity' in CL153.

Under harsher drought conditions (SWD), Icatu $aCO_2$-plants showed an up-regulation of DEGs involved in 'transporter activity', 'catalytic activity', and 'binding', but the latter two categories were involved in down-regulated DEG, together with molecular function regulators (Figure 2C,D). CL153 plants under SWD and $aCO_2$ showed a high number of up- and down-regulated DEGs associated with catalytic activity and binding functions. Under $eCO_2$, while Icatu plants showed a down-regulation of DEGs involved in catalytic activities and binding, CL153 plants rather showed up- and down-regulated DEGs involved in these two categories, as well as in transcription regulator activities.

### 2.3. Drought and $eCO_2$ Impact on DEGs Associated with Specific Biochemical Pathways

Drought had a relevant impact on a high number of DEGs involved in respiration, antioxidant, and lipid biochemical pathways, especially in Icatu where a total of 342 DEGs (respiration: 168, antioxidant: 106, lipid metabolism: 68 DEGs; Table S6) were found to be affected, in comparison with 91 in CL153 (respiration: 31, antioxidant: 29, lipid metabolism: 31 DEGs; Table S7). Under MWD and $aCO_2$, DEGs associated with these biochemical pathways were mostly down-regulated but they were substantially reduced under $eCO_2$ (Figure 3). The positive effect of $eCO_2$ was even more relevant under SWD in Icatu plants, which showed an increase in up-regulated DEGs.

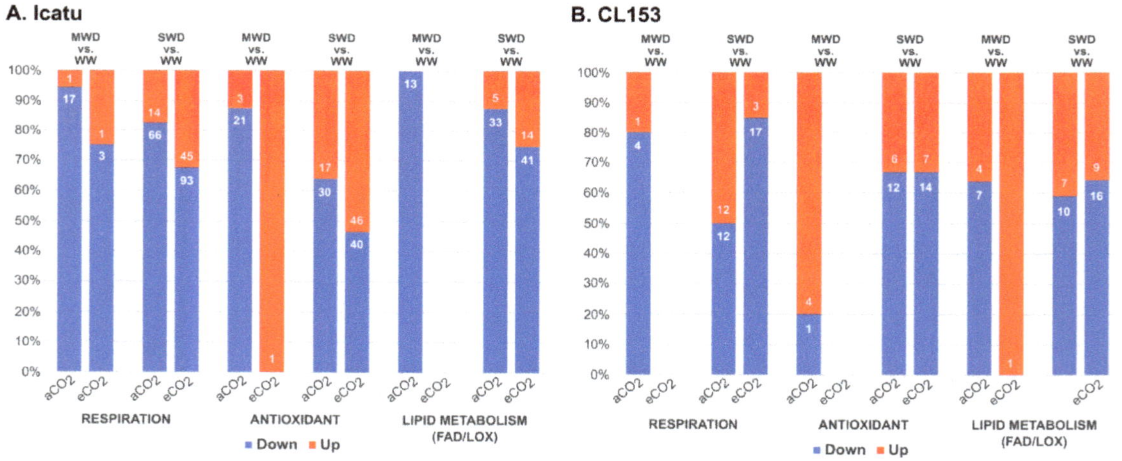


**Figure 3.** Proportion and number of significantly up- (red) and down-regulated (blue) DEGs associated with physiological and biochemical responses found in (**A**) Icatu and (**B**) CL153 plants grown under MWD or SWD in comparison with WW plants grown under $aCO_2$ or $eCO_2$, at 25/20 °C (day/night).

DEGs related to light reactions of photosynthesis, the Calvin cycle, and photorespiration visualized through MapMan showed minor effects under MWD while under SWD they were predominantly down-regulated, independently of $[CO_2]$ (Figure 4). No photosynthetic-related DEGs were found under MWD and $eCO_2$ either in Icatu (Table S8) or CL153 (Table S9). The expression of the remaining photosynthetic-related DEGs decreased with drought with a few exceptions: transkelotase and RuBisCO activase 1 were up-regulated in Icatu plants in both water deficits (Table S8); in CL153 plants, the translocase was up-regulated in both water deficits but only under $aCO_2$; while RuBisCO activase 1 was up-regulated under SWD and $aCO_2$ but down-regulated under SWD and $eCO_2$ (Table S9).

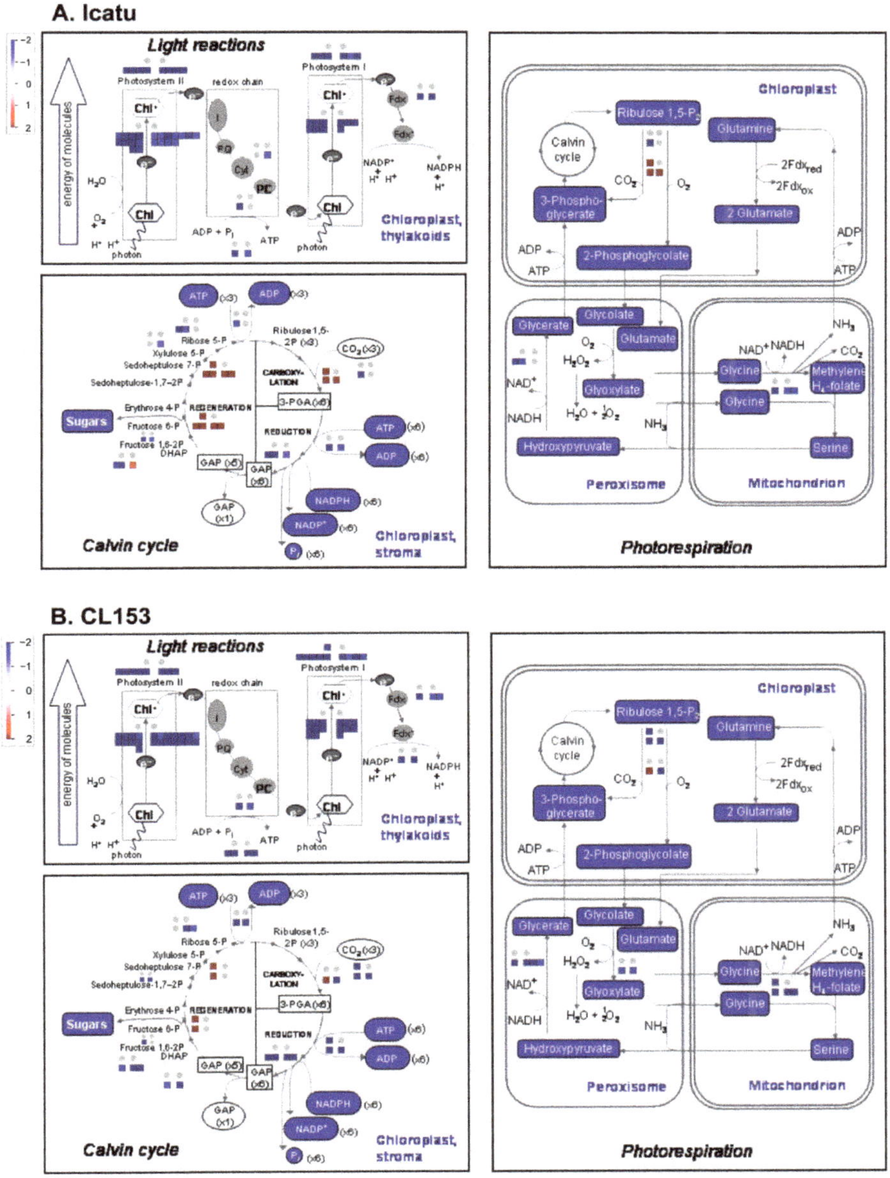


**Figure 4.** Changes in the expression of photosynthesis-related DEGs in (**A**) Icatu and (**B**) CL153 plants grown under MWD or SWD in comparison with WW plants grown under a$CO_2$ or e$CO_2$, at 25/20 °C (day/night). Pathway diagrams of light reactions of photosynthesis, the Calvin cycle, and photorespiration with superimposed color-coded squares drawn in MapMan. The color scale represents the level of regulation (blue: down-regulated DEGs; red: up-regulated DEGs). Results in the squares are presented in the following order: top left—MWD-a$CO_2$; top right—MWD-e$CO_2$; bottom left—SWD-a$CO_2$; bottom right—SWD-e$CO_2$. Grey dots represent an absence of DEGs. An absence of data in e$CO_2$ indicates the absence of DEGs.

*2.4. DEGs Involved in the Response to Water Deprivation and Desiccation*

Drought triggered a high number of DEGs involved in water deprivation and desiccation responses, especially in Icatu plants (155 DEGs) in comparison with CL153 (16 DEGs). The full list of DEGs is depicted in Table S10. DEGs involved in water-deprivation responses showed minor changes under MWD (Figure 5). For instance, no significant DEGs were detected in CL153 plants grown under MWD and $eCO_2$. Under SWD, most drought-responsive genes were down-regulated under $aCO_2$ but up-regulated under $eCO_2$, an effect mostly seen in Icatu plants (Figure 5).

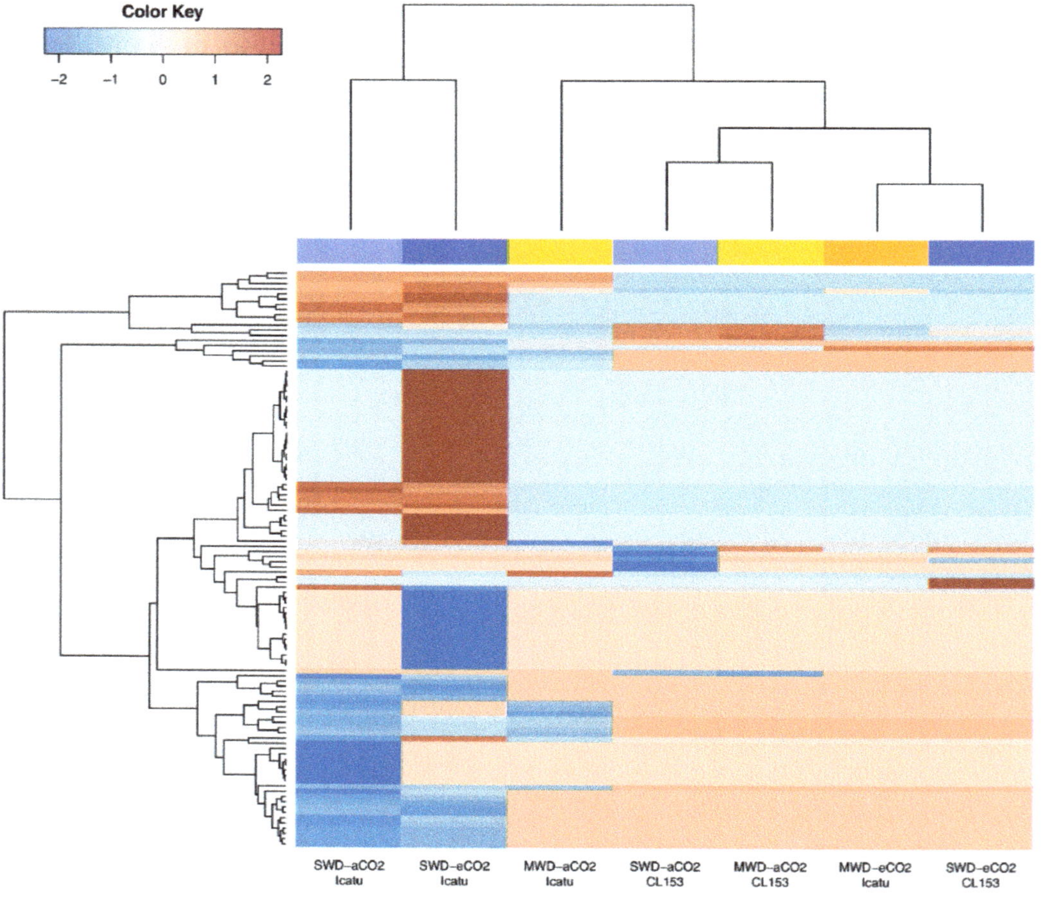


**Figure 5.** Clustered heatmaps and dendrograms of the normalized $\log_2$ fold change (FC) visualizing the expression of drought-related DEGs in Icatu and CL153 plants grown under MWD or SWD in comparison with WW plants grown under $aCO_2$ or $eCO_2$, at 25/20 °C (day/night). Values were scaled by row using Z-scores. Hot colors represent up-regulated DEGs, and cold colors represent down-regulated DEGs. Column color labels groups comparisons by water treatments (yellow/orange: MWD; light blue/dark blue: SWD; light colors represent $aCO_2$; dark colors represent $eCO_2$). An absence of data in CL153 plants under MWD-$eCO_2$ indicates the absence of significant DEGs.

Icatu plants grown under MWD and $aCO_2$ showed the most down-regulation for the PLAT domain-containing protein 3-like (FC: −7.32) and the most up-regulation for the HVA22-like protein e (FC: 4.94), while under $eCO_2$, only one DEG was found: the galactinol synthase 2-like (FC: 3.45; Table 1 and Table S10). SWD and $aCO_2$ triggered the most down-regulation of the probable xyloglucan endotransglucosylase/hydrolase protein 6 (FC: −12.31) and the most up-regulation of the protein phosphatase 2C 51-like (FC: 5.62). Under $eCO_2$ and SWD, the lowest down-regulation was also found for the probable xyloglucan endotransglucosylase/hydrolase protein 6 (FC: −11.17), while the highest up-regulation was recorded for the Late Embryogenesis Abundant protein Dc3-like (*LEA-DC3*; FC: 7.19). Notably, a high number of DEGs were involved in the antioxidant system as the galactinol synthase 2-like, sucrose synthase 2-like, the homeobox-leucine zipper *ATBH-12*, as well as aquaporins were up-regulated in Icatu under SWD and $eCO_2$ (Table S10).

**Table 1.** Regulation patterns of the top 20 up- and down-differentially expressed genes (DEGs) related to drought responses in *Coffea arabica* cv. Icatu (Icatu) plants grown under MWD or SWD in comparison with WW plants grown under $aCO_2$ or $eCO_2$, at 25/20 °C (day/night). Blue: down-regulated DEGs; red: up-regulated DEGs.

| Gene ID | Protein Name | $aCO_2$ | | $eCO_2$ | |
|---|---|---|---|---|---|
| | | **MWD** | **SWD** | **MWD** | **SWD** |
| LOC113741996 | Late Embryogenesis Abundant protein Dc3-like | | 5.51 | | 7.19 |
| LOC113704200 | Homeobox-leucine zipper protein ATHB-12-like | 4.50 | 4.83 | | 6.75 |
| LOC113740436 | Late Embryogenesis Abundant protein Dc3-like | | 3.58 | | 6.10 |
| LOC113727829 | Galactinol synthase 2-like | 2.99 | 4.29 | | 6.07 |
| LOC113706564 | Protein phosphatase 2C 51-like | | 4.43 | | 6.06 |
| LOC113733000 | Galactinol synthase 2-like | | 4.89 | 3.45 | 5.79 |
| LOC113703008 | Protein phosphatase 2C 51-like | | 5.62 | | 5.54 |
| LOC113727830 | Galactinol synthase 2-like | | 4.91 | | 5.43 |
| LOC113743599 | NAC domain-containing protein 72-like | | 3.84 | | 5.41 |
| LOC113689826 | NAC domain-containing protein 72-like | | 3.83 | | 4.86 |
| LOC113740563 | HVA22-like protein e | 4.25 | 3.96 | | 4.65 |
| LOC113695700 | Protein LE25-like | | 3.46 | | 4.65 |
| LOC113692189 | Late Embryogenesis Abundant protein 46-like | | | | 4.64 |
| LOC113740410 | HVA22-like protein e | 4.94 | 4.7 | | 4.61 |
| LOC113692190 | 18 kDa seed maturation protein-like | | | | 4.53 |
| LOC113729974 | Acidic endochitinase SE2-like | | | | 4.15 |
| LOC113740137 | Protein Aspartic Protease in Guard Cell 1-like | | | | 4.05 |
| LOC113729880 | Homeobox-leucine zipper protein ATHB-12-like | 3.79 | 3.72 | | 3.63 |
| LOC113733488 | Late Embryogenesis Abundant protein-like | | | | 3.62 |
| LOC113726525 | Zinc finger protein ZAT10-like | | 4.46 | | 3.54 |
| LOC113709223 | Aquaporin TIP1-3-like | −4.55 | −4.35 | | −3.52 |
| LOC113706149 | Probable LRR receptor-like serine/threonine-protein kinase At1g34110 | | | | −4.01 |
| LOC113702113 | Serine/threonine-protein kinase SAPK1-like isoform X3 | | −3.15 | | −4.26 |
| LOC113742299 | Protein Aspartic Protease in Guard Cell 1-like | −4.16 | −7.11 | | −4.32 |
| LOC113733193 | Probable aquaporin PIP1-2 | | −4.03 | | |
| LOC113716691 | Protein Aspartic Protease in Guard Cell 1-like | −2.96 | −4.37 | | |
| LOC113695827 | Pathogenesis-related protein PR-1-like | | −4.83 | | |
| LOC113742753 | Aquaporin TIP2-1 | −3.25 | −5.97 | | −4.38 |
| LOC113741887 | Ethylene-responsive transcription factor WIN1-like | | | | −4.46 |
| LOC113707187 | Aquaporin TIP4-1-like | | −5.45 | | −4.93 |
| LOC113740136 | Protein Aspartic Protease in Guard Cell 1-like | −3.8 | −6.26 | | −4.93 |
| LOC113732068 | Protein Eceriferum 1-like | | −4.20 | | −5.03 |
| LOC113735055 | Basic endochitinase-like | −4.17 | −7.54 | | −5.58 |
| LOC113742697 | Protein Aspartic Protease in Guard Cell 1-like | −3.81 | −6.63 | | −5.62 |
| LOC113716416 | Acidic endochitinase-like | | −4.09 | | −5.94 |
| LOC113701593 | PLAT domain-containing protein 3-like | −7.32 | −8.00 | | −6.22 |
| LOC113742441 | Protein Aspartic Protease in Guard Cell 1-like | | | | −8.98 |
| LOC113737208 | Basic endochitinase-like | −4.41 | −7.12 | | |
| LOC113739398 | Ethylene-responsive transcription factor WIN1-like | | −7.71 | | |
| LOC113697821 | Probable xyloglucan endotransglucosylase/hydrolase protein 6 | −4.72 | −12.31 | | −11.17 |

By contrast, the minor responses recorded in CL153 plants and mostly recorded under SWD and $aCO_2$ involved a high number of Aspartic Protease in Guard Cell 1-like (*ASPG1*; 6 out of 16 total DEGs; Table 2). In this genotype, the 18 kDa seed maturation was the most up-regulated under MWD, independently of $CO_2$ levels ($aCO_2$ FC: 9.18; $eCO_2$ FC: 8.79) while the lowest was the *APG1*, also independently of $CO_2$ levels ($aCO_2$ FC: −7.93; $eCO_2$ FC: −8.16; Table 2). Remarkably, under $aCO_2$, the dehydrin *DH1a* was also highly up-regulated under both water deficit levels (FC: 6.05 for MWD; FC 5.47 for SWD) while it was less expressed under $eCO_2$ and only under SWD (FC 2.37). In fact, under $eCO_2$ and MWD, no water-deprivation-related DEG was recorded. Under SWD, only one down-regulated DEG was found (the putative movement binding protein 2C; FC:−19.92) while the remaining were up-regulated, especially the *APG1* (FC: 3.96).

**Table 2.** Regulation patterns of differentially expressed genes (DEGs) related to drought responses in *Coffea canephora* cv. CL153 (CL153) grown under MWD or SWD in comparison with WW plants grown under ambient $aCO_2$ or $eCO_2$, at 25/20 °C (day/night). Blue: down-regulated DEGs; red: up-regulated DEGs.

| Gene ID | Protein Name | $aCO_2$ | | $eCO_2$ | |
|---|---|---|---|---|---|
| | | MWD | SWD | MWD | SWD |
| Cc07_g07560 | Probable xyloglucan endotransglucosylase/hydrolase protein 6 | −6.90 | −5.85 | | |
| Cc04_g09640 | Protein Aspartic Protease in Guard Cell 1-like | −5.03 | −4.92 | | |
| Cc04_g07360 | Putative protein Aspartic Protease in Guard Cell 1-like | 4.89 | 4.07 | | 3.96 |
| Cc04_g07380 | Putative protein Aspartic Protease in Guard Cell 1-like | 4.27 | 3.17 | | 2.80 |
| Cc07_g10030 | Dehydrin DH1a | 6.05 | 5.47 | | 2.37 |
| Cc01_g11790 | Mitogen-activated protein kinase 3 | | | | 1.45 |
| Cc06_g09540 | Multiprotein-bridging factor 1c | | | | 2.34 |
| Cc04_g08280 | Putative movement protein binding protein 2C | | −2.86 | | −1.92 |
| Cc06_g15980 | 18 kDa seed maturation protein | 9.18 | 8.79 | | 2.92 |
| Cc04_g07350 | Putative protein Aspartic Protease in Guard Cell 1-like | | −5.86 | | |
| Cc04_g07330 | Putative protein Aspartic Protease in Guard Cell 1-like | −7.93 | −8.16 | | |
| Cc02_g17270 | Putative Late Embryogenesis Abundant protein | 4.32 | 3.58 | | |
| Cc01_g08980 | Late Embryogenesis Abundant protein hydroxyproline-rich glycoprotein family | | −3.25 | | |
| Cc04_g07370 | Putative protein Aspartic Protease in Guard Cell 1-like | | −3.48 | | |
| Cc02_g15480 | Cellulose synthase A catalytic subunit 8 [UDP-forming] | | −5.28 | | |
| Cc07_g15660 | C2 domain-containing protein | | −1.73 | | |

### 2.5. *Enriched Gene Ontology (GO) Terms and Functional Pathways Responding to Drought and $eCO_2$*

Icatu plants under $aCO_2$ showed a moderate increase in enriched GO terms as water deficit severity increased, being all associated with down-regulated DEGs (Figure 6). $eCO_2$ attenuated MWD impact and no enriched categories were found under this drought level in Icatu plants, while under SWD, Icatu plants showed a high number of enriched GO terms mostly linked to down-regulated DEGs. From all enriched categories, the greatest effect was recorded under SWD in the 'integral component of the membrane' (GO:0016021), independently of [$CO_2$] (Figure 6; Table S11). However, under SWD and $eCO_2$, two enriched categories were linked to up-regulated DEGs: 'sequence-specific DNA binding' (GO:0043565) and the 'UDP−glycosyltransferase activity' (GO:0008194).

By contrast, drought triggered a higher number of enriched GO terms in CL153 than in Icatu (84 vs. 43) (Figure 6; Table S12). Under $aCO_2$ and MWD, CL153 plants were enriched in GO terms mostly involved up-regulated DEGs, except 'localization' (GO:0051179), 'metal ion transport' (GO:0030001), 'transport' (GO:0006810), 'transmembrane transporter activity' (GO:0022857), and 'transporter activity' (GO:0005215). The opposite was recorded in SWD plants under $aCO_2$, where all enriched categories were linked to down-regulated DEGs. By contrast, CL153 plants under $eCO_2$ showed no enriched categories under MWD (as reported also for Icatu plants), while under SWD they were mostly linked to down-regulated

DEGs, except 'defense response' (GO:0006952), 'regulation of transcription DNA-templated' (GO:0006355), 'oxidoreductase activity' (GO:0016491) and 'transcription regulator activity' (GO:0140110). CL153 plants under $aCO_2$ showed an enrichment in many down-regulated categories involving the cellular membrane, which were less recorded under $eCO_2$ (Figure 6). Remarkably, the 'catalytic activity' (GO:0003824), which was up-regulated under MWD and $aCO_2$, was down-regulated under SWD, regardless of $[CO_2]$ (Figure 6).

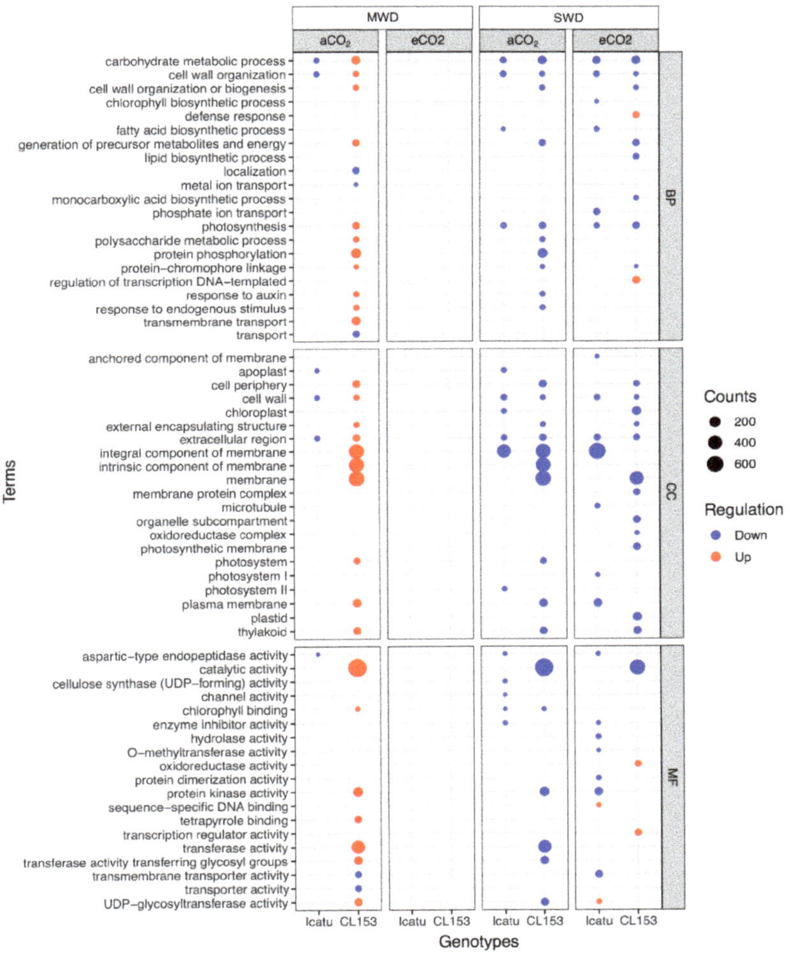


**Figure 6.** Over-representation analysis of gene ontology (GO) terms found among Icatu and CL153 DEGs performed with gProfiler against the functional annotation of *Coffea arabica* and *Coffea canephora* genomes, respectively. Significantly (g:SCS < 0.01) enriched GO was ranked by decreasing $\log_2$ fold-change (FC), considering the effect of MWD or SWD in comparison with WW plants grown under $aCO_2$ or $eCO_2$, at 25/20 °C (day/night). Dot plots with all GO terms were filtered by REVIGO with similarity = 0.5, and a count > 10 cut-offs. Terms are grouped by the main category: Biological Process (BP), Molecular Function (MF), and Cellular Component (CC). Counts (size) indicate the number of DEGs annotated with each GO term and color represents the type of regulation (blue: down-regulated DEGs; red: up-regulated DEGs). An absence of data in $eCO_2$ under MWD indicates the absence of enriched GO terms.

Based on the Kyoto Encyclopedia of Genes and Genomes (KEGG), only pathways linked to down-regulated DEGs were found to be significantly affected by drought (Figure 7). Under MWD and $aCO_2$, the 'flavonoid biosynthesis' was the only KEGG pathway affected by drought and only in Icatu plants. Under SWD, the 'photosynthesis' pathway was mostly affected in Icatu and CL153 under $aCO_2$, but only under $eCO_2$ in CL153 plants. In addition, the 'indole alkaloid biosynthesis' pathway was affected under $eCO_2$ in Icatu plants, while in CL153 plants, the 'photosynthesis antenna proteins' pathway was also affected under $aCO_2$ (Figure 7).

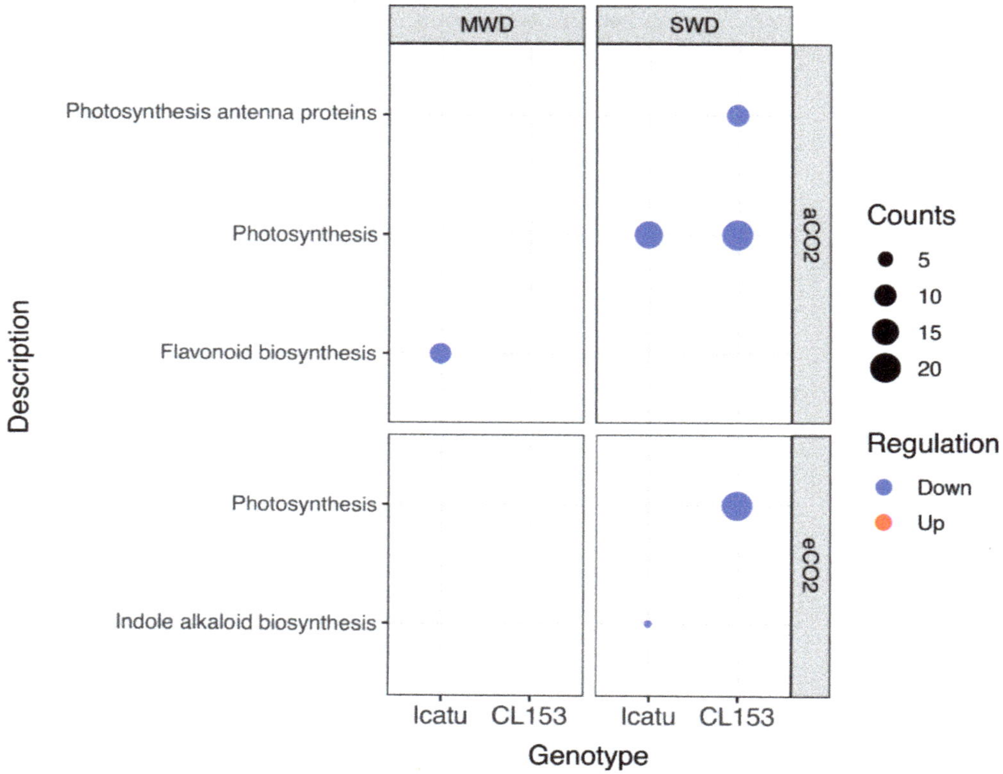


**Figure 7.** Over-representation analysis of KEGG pathways of Icatu and CL153 down-regulated DEGs performed with gProfiler, using the DEGs' *Arabidopsis thaliana* homologs mapped through blastx, against the functional annotation of its reference genome. Significantly (g:SCS < 0.01) enriched KEGG pathways of DEGs ranked by decreasing $\log_2$ fold-change (FC), considering the effect of MWD or SWD in comparison with WW plants grown under $aCO_2$ or $eCO_2$, at 25/20 °C (day/night). Counts (size) indicate the number of DEGs annotated with each GO term. An absence of data in CL153 indicates the absence of significant KEGG pathways.

### *2.6. Validation of RNA-Seq Results by qRT-PCR*

All transcripts associated with a differential gene expression (*ASPG1, GMPM1, PP2C-51, LEAD-C3, DH1a, ATHB22, SUS2, PIP2-2, XTH6, GOLS2, CuSOD1,* $APX_{Chl}$) in the RNA-seq workflow showed a change in the same direction under qRT-PCR, revealing a general agreement between RNA-seq sequencing and qRT-PCR analysis (Figure 8). Except for *XTH6*, all analyzed transcripts involved in the regulation or protection against water deficit were up-regulated under drought conditions, especially under SWD and $eCO_2$.

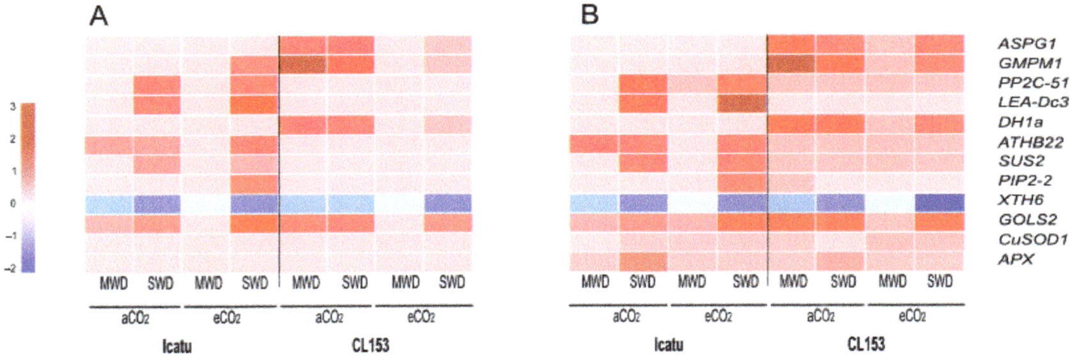


**Figure 8.** Heatmap of expression levels ($log_2$ ratio) of the selected genes obtained from RNA-seq (**A**) and validated by qRT-PCR (**B**), considering the effect of MWD or SWD in comparison with WW-Icatu and CL153 plants, grown under $aCO_2$ (light colors) or $eCO_2$ (dark colors), at 25/20 °C (day/night). *GMPM1*: 18 kDa seed maturation; *PP2C-51*: protein phosphatase 2C 51-like; *LEA-DC3*: late embryogenesis abundant protein Dc3-like; *DH1a*: dehydrin DH1a; *ATHB22*: homeobox leucine zipper; *SUS2*: sucrose synthase 2-like; *PIP2-2*: aquaporin PIP2-2-like; *XTH6*: xyloglucan endotransglucosylase/hydrolase protein 6; *GOLS2*: galactinol synthase 2-like; *CuSOD1*: superoxide dismutase [Cu-Zn]; $APX_{Chl}$: chloroplast ascorbate peroxidase.

## 3. Discussion

### 3.1. Differential Transcriptional Drought Regulation Responses in the Two Coffea Genotypes

Despite the information available, the molecular basis of stress responses in woody species is still limited in comparison with model plants. Understanding which genes may play a role under different degrees of drought severity, and which physiological traits are controlled by such genes, is of crucial importance for understanding the capabilities of trees to successfully cope with stress, especially in commercially important crops such as coffee. To maintain the global supply of this crop, it is important to promote the screening and development of tolerant varieties to face the increasingly expected impacts of drought events. In this study, we showed that the two genotypes, belonging to two different *Coffea* species, activated distinct transcriptional responses to cope with drought.

In comparison with CL153, Icatu showed (i) a higher number of expressed genes under MWD, but especially under SWD (Figure 1); (ii) a higher number of specific DEGs, especially under SWD (Figure 2); (iii) a higher number of DEGs involved in respiration, antioxidant activities, and lipid metabolism (Figure 3); (iv) a higher activation of DEGs involved in light reactions of photosynthesis, the Calvin cycle and photorespiration (Figure 4); and (v) a higher number of DEGs involved in responses to water deprivation and desiccation (Figure 5). This transcriptional result explains previous physiological studies that reported a higher photochemical performance in Icatu than in CL153, due to the preservation or reinforcement of photosynthetic components, strengthened enzymatic antioxidative system, and a greater abundance of photoprotective pigments [7,12,26,27].

Protein phosphatases (*PP2Cs*) were found to be involved in Icatu responses to drought (*PP2C 51-like*; Table 1). *PP2Cs* are a class of evolutionarily conserved serine/threonine protein phosphatases involved in stress responses and have been implicated in abscisic acid (ABA) signal transduction [28]. Transgenic studies suggest that many *PP2Cs* participate in drought responses, negatively regulating ABA signaling pathways in *Arabidopsis* [29,30], tomato [31], *Populus euphratica* [32], and *Artemisia annua* [33].

In response to drought, CL153 rather activated a high number of aspartic proteases (*ASPG1*; Table 2). Aspartic proteases play a fundamental role in the response of plants to drought, especially during ROS increments [34]. ROS can act as a positive regulator of ABA signaling in guard cells, but the excessive accumulation of ROS during stress can be toxic [35]. Thus, to balance ROS production and scavenging, either the levels are modulated through signal transduction, or the cells have mechanisms to detoxify excessive ROS during stress [36]. Hence, the up-regulation of aspartic proteases in CL153 may help to scavenge the excessive amount of ROS. The up-regulation of dehydrins (namely *DH1a*) in both water deficits may also be involved in protective reactions to dehydration [37]. Dehydrins were also found to be highly expressed in the leaves of drought-stressed *Coffea* plants such as *C. arabica* cvs. Catuaí and Mundo Novo, *C. canephora* cv. Apoatã [38], as well as in the genotypes studied here [39], where they seem to play a key role in the acclimation response of *Coffea*.

Our transcriptomic results also found a high number of DEGs involved in the antioxidant system, especially in Icatu (Figure 3). This suggests the action of three mechanisms involved in drought acclimation: (i) activation of antioxidant activities to scavenge the excessive amount of ROS and reduce oxidative damage in plants, including enzymes (e.g., catalase, superoxide dismutases, peroxidases) and non-enzymatic molecules (e.g., ascorbic acid, α-tocopherol, raffinose family oligosaccharides), in agreement with studies from other plants [40,41]; (ii) antioxidative mechanisms complemented with thermal dissipation mechanisms (e.g., photoprotective carotenoids) and changes in the cyclic electron flow (CEF) to protect the photosystem (PS) I and/or II. In fact, (i) and (ii) are transversally found in resilient *Coffea* genotypes as response mechanisms to drought [12,26], heat [42–44], cold [45,46], and high irradiance [47,48] stresses. Additionally, the activation of other molecules, such as aquaporins reported here in Icatu and in previous studies involving other *C. arabica* genotypes [49] or the heat shock protein 70 kDa as reported also in other *Coffea* studies [39,50]. Altogether, results reveal the expression of DEGs, whose products are important to limit or control water loss in Coffea, regulating stomatal closure in leaves subjected to drought conditions. Their overexpression upon stress was validated by qRT-PCRs (Figure 8), supporting the action of several antioxidant molecules in response to drought. It also corroborates the results of RNA-sequencing, providing a set of candidate genes involved in drought tolerance in coffee plants. These mainly include genes encoding antioxidant enzymes involved in the biosynthesis of small antioxidant molecules and water and ion movement, such as aquaporins and ion transporters or the biosynthesis of osmolytes. Other proteins that function directly in the protection of proteins and membranes, such as late embryogenesis abundant proteins, heat shock proteins, protein phosphatases, and aspartic proteases also play important roles in Coffea resilience to abiotic stresses. This vast information will help in the identification of target sites suitable for gene editing, which will undoubtfully make progress in the future.

### 3.2. Influence of $eCO_2$ in Drought-Responsive Genes

The level and type of transcripts found in this study support the positive effects of $eCO_2$ in mitigating some impacts of drought, especially in the functioning and components of the photosynthetic apparatus [12]. The up-regulation of antioxidant-related DEGs under $eCO_2$, mostly under SWD and in Icatu plants (Figure 3) is congruent with physiological measurements that showed an increase in Cu,Zn-SOD activity under MWD, rising further under SWD, but only under $eCO_2$ [27]. The up-regulation of antioxidant-related DEGs under $eCO_2$ might strengthen the scavenging potential for molecular $O_2$ when the photochemical use of energy is small, reducing the probability of electron flow to lower energy states [12], helping to alleviate some impacts of drought in the photosynthetic machinery (Figure 4).

The imposition of $eCO_2$ to SWD plants promoted the up-regulation of several important stress response genes, including the *LEA-DC3* in Icatu and the *APG1* in CL153 as mentioned before. Late embryogenesis abundant proteins compose the most abundant and characterized group of intrinsically disordered proteins that prevent and repair the damage caused by environmental stresses. A positive association between the accumulation of *LEA* and environmental stresses, such as drought, heat, and salinity, has been outlined in several other plant species, where they bind to enzymes to prevent the loss of activity under stressful conditions [51,52]. This is consistent with the biological functions of *LEAs* namely in oxidant scavenging activities, enzyme and nucleic acid preservations, and the membrane maintenance that occurs in genotypes/species that can better cope with environmental stresses [53].

Directly or indirectly, ABA signaling pathways, which include ABA-dependent, ABA-responsive element/ABRE-binding factors (ABRE/ABF), as well as ABA-independent genes as dehydration-responsive elements regulate *Coffea* responses to stress. ABA is involved in both drought-induced and $eCO_2$-induced stomatal closure in a dual way, including root-derived and foliar ABA. For instance, under well-watered conditions, *C. arabica* plants grown under $eCO_2$ showed lower whole-plant transpiration rates than under $aCO_2$ [21]. These changes, although unrelated to stomatal conductance or foliar ABA levels, are associated with faster stomata closure rates upon rapid increases in vapor pressure deficit under $eCO_2$ [21]. For instance, during exposure to drought, *Coffea* plants grown under $eCO_2$ can maintain higher water potentials and plant hydraulic conductance than under $aCO_2$, due to a higher transcript abundance of aquaporins [21]. In the genotypes studied here, drought alone prompted gradual ABA increases of ~46% in MWD plants in both genotypes, and 100% (CL 153) and 184% (Icatu) under SWD conditions, whereas single $eCO_2$ increased ABA levels (by 85%) but only in Icatu [12]. This is important since ABA controls stomatal closure under stress, which is one of the first defense responses to reduce water loss by restricting the transpiration flow in response to a rising air evaporative demand or a decreased soil water availability [54]. This suggests that, at least under MWD, $eCO_2$ seemed to decouple ABA action from stomatal closure in both genotypes. In agreement with this hypothesis, dehydration is postponed in MWD plants under $eCO_2$, and a delayed stomata response to soil drying under $eCO_2$ is found in some coffee genotypes [12,21,22], maintaining greater stomatal conductance than expected based on ABA concentrations. Such a greater stomatal opening under MWD would allow greater C-assimilation gains under $eCO_2$ [12], also revealing a more profound impact of $eCO_2$ than the direct stimulation of C-assimilation. A better understanding of ABA concentrations in the xylem sap is necessary to understand the sensitivity of *Coffea* stomata to [ABA] xylem.

### *3.3. Overall Regulatory Mechanisms Involved in Coffea Responses*

MWD had minor impacts under $aCO_2$ with few enriched categories being found in Icatu, and all were associated with down-regulated DEGs, while CL153 plants showed a high increase in GO categories, and were mostly associated with up-regulated DEGs (Figure 6).

Under SWD, a few enriched categories were even up-regulated in Icatu ('sequence-specific DNA binding' and the 'UDP−glycosyltransferase activity') and CL153 ('defense response', 'regulation of transcription DNA-templated', 'oxidoreductase activity' and 'transcription regulator activity'). Genes assigned to these GO terms are usually involved in a high number of developmental processes and stress responses, plant hormone activation, and the production of antioxidants in response to stresses, including drought [55]. In *Coffea*, as in other species, glycosylation catalyzed by glycosyltransferases, as well as oxidoreductase activities, play an essential role in regulating the stability, availability, and biological activity of antioxidant compounds and the integrity of cellular membranes [56]. This is crucial to cope with the effects of water deprivation in *Coffea* since tolerance is associated with the ability of tissues to withstand low water potentials and plant membrane transport systems play a significant role under water scarcity [57]. Depending on energy needs, translocation through biological membranes occurs, passively or actively, but drought can compromise the integrity of membranes and is, therefore, relevant an up-regulation of categories linked to cellular membranes in the two *Coffea* genotypes, even under MWD.

The 'flavonoid biosynthesis' KEGG pathway was affected in Icatu plants grown under MWD and $aCO_2$ (Figure 7). Plant phenolic compounds, especially flavonoids, can provide resistance to biotic and abiotic stresses, and can be enhanced upon drought [58]. As non-enzymatic antioxidants, hydroxyl groups in flavonoids participate in the scavenging of oxygen free radicals, alleviating stress-induced oxidative damage as is widely reported [59], thus, affecting acclimation responses of Icatu, at least under $aCO_2$. However, SWD affected the KEGG photosynthetic pathway, namely in Icatu plants under $aCO_2$, as well as in CL153 plants under both $[CO_2]$ levels together with the 'photosynthesis antenna proteins' pathways in CL153 plants under SWD and $aCO_2$ (Figure 7). Since these pathways are both linked to down-regulated DEGs, this would imply some effects in the photosynthetic machinery of these genotypes, especially in CL153. Indeed, a large dilution of the impacts of drought on the net photosynthesis in these genotypes was found to be promoted by $eCO_2$ under MWD, consistent with a tendency to the maintenance of the PSII efficiency and higher PSs activity in both genotypes while under SWD, the net photosynthesis and stomatal conductance were severely reduced, regardless of $[CO_2]$ [12]. Nevertheless, even under SWD, a relevant potential for C-assimilation was preserved, with the photosynthetic capacity ($A_{max}$) showing values close to 60% (CL153), or even higher than 70% (Icatu) relative to those displayed by their respective WW controls [12]. This photochemical protective mechanism results in a lower need for dissipation processes and a reduced PSII inhibition status [43] consistent with the down-regulation of photosynthetic-related DEGs found in this study (Figure 4) or the KEGG enrichment results (Figure 7) since no new molecules would be needed due to the protective mechanisms involved.

A minor inhibition in the 'indole alkaloid biosynthesis' (IAB) KEGG pathway was also recorded in Icatu under SWD and $eCO_2$. Plant peroxidases may accept alkaloids as substrates, as well as phenols and flavonoids [60], and metabolize $H_2O_2$ as an electron donor for phenol peroxidases, resulting in the formation of phenoxyl radicals, which can be regenerated by a non-enzymatic reaction with an ascorbate function as an $H_2O_2$ scavenging system [61]. Thus, either the IAB KEGG pathway is not involved/not needed in acclimation responses of Icatu or this would suggest a decrease in the defense ability of this genotype against oxidative stress. However, that contrasts with the high activity found in protective molecules and antioxidant enzymes in Icatu [12,27], and the large reinforcement of Cu,Zn-SOD, APX, and catalase (CAT) activities [7].

### 3.4. *Evidence of Post-Transcriptional Regulatory Mechanisms in Coffea Responses to Stress*

Previous findings showed that *Coffea* plants, namely Icatu, can maintain the potential photosynthetic functioning under the imposition of SWD due to a greater antioxidative response, which contrasts with the transcriptomic results shown here where photosynthetic-related DEGs were mostly down-regulated (Figures 3 and 4) and the KEGG photosynthetic pathway was highly affected under SWD (Figure 7). Physiological studies showed that increasing drought severity progressively affected gas exchange and fluorescence parameters in both genotypes, with non-stomatal limitations becoming gradually dominating, and having strong impacts on the photochemical and biochemical components and functioning of *Coffea*, especially in CL153 plants under SWD and $aCO_2$ [12]. In contrast, Icatu plants were tolerant to SWD, with minor, if any, negative impacts on the potential photosynthetic functioning and components, e.g., $A_{max}$, $F_v/F_m$, electron carriers, photosystems (PSs) and RuBisCO activity, under $aCO_2$ [12]. Under MWD, $eCO_2$ delayed stress severity and promoted photosynthetic functioning in both genotypes, with lower energy dissipation, while stomatal closure was decoupled from increases in ABA. Under SWD, most of the negative impacts felt on the photosynthetic components and their potential performance were reduced under $eCO_2$, at least considering CL153, since Icatu was barely affected in both $[CO_2]$ levels under SWD [12]. Still, strong effects were detected in RuBisCO, as the most sensitive photosynthetic component [12]. However, proteomic analyses have also shown a higher abundance of drought-responsive proteins in Icatu than in CL153, together with enriched GO terms, and enriched KEGG pathways associated with stress responses and the control of oxidative stress categories found here [12,27,62]. Thus, these contrasting results suggest the existence of important post-transcriptional regulation in *Coffea*, at least in the genotypes studied here. Other studies have also highlighted that protective metabolites often do not show a clear pattern between transcript accumulation and metabolite/physiological responses in response to stress, being likely to not be transcriptionally regulated. For instance, [63] studied the transcript response to $eCO_2$ in *Solanum lycopersicum* and its wild relative *S. pennelli*, and no clear transcriptomic pattern was found, but rather a translational regulatory mechanism, hypothetically involved in the differential ribosomal loading of transcripts in the two species. Additionally, in *Saccharomyces cerevisiae*, relatively few (~15%) of the mRNAs that were translationally up-regulated in response to $H_2O_2$ showed similar increases in transcript levels [64], revealing a complex transcriptional and translational reprogramming to stress. The existence of a complex translational program in *Coffea* would also explain the physiological and biochemical performance of these genotypes [12,26], as well the amplified acclimation responses at the proteomic level [27], namely in Icatu plants, despite the down-regulation of transcripts reported here.

## 4. Materials and Methods

### *4.1. Plant Growth Conditions*

Plants of two cropped genotypes from the two main producing coffee species, *Coffea canephora* Pierre ex A. Froehner cv. Conilon Clone 153 (CL153) and *C. arabica* L. cv. Icatu Vermelho (Icatu) were obtained, respectively, from Emcapa and IAC. A total of 26 plants were grown from the seedling stage, for seven years, in 80 L pots in walk-in growth chambers (EHHF 10000, ARALAB, Albarraque, Portugal), under controlled conditions of temperature (25/20 °C, day/night, ±1 °C), irradiance (max. ca. 750 μmol $m^{-2}$ $s^{-1}$ at the upper part of the plant, using a combination of fluorescent, metal halide and halogen lamps to provide a balanced light spectrum), relative humidity (70 ± 2%), photoperiod (12 h), and exposed to ambient ($aCO_2$, 380 ± 5 μL $L^{-1}$) or elevated ($eCO_2$, 700 ± 5 μL $L^{-1}$) atmospheric [$CO_2$] [12]. Plants were maintained without restrictions of nutrients (with fertilization provided as stated in [65]), root growing space, or water (until water deficit experiments), watering the plants every two days to maintain adequate soil moisture.

### *4.2. Imposition and Monitoring of Water Deficit Conditions and Sampling*

Plants previously maintained without water restriction were divided into three groups. The first one was maintained under well-watered (WW) conditions, with a leaf predawn water potential ($\Psi_{pd}$) above −0.35 MPa. In the other two groups, drought was imposed by a gradual reduction of irrigation, allowing plants to express their potential acclimation ability for two weeks, to promote $\Psi_{pd}$ decline to values between −1.5 and −2.5 MPa (moderate water deficit—MWD) or below −3.5 MPa (severe water deficit—SWD), representing ca. 80 (WW), 35 (MWD) and 10% (SWD) of maximal water availability in pots [6]. These MWD and SWD conditions were maintained for another two weeks by adding adequate water amounts according to each water deficit level. Samples were then collected for transcriptomic analysis. Exceptionally, Icatu $eCO_2$ plants under MWD were submitted to total water withholding in the last 5 days of the 4-week period, to force the reduction of $\Psi_{pd}$, which, even so, did not shift below −0.6 MPa. Leaf $\Psi_{pd}$ was determined immediately after leaf excision, using a pressure chamber (Model 1000, PMS Instrument Co., Albany, OR, USA).

### *4.3. RNA Extraction and Illumina Sequencing*

Newly matured leaves from plagiotropic and orthotropic branches from the upper third part (well illuminated) of each plant were collected under photosynthetic steady-state conditions after 2 h of illumination, flash frozen in liquid nitrogen, and stored at −80 °C. Total RNA was extracted from 36 samples (two genotypes × three water treatments × two [$CO_2$] × three individual plants) using the Analytik-Jena InnuSPEED Plant RNA Kit (Analytik Jena Innuscreen GmbH, Jena, Germany) following [50]. RNA quantity and quality were determined using a BioDrop Cuvette (BioDrop, UK) and an Agilent 2100 Bioanalyzer (Agilent Technologies, Santa Clara, CA, USA). RNA integrity number (RIN) for the samples ranged from 8.96 to 9.05. The mRNA libraries were constructed with the Illumina TruSeq Stranded mRNA Sample Preparation kit (Illumina, San Diego, CA, USA) and sequenced on an Illumina NovaSeq6000 at Macrogen facilities (Macrogen, Geumcheongu, Seoul, Republic of Korea).

### 4.4. Quality Analysis of Sequencing Data

Raw reads were processed using FastQC version 0.11.9 [66] to remove low-quality reads. FastQ Screen version 0.14 [67] was used to check for contaminants against the genome of the most common model organisms (e.g., *Homo sapiens*, *Mus musculus*, *Rattus norvegicus*, *Drosophila melanogaster*, *Caenorhabditis elegans*, *Saccharomyces cerevisiae*, *Escherichia coli*) and adapter databases (e.g., Mitochondria RNA, PhiX, Vector from UniVec database, FastQ Screen rRNA custom database, and FastQ Screen Adapters database). Since all reads presented an overall good quality, the trimming step was skipped. Recent studies showed that this process is redundant in the quantification of expression data from RNA-seq since most aligners can perform soft-clipping to effectively remove adapter sequences and rescue low-sequencing-quality bases that would be removed by read trimming tools, improving the accuracy in the quantification of gene expression [68].

### 4.5. Reference-Based Mapping and Assembly

The raw reads of Icatu were mapped to the reference genome of *C. arabica* downloaded from the NCBI (https://www.ncbi.nlm.nih.gov/assembly/GCF_003713225.1, accessed on 4 April 2021), while the raw reads of CL153 were mapped to the *C. canephora* genome downloaded from the Coffee Genome Hub (http://coffee-genome.org/download, accessed on 4 April 2021) [69] using STAR version 2.7.8a [70]. Htseq-count v0.11.0 [71] was then used to quantify only uniquely mapped genes. Samtools version 1.12 [72] and gffread version 0.12.1 [73] were used throughout the analysis to obtain general statistics on genome mapping. A principal component analysis (PCA) was performed on the expression data of genes, TMM (trimmed mean of means) normalized and log10-transformed using ggplot2 version 3.3.3 library [74] of R software version 4.0.2 [75]. Through visual inspection of the PCA, replicate 7B was considered an outlier and excluded from downstream analyses (Figure S1).

### 4.6. Identification of Differentially Expressed Genes (DEGs)

DEGs from the plants grown under different water treatments and different [$CO_2$] were estimated and compared as follows: MWD-a$CO_2$ vs. WWa-$CO_2$, SWD-a$CO_2$ vs. WW-a$CO_2$, MWD-e$CO_2$ vs. WW-e$CO_2$, and SWD-e$CO_2$ vs. WW-e$CO_2$. Differential expression analyses were performed using the DEGs commonly found by both DESeq2 version 3.8 [76] and edgeR version 3.26.0 [77]. The resulting values of expression were adjusted using Benjamini and Hochberg's approach to control the false discovery rate (FDR; [78]). Genes with a normalized non-zero $\log_2$ fold change (FC) expression and an FDR $< 0.01$ in both tools were defined as differentially expressed. Python's matplotlib library was used to plot Venn diagrams and bar plots [79]. Using ggplot2, heatmaps with dendrograms were plotted to visualize DEGs based on the differential expression patterns between the different comparisons. To prevent high DEGs from clustering together without considering their expression pattern, $\log_2$ FC was scaled by gene expression across treatments (row Z-score).

### 4.7. Drought and e$CO_2$ Impact on DEGs Associated with Specific Biochemical Pathways

Due to the fundamental role of photosynthesis, respiration, lipid profile changes, and the antioxidant system in the process of coffee acclimation to environmental stresses, a specific/fine-tuned search was performed among the significant DEGs associated with these processes. According to the reference genome and the UniProtKB database DEGs annotated with the following direct and child GO terms were searched to study their regulation pattern: [Photosynthesis], 'photosynthesis' (GO:0015979), 'photosystem' (GO:0009521), 'photosynthetic membrane' (GO:0034357), 'photoinhibition' (GO:0010205), 'photosynthetic phosphorylation' (GO:0009777), 'photosynthetic acclimation' (GO:0009643), 'photosynthetic state transition' (GO:0062055), 'photosynthetic NADP+ reduction' (GO:0009780), 'photosynthetic electron transport chain' (GO:0009780), 'photosynthetic electron transport chain' (GO:0009767), 'photorespiration' (GO:0009853), 'chlorophyll metabolic process' (GO:0015994), 'chlorophyll biosynthetic process' (GO:0015995), and 'chlorophyll catabolic process' (GO:0015996); [Cellular respiration], 'cellular respiration' (GO:0045333), 'oxidative phosphorylation' (GO:0006119), and 'mitochondrion' (GO:0005739), [Antioxidant activity], 'antioxidant activity' (GO:0016209), and 'response to oxidative stress' (GO:0006979); [Lipid metabolism], 'fatty acid metabolic process' (GO:0006631), 'FAD metabolic process' (GO:0046443), 'FAD transport' (GO:0015883), and LOX (GO:0004051, GO:0016702). Additionally, a more specific pathway analysis was conducted where DEGs related to the light reactions of photosynthesis, the Calvin cycle, and photorespiration were visualized through MapMan version 3.6.0RC1 [80].

### *4.8. DEGs Involved in the Response to Water Deprivation and Desiccation*

To better understand the impacts of water deficit, a specific search was performed, as described previously, among DEGs annotated with 'water transport' (GO:0006833), 'water homeostasis' (GO:0030104), 'response to water' (GO:0009415), 'response to water deprivation' (GO:0009414), and 'response to desiccation' (GO:0009269).

### *4.9. Functional Classification of Responsive DEGs*

DEGs from CL153 and Icatu comparisons were annotated following the functional annotation of the reference genomes of *C. canephora* and *C. arabica*, respectively, as stated previously. GO enrichment analyses were applied to understand the functional classification of responsive DEGs through another over-representation analysis (ORA) using gProfiler [81] under g:SCS < 0.01. Results were summarized using REVIGO [82] by removing redundant GO terms within a similarity = 0.5. Enrichment non-redundant results were plotted using ggplot2 version 3.3.3 library, using the number of DEGs annotated with each term to set a Counts > 10 cut-off [74]. Since *Coffea* genomic annotations are not complete, namely in terms of KEGG pathways, DEGs were mapped to their *Arabidopsis thaliana* homologs against a local Swissprot database, filtering gene hits by maximum e-value of $1.0 \times 10^{-3}$ and minimum identities of 40% [83], and using blastx from the Basic Local Alignment Search Tool (BLAST) version 2.10.1 command-line tool from the NCBI C++ Toolkit. These annotations were then used to perform an over-representation analysis (ORA) with gProfiler, searching for significantly (g:SCS < 0.01) enriched KEGG pathways.

### *4.10. Quantitative RT-PCR*

Twelve transcripts were randomly selected for real-time quantitative PCR (qRT-PCR) to verify the accuracy of the levels of expression obtained under RNA-seq. Genes included: *GMPM1*: 18 kDa seed maturation; *PP2C-51*: protein phosphatase 2C 51-like; *LEA-DC3*: late embryogenesis abundant protein Dc3-like; *DH1a*: dehydrin DH1a; *ATHB22*: homeobox leucine zipper; *SUS2*: sucrose synthase 2-like; *PIP2-2*: aquaporin PIP2-2-like; *XTH6*: xyloglucan endotransglucosylase/hydrolase protein 6; *GOLS2*: galactinol synthase 2-like; *CuSOD1*: Superoxide dismutase [Cu-Zn]; $APX_{Chl}$: chloroplast ascorbate peroxidase. All primer sequences are presented in Table S1. The primers were designed using Primer3 web version 4.1.0 [84] with an e-value < $2 \times 10^{-4}$ and a score >41. cDNA was synthesized from 1 µg total RNA using the SensiFASTTM cDNA Synthesis kit (Meridian BioScience, Cincinnati, OH, USA), according to the manufacturer's recommendations. The presence of a single amplification product of the expected gene size was verified by electrophoresis on a 1.5% agarose gel. PCR reactions were prepared using the SensiFASTTM SYBR No-ROX kit (Meridian BioScience, USA) according to the manufacturer's protocol. One negative sample was included for each primer pair, in which cDNA was replaced by water. Reactions were carried out in 96-well plates using a qTOWER 2.2 Thermal Cycler (Analytik, Jena, Germany) using the following parameters: hot start activation of the Taq DNA polymerase at 95 °C for 10 min, followed by 40 cycles of denaturation at 95 °C for 15 s, annealing at 60 °C for 30 s, elongation at 72 °C for 30 s. A melting curve analysis was performed at the end of the PCR run by a continuous fluorescence measurement from 55 °C to 95 °C with sequential steps of 0.5 °C for 15 s. A single peak was obtained and no signal was detected in the negative controls. Three technical replicates were used for each analyzed plant. Gene expression was quantified using malate dehydrogenase (*MDH*) and ubiquitin (*UBQ10*) as reference genes [85]. To understand the agreement of these results with the one from RNA-sequencing, heatmaps were constructed considering the levels of transcripts and their expression levels from qRT-PCRs after being $\log_2$ FC scaled by gene expression across treatments.

## 5. Conclusions

In this study, we showed how the single and combined effects of drought and $eCO_2$ triggered wide but differential responses at the transcriptional level in *Coffea*.

MWD had a minor impact on the number of transcripts differentially regulated by Icatu and CL153, contrary to SWD where a high number of DEGs were reported, being mostly down-regulated. $eCO_2$ attenuated the impacts of drought in the two genotypes, but especially in Icatu, in agreement with the contrasting physiological tolerance previously reported in these genotypes.

There was a predominance of protective and ROS-scavenging genes, directly or indirectly related to ABA signaling pathways involving *Coffea* tolerance responses. These genes were also involved in water deprivation and desiccation processes, such as *LEA* and protein phosphatases in Icatu and Aspartic Protease in Guard Cell 1-like and dehydrins in CL153, being their expression confirmed by qRT-PCR.

Enrichment analysis of GO and KEGG pathways revealed different regulatory mechanisms of Icatu and CL153 in response to drought, agreeing with the minor effects of MWD and the positive action of $eCO_2$. However, a clear effect on photosynthetic pathways was recorded, namely under SWD and $eCO_2$, contrary to previous physiological and biochemical studies.

The existence of a complex post-transcriptional regulatory mechanism is suggested to occur in *Coffea* explaining the discrepancies between transcriptional vs. proteomic and physiological data in these genotypes.

**Supplementary Materials:** The following supporting information can be downloaded at: https://www.mdpi.com/article/10.3390/ijms24043210/s1.

**Author Contributions:** Conceptualization, A.I.R.-B. and J.C.R.; Data curation, I.M., I.F. and O.S.P.; Formal analysis, I.M., I.F., O.S.P. and D.B.; Funding acquisition, J.C.R., A.I.R.-B., F.C.L. and F.M.D.; Investigation, I.M., J.C.R., A.I.R.-B., F.C.L., F.M.D. and F.P.; Methodology, I.M., A.I.R.-B., O.S.P. and J.C.R.; Project administration, J.C.R. and A.I.R.-B.; Supervision, I.M., O.S.P., A.I.R.-B. and J.C.R.; Writing—Original draft, I.M., I.F., O.S.P., J.C.R. and A.I.R.-B.; Writing—Review and editing, all authors. All authors have read and agreed to the published version of the manuscript.

**Funding:** This work received funding support from the European Union's Horizon 2020 (H2020) research and innovation program (grant agreement No 727934, project BreedCAFS—Breeding Coffee for Agroforestry Systems), and from national funds from Fundação para a Ciência e a Tecnologia, I.P. (FCT), Portugal, through the project PTDC/ASP-AGR/31257/2017, through the Scientific Employment Stimulus—Individual Call (CEEC Individual)—2021.01107.CEECIND/CP1689/CT0001 (IM), and through the research units UIDB/00239/2020 (CEF) UID/AGR/04129/2020 (LEAF) and UIDP/04035/2020 (GeoBioTec). Fellowships from the Conselho Nacional de Desenvolvimento Cientifico e Tecnologico, Brazil (CNPq) for F.L.P. and F.M.D. and the Fundação de Amparo á Pesquisa do Estado de Minas Gerais, Brazil (FAPEMIG, project CRA-RED-00053-16; APQ01512-18, to F.M.D.) are also greatly acknowledged.

**Institutional Review Board Statement:** Not applicable.

**Informed Consent Statement:** Not applicable.

**Data Availability Statement:** Raw reads have been deposited in the NCBI Sequence Read Archive, BioProject accession PRJNA787748.

**Acknowledgments:** The authors would like to thank Novadelta–Comércio e Indústria de Cafés Lda., as well as Paula Alves for technical assistance.

**Conflicts of Interest:** The authors declare no conflict of interest.

## References

1. Solomon, S.; Plattner, G.K.; Knutti, R.; Friedlingstein, P. Irreversible climate change due to carbon dioxide emissions. *Proc. Natl. Acad. Sci. USA* **2009**, *106*, 1704–1709. [CrossRef] [PubMed]
2. Raza, A.; Mubarik, M.S.; Sharif, R.; Habib, M.; Jabeen, W.; Zhang, C.; Chen, H.; Chen, Z.-H.; Siddique, K.H.M.; Zhuang, W.; et al. The Plant Genome Developing drought-smart, ready-to-grow future crops. *Plant Genome* **2022**, e20279. [CrossRef] [PubMed]
3. Brodribb, T.J.; McAdam, S.A.M. Evolution of the Stomatal Regulation of Plant Water Content. *Plant Physiol.* **2017**, *174*, 639–649. [CrossRef]
4. Chaves, M.M.; Maroco, J.P.; Pereira, J.S. Understanding plant responses to drought-From genes to the whole plant. *Funct. Plant Biol.* **2003**, *30*, 239–264. [CrossRef] [PubMed]
5. Muller, B.; Pantin, F.; Génard, M.; Turc, O.; Freixes, S.; Piques, M.; Gibon, Y. Water deficits uncouple growth from photosynthesis, increase C content, and modify the relationships between C and growth in sink organs. *J. Exp. Bot.* **2011**, *62*, 1715–1729. [CrossRef]
6. Ramalho, J.C.; DaMatta, F.M.; Rodrigues, A.P.; Scotti-Campos, P.; Pais, I.; Batista-Santos, P.; Partelli, F.L.; Ribeiro, A.; Lidon, F.C.; Leitão, A.E. Cold impact and acclimation response of *Coffea* spp. plants. *Theor. Exp. Plant Physiol.* **2014**, *26*, 5–18. [CrossRef]
7. Ramalho, J.C.; Rodrigues, A.P.; Lidon, F.C.; Marques, L.M.C.; Leitão, A.E.; Fortunato, A.S.; Pais, I.P.; Silva, M.J.; Scotti-Campos, P.; Lopes, A.; et al. Stress cross-response of the antioxidative system promoted by superimposed drought and cold conditions in *Coffea* spp. *PLoS ONE* **2018**, *13*, e0198694. [CrossRef]
8. Fahad, S.; Bajwa, A.A.; Nazir, U.; Anjum, S.A.; Farooq, A.; Zohaib, A.; Sadia, S.; Nasim, W.; Adkins, S.; Saud, S.; et al. Crop production under drought and heat stress: Plant responses and management options. *Front. Plant Sci.* **2017**, *8*, 1147. [CrossRef]
9. Chaves, M.M.; Flexas, J.; Pinheiro, C. Photosynthesis under drought and salt stress: Regulation mechanisms from whole plant to cell. *Ann. Bot.* **2009**, *103*, 551–560. [CrossRef]
10. Ainsworth, E.A.; Rogers, A. The response of photosynthesis and stomatal conductance to rising [$CO_2$]: Mechanisms and environmental interactions. *Plant Cell Environ.* **2007**, *30*, 258–270. [CrossRef]
11. Leakey, A.D.B.; Ainsworth, E.A.; Bernacchi, C.J.; Rogers, A.; Long, S.P.; Ort, D.R. Elevated $CO_2$ effects on plant carbon, nitrogen, and water relations: Six important lessons from FACE. *J. Exp. Bot.* **2009**, *60*, 2859–2876. [CrossRef]
12. Semedo, J.N.; Rodrigues, A.P.; Lidon, F.C.; Pais, I.P.; Marques, I.; Gouveia, D.; Armengaud, J.; Silva, M.J.; Martins, S.; Semedo, M.C.; et al. Intrinsic non-stomatal resilience to drought of the photosynthetic apparatus in *Coffea* spp. is strengthened by elevated air [$CO_2$]. *Tree Physiol.* **2021**, *41*, 708–727. [CrossRef]
13. Tausz-Posch, S.; Tausz, M.; Bourgault, M. Elevated [$CO_2$] effects on crops: Advances in understanding acclimation, nitrogen dynamics and interactions with drought and other organisms. *Plant Biol.* **2020**, *22*, 38–51. [CrossRef]
14. De Souza, A.P.; Cocuron, J.-C.; Garcia, A.C.; Alonso, A.P.; Buckeridge, M.S. Changes in Whole-Plant Metabolism during the Grain-Filling Stage in Sorghum Grown under Elevated $CO_2$ and Drought. *Plant Physiol.* **2015**, *169*, 1755–1765. [CrossRef]
15. Gray, S.B.; Dermody, O.; Klein, S.P.; Locke, A.M.; McGrath, J.M.; Paul, R.E.; Rosenthal, D.M.; Ruiz-Vera, U.M.; Siebers, M.H.; Strellner, R.; et al. Intensifying drought eliminates the expected benefits of elevated carbon dioxide for soybean. *Nat. Plants* **2016**, *2*, 16132. [CrossRef]
16. ICO International Coffee Organization-What's New. Available online: https://www.ico.org/ (accessed on 22 November 2022).
17. DaMatta, F.M.; Rahn, E.; Läderach, P.; Ghini, R.; Ramalho, J.C. Why could the coffee crop endure climate change and global warming to a greater extent than previously estimated? *Clim. Change* **2019**, *152*, 167–178. [CrossRef]
18. Cassamo, C.T.; Mangueze, A.V.J.; Leitão, A.E.; Pais, I.P.; Moreira, R.; Campa, C.; Chiulele, R.; Reis, F.O.; Marques, I.; Scotti-Campos, P.; et al. Shade and Altitude Implications on the Physical and Chemical Attributes of Green Coffee Beans from Gorongosa Mountain, Mozambique. *Agronomy* **2022**, *12*, 2540. [CrossRef]
19. DaMatta, F.M.; Cochicho Ramalho, J.D. Impacts of drought and temperature stress on coffee physiology and production: A review. *Braz. J. Plant Physiol.* **2006**, *18*, 55–81. [CrossRef]
20. Vinecky, F.; Davrieux, F.; Mera, A.C.; Alves, G.S.C.; Lavagnini, G.; Leroy, T.; Bonnot, F.; Rocha, O.C.; Bartholo, G.F.; Guerra, A.F.; et al. Controlled irrigation and nitrogen, phosphorous and potassium fertilization affect the biochemical composition and quality of Arabica coffee beans. *J. Agric. Sci.* **2017**, *155*, 902–918. [CrossRef]
21. Avila, R.T.; Cardoso, A.A.; de Almeida, W.L.; Costa, L.C.; Machado, K.L.G.; Barbosa, M.L.; de Souza, R.P.B.; Oliveira, L.A.; Batista, D.S.; Martins, S.C.V.; et al. Coffee plants respond to drought and elevated [$CO_2$] through changes in stomatal function, plant hydraulic conductance, and aquaporin expression. *Environ. Exp. Bot.* **2020**, *177*, 104148. [CrossRef]
22. Avila, R.T.; de Almeida, W.L.; Costa, L.C.; Machado, K.L.G.; Barbosa, M.L.; de Souza, R.P.B.; Martino, P.B.; Juárez, M.A.T.; Marçal, D.M.S.; Martins, S.C.V.; et al. Elevated air [$CO_2$] improves photosynthetic performance and alters biomass accumulation and partitioning in drought-stressed coffee plants. *Environ. Exp. Bot.* **2020**, *177*, 104137. [CrossRef]
23. Sanches, R.F.E.; da Cruz Centeno, D.; Braga, M.R.; da Silva, E.A. Impact of high atmospheric $CO_2$ concentrations on the seasonality of water-related processes, gas exchange, and carbohydrate metabolism in coffee trees under field conditions. *Clim. Chang.* **2020**, *162*, 1231–1248. [CrossRef]

24. Catarino, I.C.A.; Monteiro, G.B.; Ferreira, M.J.P.; Torres, L.M.B.; Domingues, D.S.; Centeno, D.C.; Lobo, A.K.M.; Silva, E.A. Elevated [$CO_2$] Mitigates Drought Effects and Increases Leaf 5-O-Caffeoylquinic Acid and Caffeine Concentrations during the Early Growth of Coffea Arabica Plants. *Front. Sustain. Food Syst.* **2021**, *5*, 266. [CrossRef]
25. Rodrigues, A.M.; Jorge, T.; Osorio, S.; Pott, D.M.; Lidon, F.C.; DaMatta, F.M.; Marques, I.; Ribeiro-barros, A.I.; Ramalho, J.C.; António, C. Primary Metabolite Profile Changes in *Coffea* spp. Promoted by Single and Combined Exposure to Drought and Elevated $CO_2$ Concentration. *Metabolites* **2021**, *11*, 427. [CrossRef]
26. Dubberstein, D.; Lidon, F.C.; Rodrigues, A.P.; Semedo, J.N.; Marques, I.; Rodrigues, W.P.; Gouveia, D.; Armengaud, J.; Semedo, M.C.; Martins, S.; et al. Resilient and Sensitive Key Points of the Photosynthetic Machinery of *Coffea* spp. to the Single and Superimposed Exposure to Severe Drought and Heat Stresses. *Front. Plant Sci.* **2020**, *11*, 1049. [CrossRef]
27. Marques, I.; Rodrigues, A.P.; Gouveia, D.; Lidon, F.C.; Martins, S.; Semedo, M.C.; Gaillard, J.C.; Pais, I.P.; Semedo, J.N.; Scotti-Campos, P.; et al. High-resolution shotgun proteomics reveals that increased air [$CO_2$] amplifies the acclimation response of *Coffea* species to drought regarding antioxidative, energy, sugar, and lipid dynamics. *J. Plant Physiol.* **2022**, *276*, 153788. [CrossRef]
28. Vranová, E.; Langebartels, C.; Van Montagu, M.; Inzé, D.; Van Camp, W. Oxidative stress, heat shock and drought differentially affect expression of a tobacco protein phosphatase 2C. *J. Exp. Bot.* **2000**, *51*, 1763–1764. [CrossRef]
29. Merlot, S.; Gosti, F.; Guerrier, D.; Vavasseur, A.; Giraudat, J. The ABI1 and ABI2 protein phosphatases 2C act in a negative feedback regulatory loop of the abscisic acid signalling pathway. *Plant J.* **2001**, *25*, 295–303. [CrossRef]
30. Komatsu, K.; Suzuki, N.; Kuwamura, M.; Nishikawa, Y.; Nakatani, M.; Ohtawa, H.; Takezawa, D.; Seki, M.; Tanaka, M.; Taji, T.; et al. Group A PP2Cs evolved in land plants as key regulators of intrinsic desiccation tolerance. *Nat. Commun.* **2013**, *4*, 2219. [CrossRef]
31. Sun, L.; Wang, Y.P.; Chen, P.; Ren, J.; Ji, K.; Li, Q.; Li, P.; Dai, S.J.; Leng, P. Transcriptional regulation of SlPYL, SlPP2C, and SlSnRK2 gene families encoding ABA signal core components during tomato fruit development and drought stress. *J. Exp. Bot.* **2011**, *62*, 5659–5669. [CrossRef]
32. Chen, J.; Zhang, D.; Zhang, C.; Xia, X.; Yin, W.; Tian, Q. A Putative PP2C-encoding gene negatively regulates ABA signaling in populus euphratica. *PLoS ONE* **2015**, *10*, e0139466. [CrossRef]
33. Zhang, F.; Fu, X.; Lv, Z.; Shen, Q.; Yan, T.; Jiang, W.; Wang, G.; Sun, X.; Tang, K. Type 2C phosphatase 1 of *Artemisia annua* L. is a negative regulator of ABA signaling. *Biomed Res. Int.* **2014**, *2014*, 521794. [CrossRef]
34. D'Ippólito, S.; Rey-Burusco, M.F.; Feingold, S.E.; Guevara, M.G. Role of proteases in the response of plants to drought. *Plant Physiol. Biochem.* **2021**, *168*, 1–9. [CrossRef]
35. Meinhard, M.; Rodriguez, P.L.; Grill, E. The sensitivity of ABI2 to hydrogen peroxide links the abscisic acid-response regulator to redox signalling. *Planta* **2002**, *214*, 775–782. [CrossRef]
36. Mittler, R. Oxidative stress, antioxidants and stress tolerance. *Trends Plant Sci.* **2002**, *7*, 405–410. [CrossRef]
37. Liu, Y.; Song, Q.; Li, D.; Yang, X.; Li, D. Multifunctional roles of plant dehydrins in response to environmental stresses. *Front. Plant Sci.* **2017**, *8*, 1018. [CrossRef]
38. Santos, A.B.; Mazzafera, P. Dehydrins Are Highly Expressed in Water-Stressed Plants of Two Coffee Species. *Trop. Plant Biol.* **2012**, *5*, 218–232. [CrossRef]
39. Fernandes, I.; Marques, I.; Paulo, O.S.; Batista, D.; Partelli, F.L.; Lidon, F.C.; DaMatta, F.M.; Ramalho, J.C.; Ribeiro-Barros, A.I. Understanding the impact of drought in *Coffea* genotypes: Transcriptomic analysis supports a common high resilience to moderate water deficit but a genotype dependent sensitivity to severe water deficit. *Agronomy* **2021**, *11*, 2255. [CrossRef]
40. Noctor, G.; Reichheld, J.P.; Foyer, C.H. ROS-related redox regulation and signaling in plants. *Semin. Cell Dev. Biol.* **2018**, *80*, 3–12. [CrossRef]
41. Dumanović, J.; Nepovimova, E.; Natić, M.; Kuča, K.; Jaćević, V. The Significance of Reactive Oxygen Species and Antioxidant Defense System in Plants: A Concise Overview. *Front. Plant Sci.* **2021**, *11*, 552969. [CrossRef]
42. Martins, M.Q.; Rodrigues, W.P.; Fortunato, A.S.; Leitão, A.E.; Rodrigues, A.P.; Pais, I.P.; Martins, L.D.; Silva, M.J.; Reboredo, F.H.; Partelli, F.L.; et al. Protective response mechanisms to heat stress in interaction with high [$CO_2$] conditions in *Coffea* spp. *Front. Plant Sci.* **2016**, *7*, 947. [CrossRef] [PubMed]
43. Rodrigues, W.P.; Martins, M.Q.; Fortunato, A.S.; Rodrigues, A.P.; Semedo, J.N.; Simões-Costa, M.C.; Pais, I.P.; Leitão, A.E.; Colwell, F.; Goulao, L.; et al. Long-term elevated air [$CO_2$] strengthens photosynthetic functioning and mitigates the impact of supra-optimal temperatures in tropical *Coffea arabica* and *C. canephora* species. *Glob. Chang. Biol.* **2016**, *22*, 415–431. [CrossRef] [PubMed]
44. Vinci, G.; Marques, I.; Rodrigues, A.P.; Martins, S.; Leitão, A.E.; Semedo, M.C.; Silva, M.J.; Lidon, F.C.; DaMatta, F.M.; Ribeiro-Barros, A.I.; et al. Protective Responses at the Biochemical and Molecular Level Differ between a *Coffea arabica* L. Hybrid and Its Parental Genotypes to Supra-Optimal Temperatures and Elevated Air [$CO_2$]. *Plants* **2022**, *11*, 2702. [CrossRef]
45. Fortunato, A.S.; Lidon, F.C.; Batista-Santos, P.; Eduardo Leitão, A.; Pais, I.P.; Ribeiro, A.I.; Cochicho Ramalho, J. Biochemical and molecular characterization of the antioxidative system of *Coffea* sp. under cold conditions in genotypes with contrasting tolerance. *J. Plant Physiol.* **2010**, *167*, 333–342. [CrossRef]
46. Ramalho, J.C.; Fortunato, A.S.; Goulao, L.F.; Lidon, F.C. Cold-induced changes in mineral content in leaves of *Coffea* spp. Identification of descriptors for tolerance assessment. *Biol. Plant.* **2013**, *57*, 495–506. [CrossRef]

47. Ramalho, J.C.; Campos, P.S.; Teixeira, M.; Nunes, M.A. Nitrogen dependent changes in antioxidant system and in fatty acid composition of chloroplast membranes from *Coffea arabica* L. plants submitted to high irradiance. *Plant Sci.* **1998**, *135*, 115–124. [CrossRef]
48. Ramalho, J.C.; Pons, T.L.; Groeneveld, H.W.; Nunes, M.A. Photosynthetic responses of *Coffea arabica* leaves to a short-term high light exposure in relation to N availability. *Physiol. Plant.* **1997**, *101*, 229–239. [CrossRef]
49. Miniussi, M.; Del Terra, L.; Savi, T.; Pallavicini, A.; Nardini, A. Aquaporins in Coffea arabica L.: Identification, expression, and impacts on plant water relations and hydraulics. *Plant Physiol. Biochem.* **2015**, *95*, 92–102. [CrossRef]
50. Marques, I.; Fernandes, I.; Paulo, O.S.; Lidon, F.C.; DaMatta, F.M.; Ramalho, J.C.; Ribeiro-barros, A.I. A transcriptomic approach to understanding the combined impacts of supra-optimal temperatures and $CO_2$ revealed different responses in the polyploid *Coffea arabica* and its diploid progenitor *C. canephora*. *Int. J. Mol. Sci.* **2021**, *22*, 3125. [CrossRef]
51. Devnarain, N.; Crampton, B.G.; Olivier, N.; van der Westhuyzen, C.; Becker, J.V.W.; O'Kennedy, M.M. Transcriptomic analysis of a *Sorghum bicolor* landrace identifies a role for beta-alanine betaine biosynthesis in drought tolerance. *S. Afr. J. Bot.* **2019**, *127*, 244–255. [CrossRef]
52. Aziz, M.A.; Sabeem, M.; Mullath, S.K.; Brini, F.; Masmoudi, K. Plant Group II LEA Proteins: Intrinsically Disordered Structure for Multiple Functions in Response to Environmental Stresses. *Biomolecules* **2021**, *11*, 1662. [CrossRef]
53. Magwanga, R.O.; Lu, P.; Kirungu, J.N.; Lu, H.; Wang, X.; Cai, X.; Zhou, Z.; Zhang, Z.; Salih, H.; Wang, K.; et al. Characterization of the late embryogenesis abundant (LEA) proteins family and their role in drought stress tolerance in upland cotton. *BMC Genet.* **2018**, *19*, 6. [CrossRef]
54. Buckley, T.N. How do stomata respond to water status? *New Phytol.* **2019**, *224*, 21–36. [CrossRef]
55. Cui, L.; Yao, S.; Dai, X.; Yin, Q.; Liu, Y.; Jiang, X.; Wu, Y.; Qian, Y.; Pang, Y.; Gao, L.; et al. Identification of UDP-glycosyltransferases involved in the biosynthesis of astringent taste compounds in tea (*Camellia sinensis*). *J. Exp. Bot.* **2016**, *67*, 2285–2297. [CrossRef]
56. Dong, N.-Q.; Sun, Y.; Guo, T.; Shi, C.-L.; Zhang, Y.-M.; Kan, Y.; Xiang, Y.-H.; Zhang, H.; Yang, Y.-B.; Li, Y.-C.; et al. UDP-glucosyltransferase regulates grain size and abiotic stress tolerance associated with metabolic flux redirection in rice. *Nat. Commun.* **2020**, *11*, 2629. [CrossRef]
57. Jarzyniak, K.M.; Jasiński, M. Membrane transporters and drought resistance-A complex issue. *Front. Plant Sci.* **2014**, *5*, 687. [CrossRef]
58. Ahmed, U.; Rao, M.J.; Qi, C.; Xie, Q.; Noushahi, H.A.; Yaseen, M.; Shi, X.; Zheng, B. Expression profiling of flavonoid biosynthesis genes and secondary metabolites accumulation in populus under drought stress. *Molecules* **2021**, *26*, 5546. [CrossRef]
59. Watkins, J.M.; Hechler, P.J.; Muday, G.K. Ethylene-induced flavonol accumulation in guard cells suppresses reactive oxygen species and moderates stomatal aperture. *Plant Physiol.* **2014**, *164*, 1707–1717. [CrossRef]
60. Takahama, U. Oxidation of vacuolar and apoplastic phenolic substrates by peroxidase: Physiological significance of the oxidation reactions. *Phytochem. Rev.* **2004**, *3*, 207–219. [CrossRef]
61. Ferreres, F.; Figueiredo, R.; Bettencourt, S.; Carqueijeiro, I.; Oliveira, J.; Gil-Izquierdo, A.; Pereira, D.M.; Valentão, P.; Andrade, P.B.; Duarte, P.; et al. Identification of phenolic compounds in isolated vacuoles of the medicinal plant *Catharanthus roseus* and their interaction with vacuolar class III peroxidase: An $H_2O_2$ affair? *J. Exp. Bot.* **2011**, *62*, 2841–2854. [CrossRef]
62. Marques, I.; Gouveia, D.; Gaillard, J.C.; Martins, S.; Semedo, M.C.; Lidon, F.C.; DaMatta, F.M.; Ribeiro-Barros, A.I.; Armengaud, J.; Ramalho, J.C. Next-Generation Proteomics Reveals a Greater Antioxidative Response to Drought in *Coffea arabica* than in *Coffea Canephora*. *Agronomy* **2022**, *12*, 148. [CrossRef]
63. Gray, S.B.; Rodriguez-Medina, J.; Rusoff, S.; Toal, T.W.; Kajala, K.; Runcie, D.E.; Brady, S.M. Translational regulation contributes to the elevated $CO_2$ response in two *Solanum* species. *Plant J.* **2020**, *102*, 383–397. [CrossRef] [PubMed]
64. Shenton, D.; Smirnova, J.B.; Selley, J.N.; Carroll, K.; Hubbard, S.J.; Pavitt, G.D.; Ashe, M.P.; Grant, C.M. Global Translational Responses to Oxidative Stress Impact upon Multiple Levels of Protein Synthesis. *J. Biol. Chem.* **2006**, *281*, 29011–29021. [CrossRef]
65. Ramalho, J.C.; Rodrigues, A.P.; Semedo, J.N.; Pais, I.P.; Martins, L.D.; Simões-Costa, M.C.; Leitão, A.E.; Fortunato, A.S.; Batista-Santos, P.; Palos, I.M.; et al. Sustained photosynthetic performance of *Coffea* spp. under long-term enhanced [$CO_2$]. *PLoS ONE* **2013**, *8*, 82712. [CrossRef] [PubMed]
66. Andrews, S. FastQC: A Quality Control Tool for High Throughput Sequence Data. 2010. Available online: http://www.bioinformatics.babraham.ac.uk/projects/fastqc (accessed on 4 April 2021).
67. Wingett, S.W.; Andrews, S. FastQ Screen: A tool for multi-genome mapping and quality control. *F1000Research* **2018**, *7*, 1338. [CrossRef]
68. Liao, Y.; Shi, W. Read trimming is not required for mapping and quantification of RNA-seq reads at the gene level. *NAR Genom. Bioinforma.* **2020**, *2*, lqaa068. [CrossRef]
69. Denoeud, F.; Carretero-Paulet, L.; Dereeper, A.; Droc, G.; Guyot, R.; Pietrella, M.; Zheng, C.; Alberti, A.; Anthony, F.; Aprea, G.; et al. The coffee genome provides insight into the convergent evolution of caffeine biosynthesis. *Science* **2014**, *345*, 1181–1184. [CrossRef]
70. Dobin, A.; Davis, C.A.; Schlesinger, F.; Drenkow, J.; Zaleski, C.; Jha, S.; Batut, P.; Chaisson, M.; Gingeras, T.R. STAR: Ultrafast universal RNA-seq aligner. *Bioinformatics* **2013**, *29*, 15–21. [CrossRef]
71. Anders, S.; Pyl, P.T.; Huber, W. HTSeq—A Python framework to work with high-throughput sequencing data. *Bioinformatics* **2015**, *31*, 166. [CrossRef]

72. Li, H.; Handsaker, B.; Wysoker, A.; Fennell, T.; Ruan, J.; Homer, N.; Marth, G.; Abecasis, G.; Durbin, R. The Sequence Alignment/Map format and SAMtools. *Bioinformatics* **2009**, *25*, 2078–2079. [CrossRef]
73. Pertea, G.G. gffread: GFF/GTF Utility Providing Format Conversions, Region Filtering, FASTA Sequence Extraction and More. 2015. Available online: https://github.com/gpertea/gffread (accessed on 15 January 2021).
74. Wickham, H. *ggplot2: Elegant Graphics for Data Analysis*; Springer: New York, NY, USA, 2016; ISBN 978-3-319-24277-4.
75. Team, R.C. *R: A Language and Environment for Statistical Computing*; R Foundation for Statistical Computing: Vienna, Austria, 2018.
76. Love, M.I.; Huber, W.; Anders, S. Moderated estimation of fold change and dispersion for RNA-seq data with DESeq2. *Genome Biol.* **2014**, *15*, 550. [CrossRef]
77. Robinson, M.D.; McCarthy, D.J.; Smyth, G.K. edgeR: A Bioconductor package for differential expression analysis of digital gene expression data. *Bioinformatics* **2009**, *26*, 139–140. [CrossRef]
78. Benjamini, Y.; Hochberg, Y. On the Adaptive Control of the False Discovery Rate in Multiple Testing With Independent Statistics. *J. Educ. Behav. Stat.* **2016**, *25*, 60–83. [CrossRef]
79. Hunter, J.D. Matplotlib: A 2D graphics environment. *Comput. Sci. Eng.* **2007**, *9*, 90–95. [CrossRef]
80. Thimm, O.; Bla È Sing, O.; Gibon, Y.; Nagel, A.; Meyer, S.; Kru È Ger, P.; Selbig, J.; Mu È Ller, L.A.; Rhee, S.Y.; Stitt, M. MAPMAN: A user-driven tool to display genomics data sets onto diagrams of metabolic pathways and other biological processes. *Plant J.* **2004**, *37*, 914–939. [CrossRef]
81. Raudvere, U.; Kolberg, L.; Kuzmin, I.; Arak, T.; Adler, P.; Peterson, H.; Vilo, J. G: Profiler: A web server for functional enrichment analysis and conversions of gene lists (2019 update). *Nucleic Acids Res.* **2019**, *47*, W191–W198. [CrossRef]
82. Supek, F.; Bošnjak, M.; Škunca, N.; Šmuc, T. Revigo summarizes and visualizes long lists of gene ontology terms. *PLoS ONE* **2011**, *6*, 21800. [CrossRef]
83. Chen, C.; Huang, H.; Wu, C.H. Protein bioinformatics databases and resources. In *Methods in Molecular Biology*; Humana Press Inc.: Totowa, NJ, USA, 2017; Volume 1558, pp. 3–39. [CrossRef]
84. Untergasser, A.; Cutcutache, I.; Koressaar, T.; Ye, J.; Faircloth, B.C.; Remm, M.; Rozen, S.G. Primer3-new capabilities and interfaces. *Nucleic Acids Res.* **2012**, *40*, e115. [CrossRef]
85. Martins, M.Q.; Fortunato, A.S.; Rodrigues, W.P.; Partelli, F.L.; Campostrini, E.; Lidon, F.C.; DaMatta, F.M.; Ramalho, J.C.; Ribeiro-Barros, A.I. Selection and Validation of Reference Genes for Accurate RT-qPCR Data Normalization in *Coffea* spp. under a Climate Changes Context of Interacting Elevated [$CO_2$] and Temperature. *Front. Plant Sci.* **2017**, *8*, 307. [CrossRef]

International Journal of
*Molecular Sciences*

MDPI

*Article*

# Mechanisms of Kale (*Brassica oleracea* var. *acephala*) Tolerance to Individual and Combined Stresses of Drought and Elevated Temperature

**Nataša Bauer [1,†], Mirta Tkalec [1,†], Nikola Major [2,3], Ana Talanga Vasari [1], Mirta Tokić [1], Sandra Vitko [1], Dean Ban [2,3], Smiljana Goreta Ban [2,3,*] and Branka Salopek-Sondi [4,*]**

1 Department for Molecular Biology, Faculty of Science, University of Zagreb, 10000 Zagreb, Croatia
2 Institute of Agriculture and Tourism, 52440 Poreč, Croatia
3 Centre of Excellence for Biodiversity and Molecular Plant Breeding, 10000 Zagreb, Croatia
4 Laboratory for Chemical Biology, Department for Molecular Biology, Ruđer Bošković Institute, 10000 Zagreb, Croatia
* Correspondence: smilja@iptpo.hr (S.G.B.); salopek@irb.hr (B.S.-S.); Tel.: +385-99-338-1304 (S.G.B.); +385-14-561-143 (B.S.-S.)
† These authors contributed equally to this work.

**Abstract:** Rising temperatures and pronounced drought are significantly affecting biodiversity worldwide and reducing yields and quality of Brassica crops. To elucidate the mechanisms of tolerance, 33 kale accessions (*B. oleracea* var. *acephala*) were evaluated for individual (osmotic and elevated temperature stress) and combined stress (osmotic + temperature). Using root growth, biomass and proline content as reliable markers, accessions were evaluated for stress responses. Four representatives were selected for further investigation (photosynthetic performance, biochemical markers, sugar content, specialized metabolites, transcription level of transcription factors *NAC*, *HSF*, *DREB* and expression of heat shock proteins HSP70 and HSP90): very sensitive (392), moderately sensitive (395), tolerant (404) and most tolerant (411). Accessions more tolerant to stress conditions were characterized by higher basal content of proline, total sugars, glucosinolates and higher transcription of *NAC* and *DREB*. Under all stress conditions, 392 was characterized by a significant decrease in biomass, root growth, photosynthesis performance, fructan content, especially under osmotic and combined stress, a significant increase in *HSF* transcription and HSP accumulation under temperature stress and a significant decrease in *NAC* transcription under all stresses. The most tolerant accession under all applied stresses, 411 showed the least changes in all analyzed parameters compared with the other accessions.

**Keywords:** growth performance; heat shock proteins; photosynthetic efficiency; stress markers; *NAC*; *HSFs*; *DREB*

**Citation:** Bauer, N.; Tkalec, M.; Major, N.; Talanga Vasari, A.; Tokić, M.; Vitko, S.; Ban, D.; Ban, S.G.; Salopek-Sondi, B. Mechanisms of Kale (*Brassica oleracea* var. *acephala*) Tolerance to Individual and Combined Stresses of Drought and Elevated Temperature. *Int. J. Mol. Sci.* **2022**, *23*, 11494. https://doi.org/10.3390/ijms231911494

Academic Editors: José C. Ramalho, Isabel Marques and Ana I Ribeiro-Barros

Received: 6 September 2022
Accepted: 23 September 2022
Published: 29 September 2022



## 1. Introduction

The Brassicaceae family is cultivated worldwide and includes many economically important species. An important agricultural genus in the Brassicaceae family is *Brassica*, which includes oilseeds (canola, mustard) and vegetables (broccoli, cabbage, cauliflower, kale, kohlrabi) that have long been used for agriculture on all continents [1]. Brassica vegetables have attracted public and scientific attention because of their health potential due to their richness in "healthy phytochemicals" (carotenoids, phenols, glucosinolates, vitamins) and their great adaptability to climatic conditions that allow for their cultivation all over the world. Various epidemiological and meta-analyses suggest that consumption of *Brassica* vegetables plays a preventive role against a number of chronic diseases due to its antioxidant, antimicrobial and anticancer effects and these findings allow for the recognition of *Brassica* vegetables as a functional food [2,3]. One cruciferous vegetable that frequently appears on lists of 'healthiest foods' or 'superfoods' is kale (*Brassica oleracea* var. *acephala)*, a leafy, non-headed cabbage. Although kale has been cultivated for several centuries and is found in many traditional dishes, especially in the Mediterranean region, it became very popular in the United States after 2010 [3].

However, the cultivation of brassicas, like that of other crops, is strongly affected by climate change. Increasing temperatures and decreasing precipitation are causing abiotic stress in plants, which can significantly affect biodiversity worldwide and reduce crop yields and quality of agriculturally important *Brassica* species, which, in turn, affects food chain resilience. Approximately 40% of the world's land area is affected by drought. Average surface temperatures have already increased by about 1 °C from 1990 to the present and global average surface temperatures are expected to increase by 2.6 to 4.8 °C by the end of the century, resulting in a warmer and drier climate [4]. Drought and heat stress usually occur together and are the major obstacles to crop production [5]. The problem is particularly severe in Mediterranean, semi-arid and arid regions and is expected to worsen in the future [6]. High temperatures affect plant growth and development of Brassica species and lead to yield losses. For example, a yield reduction of 17% was observed in Brassica plants exposed to a temperature increase of 1 °C [7].

Plant response to stress caused by abiotic factors is a complex trait regulated by many factors (gene expression, plant hormones, antioxidants, osmoprotectants, etc.) [8]. Transcriptional regulation of stress-responsive genes is an essential step in determining the mechanisms of tolerance to abiotic stress. Transcription factors (TFs), as major regulators of gene expression, may be important targets for the development of crops with enhanced abiotic stress tolerance. Several transcription factors, such as heat shock transcription factors (HSFs), dehydration-responsive element binding protein (DREB) and NAC (NAM, ATAF and CUC domain) are reported as key regulators under stress conditions that control the transcription of stress-related genes and the synthesis of stress-responsive proteins that ensure plant survival under stress conditions [9,10].

Our recent comparative studies of *Brassica* species have shown that kale is more tolerant to salinity and drought compared to Chinese cabbage and white cabbage [11,12]. Furthermore, kale shows good tolerance to extreme temperature fluctuations [3]. Due to its good tolerance to abiotic stresses caused by climate change, kale has become very popular among farmers in recent decades. Kale production significantly increased in the US, from 3994 to 6256 harvested acres, in the period from 2007 to 2012, respectively [13]. In the last few decades, morphological, agronomical, genetical and phytochemical characteristics of local kale varieties have been considered and investigated in many countries [3].

Kale is commonly grown along the coastal region of Croatia on small family farms and with low commercial importance. Since each farmer is preserving their own seeds, a significant morphological diversity among kale populations is present in the region [14]. There is a possibility that more tolerant accessions to abiotic stress are present in the area and, therefore, could provide important genetic resources for selection of the cultivars more adapted to climate changes.

Although many studies have been focused to shed light on tolerance mechanisms [8], there is still a large gap in understanding the complex network of multiple processes involved in the mechanisms of stress tolerance. Therefore, research on the mechanisms of tolerance is of great importance from an ecological and agronomic point of view [15].

The main objective of this work was to investigate the tolerance of 33 local kale accessions (*B. oleracea* var. *acephala*) to individual (osmotic and heat stress) and combined stress (osmotic + heat). Based on growth performance (root growth and biomass) and proline level as a reliable stress marker, the most tolerant and the most sensitive accessions were selected to shed light on the mechanisms of abiotic stress tolerance.

## 2. Results

### 2.1. Screening of Kale Accessions to Abiotic Stresses

To shed light on the mechanisms of abiotic stress tolerance, 33 kale accessions (Table 1) were analyzed for elevated temperature (T), osmotic stress (M) and combined stress (elevated temperature + osmotic stress; TM) (Figure 1A, Table S1).

**Table 1.** List of kale (*B. oleracea* var. *acephala*) accessions used in tolerance evaluation. Kale accessions are part of the seed collection of Institute for Agriculture and Tourism Poreč.

| Acc. | City | Country | Coordinates | |
|---|---|---|---|---|
| 202 | Kaštelir | Croatia | 45°18′21.9708″ N | 13°41′11.7018″ E |
| 206 | Labinci | Croatia | 45°17′38.3244″ N | 13°41′12.6636″ E |
| 379 | Fuškulin | Croatia | 45°10′57.3492″ N | 13°38′32.5608″ E |
| 380 | Island Iž 2. | Croatia | 44°0′56.74″ N | 15°8′21.19″ E |
| 381 | Preko, island Ugljan | Croatia | 44°4′45.5916″ N | 15°11′14.5212″ E |
| 383 | Vrgorac | Croatia | 43°12′4.3806″ N | 17°23′16.0326″ E |
| 384 | Konavle, Pavlje Brdo | Croatia | 42°30′55.0296″ N | 18°20′50.5494″ E |
| 385 | Vitina, Mostar | Bosnia and Herzegovina | 43°14′33.5178″ N | 17°28′29.3694″ E |
| 386 | Vrgorac | Croatia | 43°11′28.1904″ N | 17°21′53.643″ E |
| 387 | Ponikve, Pelješac | Croatia | 42°50′52.983″ N | 17°37′36.786″ E |
| 388 | Topići, Baška voda | Croatia | 43°22′0.0912″ N | 16°57′41.2698″ E |
| 390 | Island Lošinj | Croatia | 44°34′53.097″ N | 14°24′21.9018″ E |
| 391 | Ugljan, islland Ugljan | Croatia | 44°8′0.1392″ N | 15°6′8.7804″ E |
| 392 | Dubrovnik | Croatia | 42°38′40.8264″ N | 18°7′21.2592″ E |
| 393 | Mostar | Bosnia and Herzegovina | 43°21′35.7186″ N | 17°49′7.269″ E |
| 394 | Oključina, island Vis | Croatia | 43°4′17.7348″ N | 16°6′36.1296″ E |
| 395 | Mostar | Bosnia and Herzegovina | 43°20′25.2564″ N | 17°43′28.1274″ E |
| 396 | Island Lošinj | Croatia | 44°41′9.942″ N | 14°22′12.864″ E |
| 397 | Vrgorac | Croatia | 43°12′25.7106″ N | 17°21′36.8496″ E |
| 398 | Vrgorac, Prapatnice | Croatia | 43°13′43.85″ N | 17°20′59.03″ E |
| 399 | Čarsko polje, island Korčula | Croatia | 42°55′59.8362″ N | 16°56′6.0828″ E |
| 401 | Zavalatica, island Korčula | Croatia | 42°55′7.0968″ N | 16°56′9.4272″ E |
| 402 | Katuni | Croatia | 43°27′45.39″ N | 16°53′10.39″ E |
| 403 | Drinovci | Croatia | 43°21′33.0696″ N | 17°19′23.9874″ E |
| 404 | Blato na Cetini, Omiš | Croatia | 43°28′34.87″ N | 16°49′7.39″ E |
| 405 | Opuzen | Croatia | 43°0′19.28″ N | 17°33′34.69″ E |
| 406 | Babino polje, isl. Mljet | Croatia | 42°44′22.63″ N | 17°30′2.93″ E |
| 408 | Poreč-Pištan | Croatia | 45°13′29.33″ N | 13°37′39.76″ E |
| 411 | Island Iž 1. | Croatia | 44°2′37.248″ N | 15°6′43.671″ E |
| 413 | Tomislav grad | Bosnia and Herzegovina | 43°43′7.39″ N | 17°13′25.07″ E |
| 414 | Srijane | Croatia | 43°31′20.94″ N | 16°41′14.45″ E |
| 416 | Vinjani donji | Croatia | 43°26′28.23″ N | 17°14′25.66″ E |
| 423 | Kreševo | Croatia | 43°29′14.65″ N | 16°52′35.42″ E |

As can be seen, kale accessions were grouped based on their growth performance (biomass accumulation and root growth), water content and proline content as reliable stress markers (Figure 1). The more tolerant accessions were characterized by higher (more intense red color) biomass production, greater root length, higher water content and lower (more intense green color) proline content under stress conditions. The most severe growth inhibition (biomass accumulation and root growth), reduction in water content and increase in proline content were associated with osmotic stress (M) and combined stress (TM), while temperature stress alone (T) caused less noticeable changes compared to controls (C) (Table S1).

### 2.2. Selection of Tolerant and Sensitive Accessions

Based on growth parameters (biomass accumulation and root length), water content and proline level, four kale accessions with different stress responses were selected: 392, 395, 404 and 411 (Figure 2). These accessions were selected as representative models for a group of accessions that responded similarly to the stress factors studied. Accessions 392 and 395 were considered to be more sensitive representatives among examined accessions, while accessions 404 and 411 were considered to be more tolerant representatives of the 33 studied accessions to the applied stresses compared to the corresponding controls (Figure 2A). As can be seen, root length and proline level were the most distinctive parameters between tolerant and sensitive accessions (Figure 2B). Root lengths reached 22%, 61% and 7% of length in accession 392 and 32%, 58% and 27% of length in accession 395 under osmotic, temperature and combined stress, respectively, compared to the corresponding controls. On the other hand, in more tolerant accessions, root length reached 70%, 75% and 41% of length in accession 404 and 59%, 82% and 56% of length in accession 411 under osmotic, temperature and combined stress, respectively, compared to the corresponding controls. Proline content was significantly increased under stress conditions, but more prominently in the sensitive accessions (up to 14.2-fold in accession 392 under osmotic stress and up to 17.4-fold in accession 395 under osmotic stress compared to the controls) than in the more tolerant ones (up to 8.6-fold in accession 404 under combined stress and 6.9-fold in accession 411 under osmotic stress compared with controls) (Table S1).

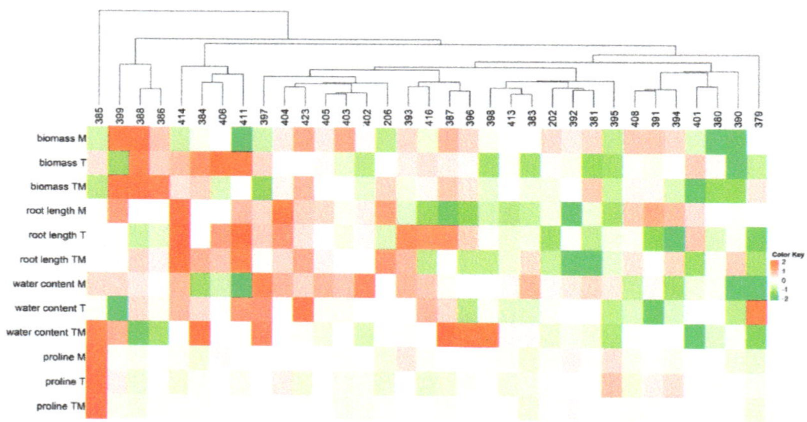


**Figure 1.** Heatmap and cluster analysis of biomass, root length, water and proline content in 33 *Brassica oleracea* var. *acephala* accessions. Abbreviations M, T and TM represent plants treated with mannitol, elevated temperature and a combination of mannitol and elevated temperature, respectively. The relative value for each parameter and accession is normalized to the respective control and is represented by color intensity, with red indicating higher values and green indicating lower values.

Interestingly, the basal level of proline under control conditions was lower in the sensitive accessions (0.8 µg $mg^{-1}$ DW in 392 and 0.6 µg $mg^{-1}$ DW in 395) than in the more tolerant accessions (2.4 µg $mg^{-1}$ DW in 404 and 2.7 µg $mg^{-1}$ DW in 411) (Table S1).

Water content was most reduced in seedlings treated with mannitol and combined stress in 392 (16% and 18%, respectively) and 395 (15% and 27%, respectively) accessions. A similar decrease in water content was seen in 404 under mannitol and combined stress (13% and 15%, respectively). Temperature stress caused a 2–4% decrease in water content for accessions 392, 395 and 404. Accession 411 performed well under temperature stress and showed no loss of water content under temperature stress compared to the control. However, in this accession, mannitol caused a significant decrease in water content (30%), while the combined stress resulted in a 15% decrease in water content compared to the control.

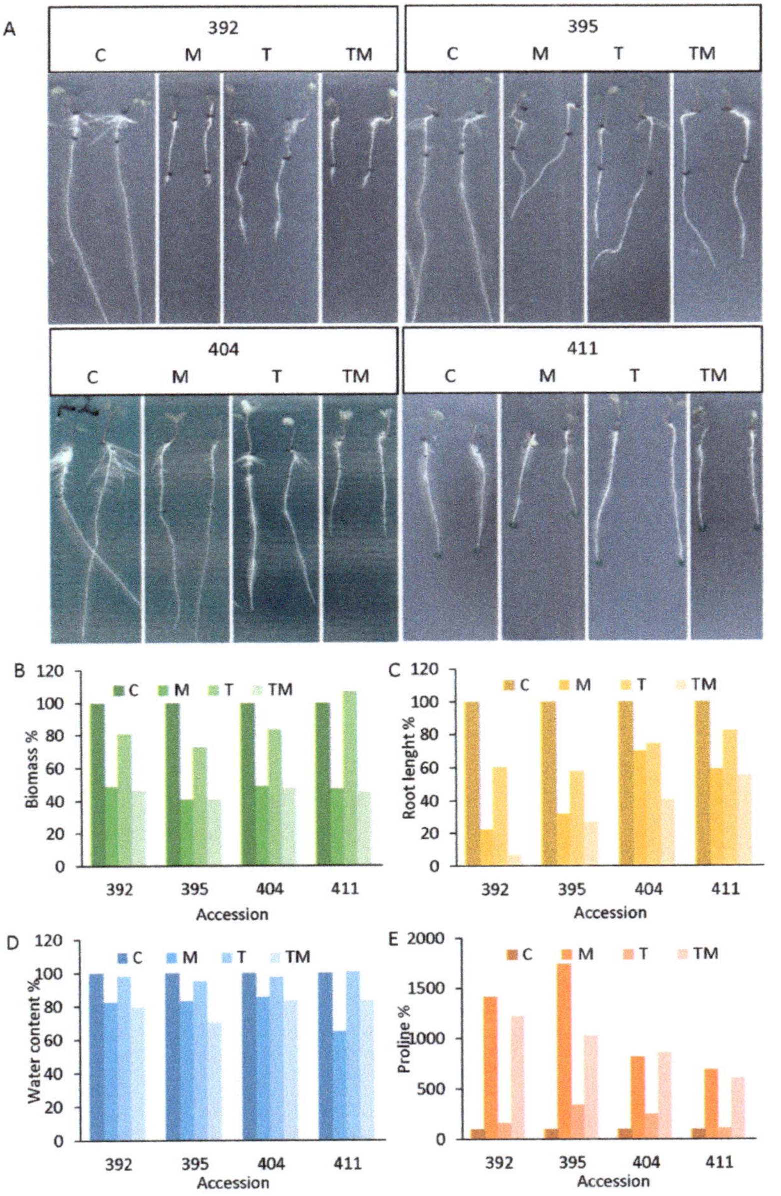


**Figure 2.** Accessions 392, 395, 404 and 411 selected for further analysis: (**A**) photographs of kale seedlings from accessions 392, 395, 404 and 411 under stress conditions (abbreviations M, T and TM represent plants treated with mannitol, elevated temperature and a combination of mannitol and elevated temperature, respectively) compared to the corresponding controls (**C**). Stress markers as % of the corresponding controls (C = 100%): (**B**) biomass, (**C**) root length, (**D**) water content and (**E**) proline level for accessions selected as sensitive (392 and 395) and tolerant (404 and 411) to the stresses studied. Raw data are presented in Table S1.

### 2.3. Selection of Tolerant and Sensitive Accessions

Photosynthetic performance parameters $PI_{abs}$ and $F_v/F_m$ were chosen to present the influence of abiotic stresses on the status of the photosynthetic apparatus of selected kale accessions (Figure 3).

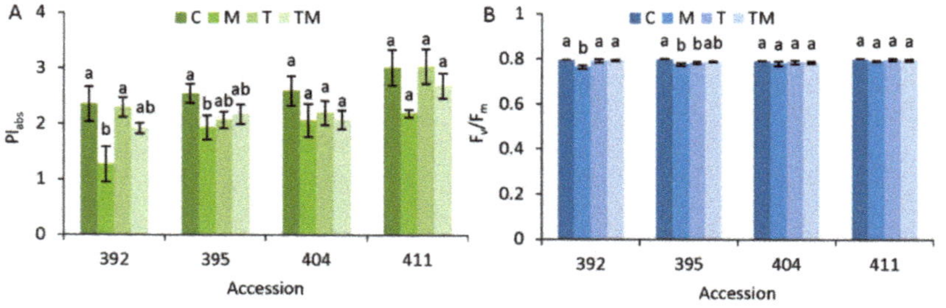


**Figure 3.** Photosynthetic parameters (**A**) $PI_{abs}$ and (**B**) $F_v/F_m$ measured in the selected kale accessions: 392, 395, 404 and 411. Abbreviations M, T and TM represent plants treated with mannitol, elevated temperature and a combination of mannitol and elevated temperature, respectively, compared to the corresponding controls, C. Data are average values of n = 5 ± SE. Different letters denote significant difference between treatments of each accession ($p < 0.05$, one-way ANOVA, LSD test).

The $PI_{abs}$ parameter was more sensitive to the applied abiotic stressors than the $F_v/F_m$ parameter. As can be seen, osmotic stress (M) caused a significant decrease in $PI_{abs}$ compared to the control in accessions 392 (1.9-fold) and 395 (1.3-fold) (Figure 3A). However, this was not observed in accessions 404 and 411. Combined stress (TM) resulted in an approximately 1.2-fold decrease in $PI_{abs}$ in all accessions compared with the corresponding control, but this was not statistically significant. Temperature stress (T) did not result in a significant change in $PI_{abs}$ in any of the accessions tested. Compared to the control, the $F_v/F_m$ parameter changed significantly in accessions 392 and 395 under osmotic stress and in 395 also under temperature stress.

### 2.4. Stress Markers

To investigate the oxidative stress level caused by the application of mannitol, increased temperature and combined stress, the hydrogen peroxide level ($H_2O_2$), catalase activity (CAT), lipid peroxidation level (MDA) and reduced glutathione level (GSH) were analyzed (Figure 4).

As can be seen, $H_2O_2$ content was differently changed in examined accessions under diverse stress treatments (Figure 4A). $H_2O_2$ was significantly increased in 392 accessions only under combined stress compared to the corresponding control. It was significantly increased in 404 accessions under temperature and combined stress and in 411 accessions under osmotic and combined stress compared with the control (Figure 4A).

Content of $H_2O_2$ was not changed in the 395 accession under stress conditions. Interestingly, the control sample of 395 accession showed higher CAT activity compared to the controls of the other accessions, which decreased (Figure 4B). CAT activity increased it significantly under osmotic and combined stress (significantly, in accessions 392, 404 and 411 under combined or temperature stress, respectively, compared with controls). MDA content (Figure 4C) was significantly increased under osmotic and combined stress in 392 accession and under combined stress in accessions 395 and 411 compared with their corresponding controls. GSH content was higher in accessions 392 and 395 under control conditions than in accessions 404 and 411 (Figure 4D). A tendency for GSH content to decrease was observed in all accessions under the applied stress conditions. In accessions 395 and 411, GSH content was significantly decreased under all stress conditions compared to the control. Additionally, it was significantly decreased under osmotic and combined stress in the 404 accession and only under osmotic stress in the 392 accession compared to the control.

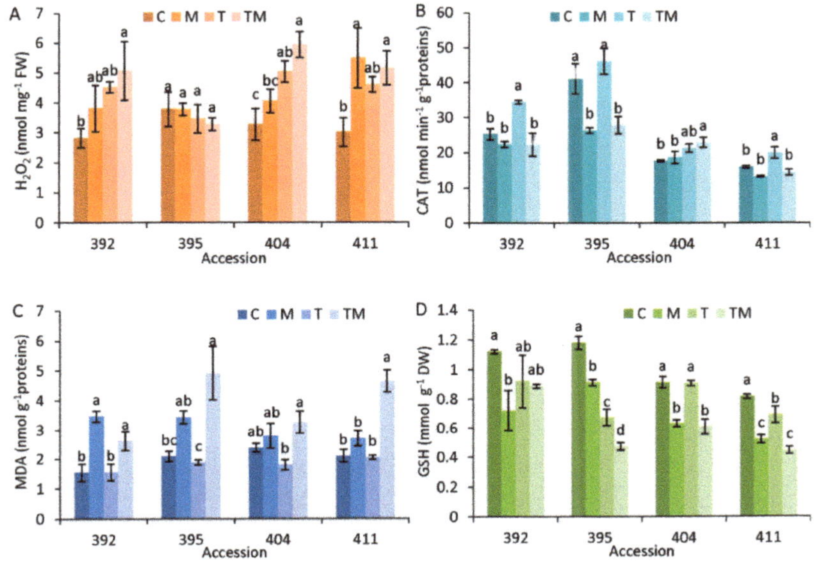


**Figure 4.** Oxidative stress markers (**A**) hydrogen peroxide ($H_2O_2$), (**B**) catalase (CAT) activity, (**C**) level of lipid peroxidation (MDA) and (**D**) reduced glutathione (GSH) content measured in the selected kale accessions: 392, 395, 404, 411. Abbreviations M, T and TM represent plants treated with mannitol, elevated temperature and a combination of mannitol and elevated temperature, respectively, compared to the corresponding controls, C. Data are average values of $n$ = 3–5 ± SE. Different letters denote significant difference between treatments of each accession ($p < 0.05$, one-way ANOVA, LSD test).

### *2.5. Specialized Metabolites and Antioxidant Activity*

Specialized metabolites: total phenols, total flavonoids and total glucosinolates were measured spectrophotometrically (Figure 5).

As shown, total phenols (Figure 5A) were decreased significantly only in the 411 accession under osmotic and combined stress compared with the control. Flavonoids (Figure 5B) decreased under all stress conditions compared with the control, except for temperature stress in accession 411 and the decrease was more pronounced under osmotic and combined stress. Accessions 395, 404 and 411 had higher total glucosinolate content than 392 under control conditions. Compared to the control, a significant decrease in glucosinolates was observed in all accessions under osmotic and combined stress. Moreover, accessions 404 and 411 showed a significant decrease in total glucosinolates under temperature stress. Antioxidant activity measured by the DPPH method did not show significant changes in the 411 accession, regardless of the stress applied. However, it was significantly increased compared to control in accessions 392, 395 and 404 under osmotic and combined stress.

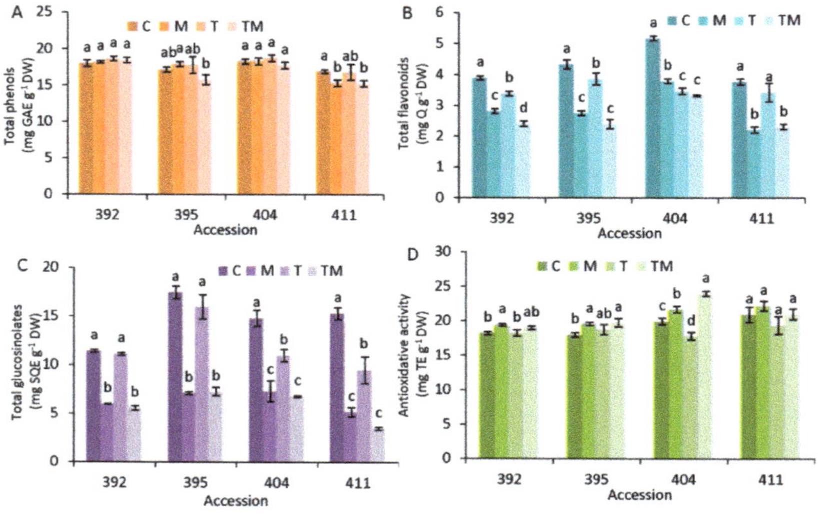


**Figure 5.** (**A**) Total phenols, (**B**) total flavonoids, (**C**) total glucosinolates and (**D**) antioxidative activity measured by DPPH in accessions 392, 395, 404 and 411 under osmotic stress, M, temperature stress, T, and combined stress, TM, compared to the control, C. Data are average of $n = 3 \pm$ SE. Different letters denote significant difference between treatments of each accession ($p < 0.05$, one-way ANOVA, LSD test).

### 2.6. Sugar Analysis

Total sugars were measured by spectrophotometry (Figure 6A). Control samples of accessions 392 and 395 had lower values of total sugars (4.6 and 4.8 mg $g^{-1}$ DW, respectively), while accessions 404 and 411 had a higher level of total sugars (6.7 and 5.6 mg $g^{-1}$ DW, respectively). The level of total sugars increased significantly under temperature (1.7-fold) and combined (1.4-fold) stress only in the 392 accession. Selected sugars, trehalose, sucrose, fructose and total fructans, were measured by HPLC. Under our chromatography conditions, trehalose and sucrose peaks overlapped; thus, we presented both sugars as a joint value (T + S). As can be seen in Figure 6B, osmotic stress decreased T+S significantly compared to corresponding controls in all accessions. Temperature stress did not change T+S level in accessions 392 and 395, while it caused a significant increase in T+S level in accessions 404 and 411. Fructose content (F) was increased in accessions 392 and 395, mostly under osmotic stress (Figure 6C).

In the 404 accession, fructose content significantly decreased in all applied treatments, while it was unchanged in the 411 accession. Glucose content (G) was not changed in the 392 accession under stress conditions, while in accessions 395 and 404, it decreased under osmotic and combined stress (Figure 6D). Fructans were present at a high level in accessions 392 and 395 under control conditions and decreased significantly under osmotic and combined stress, while elevated temperature did not significantly change fructan level in 395 (Figure 6E). On the other hand, fructans increased significantly in the 404 accession under all stress conditions, particularly under high temperature compared to the control. In the 411 accession, fructans did not change under stress conditions compared to the corresponding control.

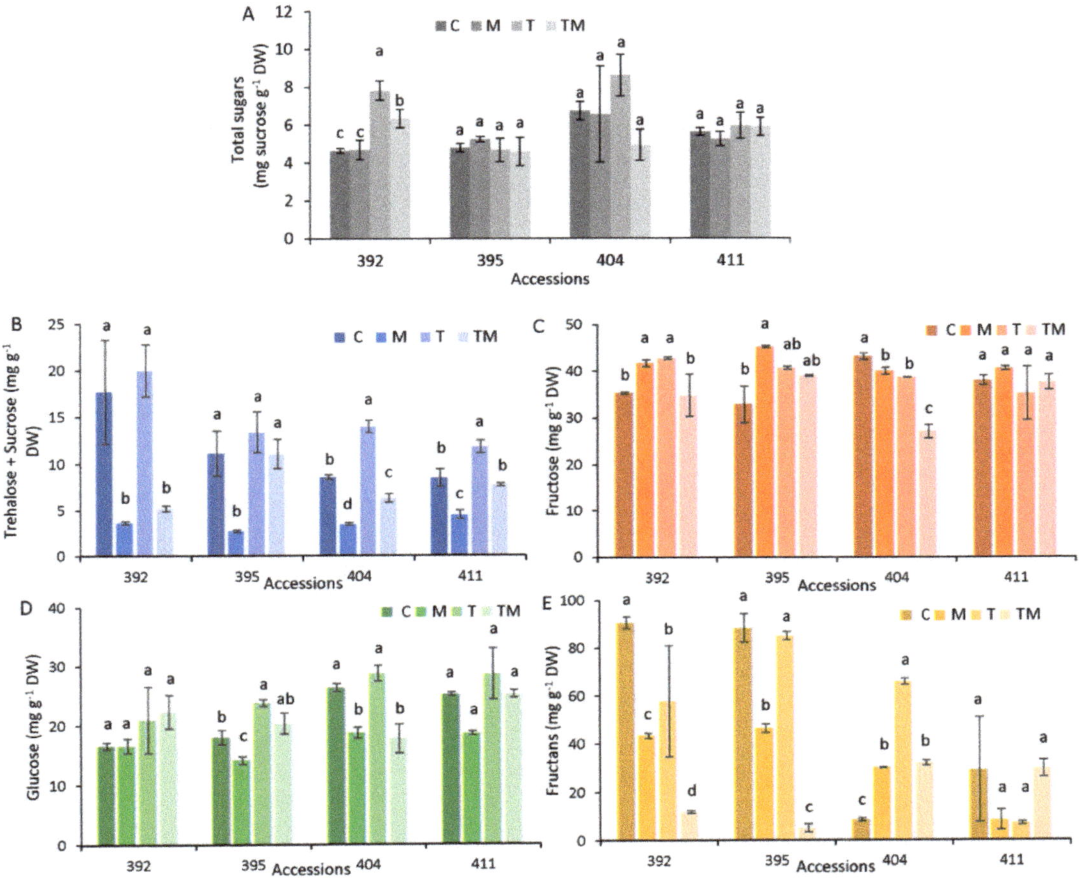

**Figure 6.** Sugar analysis in accessions 392, 395, 404 and 411 under stress conditions (M—mannitol, T—elevated temperature, TM—combined stress, C—control): **(A)** total sugars measured by spectrophotometry; **(B)** trehalose + sucrose, T+S, **(C)** fructose, F, **(D)** glucose, G and **(E)** fructans, FS, measured by HPLC. Data are average value ± SE, $n = 3$. Different letters denote significant difference between treatments of each accession ($p < 0.05$, one-way ANOVA, LSD test).

### 2.7. Gene Expression Analysis

Transcript level analysis was performed for heat shock transcription factors *HSFA2* and *HSFA7*, for *DREB2A* and transcription factors *NAC041* and *NAC084*. In control conditions, basal gene expression of transcription factors varied among accessions (Figure 7).

It is interesting that the most tolerant accession 411 had significantly higher basal gene expression of *NAC* factors and *DREB2A* compared to more sensitive accessions 392 and 395. Accession 395 showed the lowest basal gene expression of *NACs* and *HSFA2* and *DREB2A* transcription factors.

Changes in transcript level of selected genes under stress conditions are shown in Figure 8. As can be seen, a significant increase in *HSF* transcript level was observed in all accessions under temperature stress (Figure 8A, Table S3). The highest increase in *HSFA2* was observed in accession 395. Temperature stress induced increases of 345-, 560-, 148- and 124-fold in *HSFA2* expression in 392, 395, 404 and 411 accession, respectively, compared to the control. There was a tendency of increasing the expression of *HSFA2* in osmotic and combined stress, although the change was statistically significant only in accession 404 under combined stress. *HSFA7* expression was significantly induced under temperature stress in all accessions (488-, 17-, 202- and 88-fold in accessions 392, 395, 404 and 411, respectively) and in accession 392 under osmotic stress (8-fold), while osmotic stress reduced its expression in accessions 395 and 404, although changes were statistically non-significant.

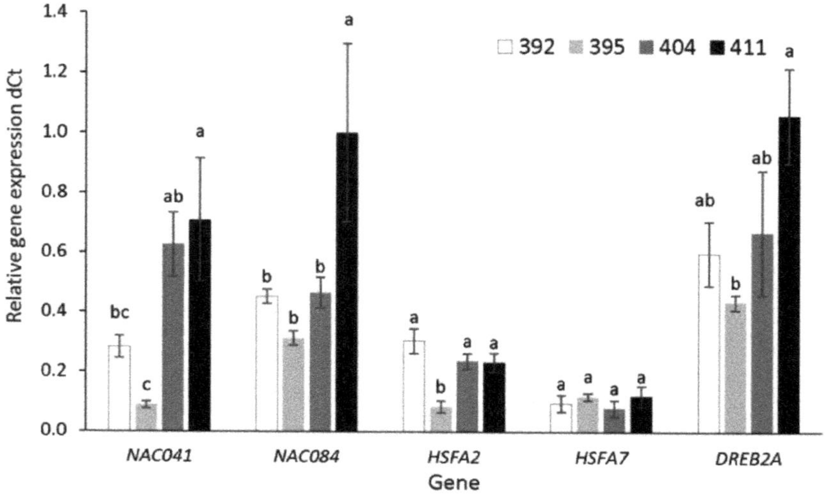


**Figure 7.** Basal relative gene expression of transcription factors *NAC041* and *NAC084*, *HSFA2*, *HSFA7* and *DREB2A* in accessions 392, 395, 404 and 411, under control condition measured by qPCR compared to referent genes (dCt). Data are average ± SE, $n = 3$. Different letters present significant difference between accessions for each individual gene (one-way ANOVA, LSD test).

*DREB2A* transcript was significantly increased under temperature stress in accession 392 (35-fold) while its transcription was increased in accessions 395 (11-fold), 404 (22-fold) and 411 (9-fold), although these increases were statistically non-significant (Figure 8A, Table S3). Combined stress caused a significant increase in *DREB2A* transcript level in 395, 404 and 411 (21-, 38- and 14-fold, respectively). Under mannitol stress, DREB2A showed an intention to increase in all accessions, although changes were statistically non-significant (Figure 8A, Table S3).

Considering NAC transcription factors, there was significant downregulation of *NAC041* and *NAC084* genes observed in the 392 accession under all stress conditions (Figure 8B). *NAC041* expression was reduced in accession 404 under osmotic and combined stress and induced in accession 395 under temperature stress. On the other hand, *NAC041* and *NAC084* were significantly overexpressed, 4.8-fold and 1.8-fold for temperature stress, respectively, and expression of *NAC084* increased 1.6-fold for combined stress in 395 accession. *NAC084* transcript was also significantly increased (1.7-fold) under combined stress in 404 accession and under mannitol stress (1.9-fold) in accession 411, while *NAC041* showed a tendency to decrease in both tolerant accessions, with significant changes only in accession 404.

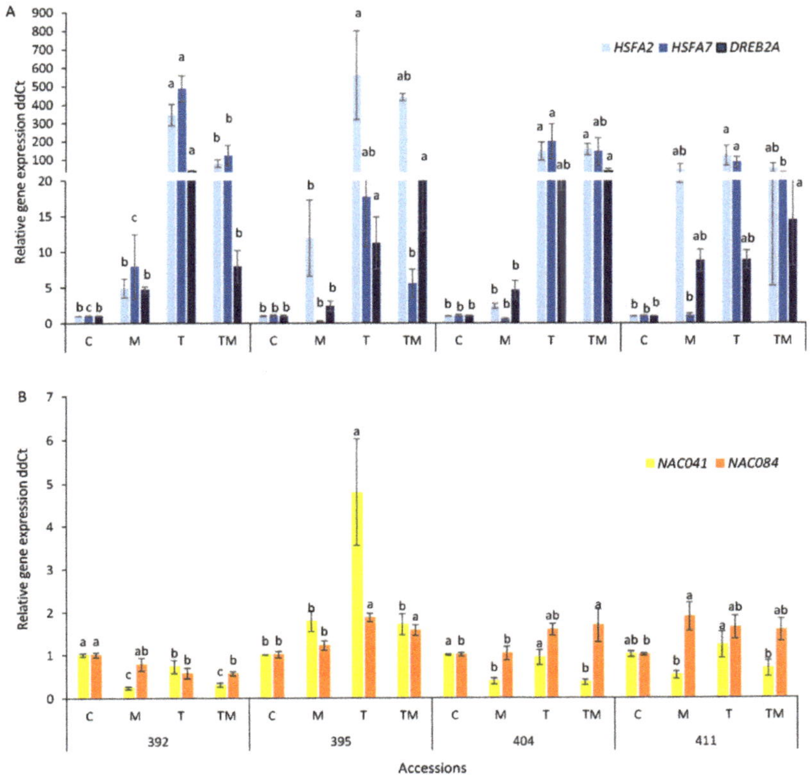


**Figure 8.** Relative gene expression of transcription factors (**A**) *HSFA2*, *HSFA7* and *DREB2A* and (**B**) *NAC041* and *NAC084*, in accessions 392, 395, 404 and 411, 24 h under control (C = 1) and osmotic stress, M, temperature stress, T and combined stress, TM, conditions measured by qPCR (ddCt). Gene expression was presented as fold change compared to the controls. Data are average ± SE, $n = 3$. Different letters present significant difference among treatments of individual accessions (one-way ANOVA, LSD test).

### 2.8. Protein Immmunodetection

Heat shock proteins HSP70 and HSP90 were analyzed by immunodetection (Western blot) and protein signals were semi-quantified and expressed as % compared to the control (C = 100 %) (Figure 9).

Representative Western blot images are presented in Figure S1. It is noticeable that more tolerant accessions 404 and 411 contained a higher level of HSPs in the control conditions compared to more sensitive accessions 392 and 395 (Figure S1A–C). As can be seen, HSP70 and HSP90 were downregulated in osmotic stress in all accessions, but particularly in accessions 392, 404 and 411. The smallest decrease in HSP70 under osmotic stress was in the 395 accession. HSP70 was markedly upregulated under temperature stress in accessions 392 and 395 (up to 140% and 180%, respectively), while it was downregulated or unchanged in accessions 404 and 411, respectively, compared to the control. Under combined stress, HSP70 was unchanged in the 392 accession, increasing up to 180% in the 395 accession and downregulated in accessions 404 and 411 (50% and 40% lower compared to the control). HSP90 was increased up to 300% in accessions 392, 395 and 404 under temperature stress, while it was unchanged in the 411 accession. A similar trend was observed under combined stress—HSP90 was increased slightly in accessions 392, 395 and 404, but decreased significantly in the 411 accession.

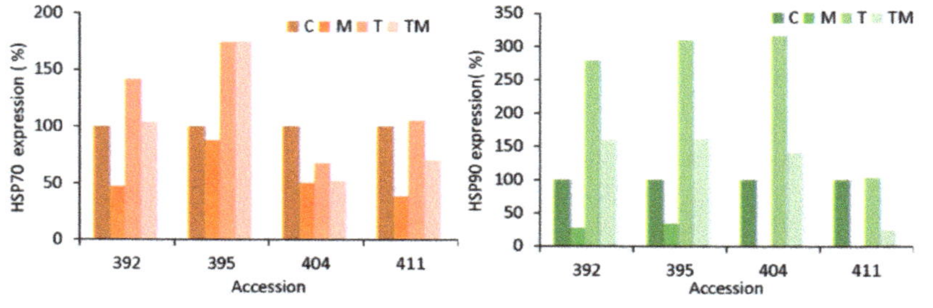


**Figure 9.** Immunodetection of heat shock proteins HSP70 and HSP90 in kale accessions 392, 395, 404 and 411 under osmotic (M), temperature (T) and combined stress (TM) compared to the corresponding controls (C). Data are expressed as % compared to the control (C = 100%).

### *2.9. Stress Response: Interactions of Accession and Stress Treatment*

Using the two-way ANOVA method, we analyzed the impact of accession (A), stress treatment (T) and their interaction (A×T) on the stress response of the selected kale accessions (Table 2). As shown, the majority of the measured parameters was highly influenced by the interaction of accession and treatment.

**Table 2.** Two-way ANOVA summary table represents effect of accession (A), treatment (T) or accession × treatment (A×T) on total phenols, total flavonoids, antioxidative capacity, total glucosinolates, MDA content, CAT activity, $H_2O_2$ content, glutathione content (GSH), photosynthetic parameters $F_v/F_m$ and $PI_{abs}$, total sugars, proline content, root growth and transcription levels of *NAC041*, *NAC084*, *HSFA2*, *HSFA7* and *DREB2A* genes. Results were considered as non-significant (ns) at $p > 0.05$, as significant (*) at $p < 0.05$, as very significant (**) at $p < 0.01$ and as highly significant (***) at $p < 0.001$.

| Parameter | Source | SS | Df | MS | F | p | Significance Level |
|---|---|---|---|---|---|---|---|
| Total phenols | A | 61.02 | 3 | 20.34 | 14.07 | 0.00000 | *** |
| | T | 12.80 | 3 | 4.27 | 2.95 | 0.03868 | * |
| | A×T | 18.82 | 9 | 2.09 | 1.45 | 0.18598 | ns |
| Total flavonoids | A | 9.42 | 3 | 3.14 | 42.69 | 0.00000 | *** |
| | T | 34.45 | 3 | 11.48 | 156.08 | 0.00000 | *** |
| | A×T | 5.52 | 9 | 0.61 | 8.34 | 0.00000 | *** |
| Antioxidative activity | A | 76.93 | 3 | 25.64 | 13.92 | 0.00000 | *** |
| | T | 73.82 | 3 | 24.61 | 13.35 | 0.00000 | *** |
| | A×T | 51.94 | 9 | 5.77 | 3.13 | 0.00327 | ** |
| Total glucosinolates | A | 183.15 | 3 | 61.05 | 30.19 | 0.00000 | *** |
| | T | 1137.48 | 3 | 379.16 | 187.5 | 0.00000 | *** |
| | A×T | 111.58 | 9 | 12.40 | 6.13 | 0.00000 | *** |
| MDA | A | 4.23 | 3 | 1.41 | 4.12 | 0.01404 | * |
| | T | 32.87 | 3 | 10.96 | 31.99 | 0.00000 | *** |
| | A×T | 9.58 | 9 | 1.06 | 3.11 | 0.00847 | ** |
| CAT activity | A | 2561.07 | 3 | 853.69 | 71.86 | 0.00000 | *** |
| | T | 742.76 | 3 | 247.59 | 20.84 | 0.00000 | *** |
| | A×T | 545.15 | 9 | 60.57 | 5.10 | 0.00027 | *** |

**Table 2.** *Cont.*

| Parameter | Source | SS | Df | MS | F | $p$ | Significance Level |
|---|---|---|---|---|---|---|---|
| $H_2O_2$ content | A | 8.28 | 3 | 2.76 | 3.14 | 0.03864 | * |
| | T | 17.37 | 3 | 5.79 | 6.59 | 0.00136 | ** |
| | A×T | 14.83 | 9 | 1.65 | 1.88 | 0.09231 | ns |
| GSH | A | 0.85 | 3 | 0.28 | 10.45 | 0.00001 | *** |
| | T | 1.87 | 3 | 0.62 | 22.99 | 0.00000 | *** |
| | A×T | 0.99 | 9 | 0.11 | 4.06 | 0.00034 | *** |
| $F_v/F_m$ | A | 0.00 | 3 | 0.00 | 3.40 | 0.02667 | * |
| | T | 0.00 | 3 | 0.00 | 6.10 | 0.00156 | ** |
| | A×T | 0.00 | 9 | 0.00 | 1.00 | 0.42690 | ns |
| $PI_{abs}$ | A | 4.56 | 3 | 1.52 | 7.68 | 0.00034 | *** |
| | T | 3.51 | 3 | 1.17 | 5.92 | 0.00184 | ** |
| | A×T | 1.40 | 9 | 0.16 | 0.79 | 0.62964 | ns |
| Total sugars | A | 35.56 | 3 | 11.85 | 5.45 | 0.00204 | ** |
| | T | 25.22 | 3 | 8.41 | 3.87 | 0.01296 | * |
| | A×T | 44.17 | 9 | 4.91 | 2.26 | 0.02835 | * |
| Proline content | A | 643.29 | 3 | 214.43 | 38.88 | 0.00000 | *** |
| | T | 2011.56 | 3 | 670.52 | 121.59 | 0.00000 | *** |
| | A×T | 284.31 | 9 | 31.59 | 5.73 | 0.00003 | *** |
| Root growth | A | 41.05 | 3 | 13.68 | 7.04 | 0.00015 | *** |
| | T | 598.51 | 3 | 199.50 | 102.68 | 0.00000 | *** |
| | A×T | 104.58 | 9 | 11.62 | 5.98 | 0.00000 | *** |
| *NAC041* | A | 2.63 | 3 | 0.88 | 12.63 | 0.00000 | *** |
| | T | 1.46 | 3 | 0.49 | 7.00 | 0.00031 | *** |
| | A×T | 0.68 | 9 | 0.08 | 1.09 | 0.37708 | ns |
| *NAC084* | A | 19.42 | 3 | 6.47 | 15.64 | 0.00000 | *** |
| | T | 0.79 | 3 | 0.26 | 0.64 | 0.59454 | ns |
| | A×T | 3.05 | 9 | 0.34 | 0.82 | 0.60024 | ns |
| *HSFA2* | A | 4363.49 | 3 | 1454.5 | 4.75 | 0.00424 | ** |
| | T | 38,958.62 | 3 | 12,986.21 | 42.41 | 0.00000 | *** |
| | A×T | 14,531.79 | 9 | 1614.64 | 5.27 | 0.00001 | *** |
| *HSFA7* | A | 3390.37 | 3 | 1130.12 | 26.89 | 0.00000 | *** |
| | T | 7988.41 | 3 | 2662.80 | 63.36 | 0.00000 | *** |
| | A×T | 8587.73 | 9 | 954.19 | 22.70 | 0.00000 | *** |
| *DREB2A* | A | 50.25 | 3 | 16.75 | 0.69 | 0.55866 | ns |
| | T | 1884.68 | 3 | 628.23 | 26.01 | 0.00000 | *** |
| | A×T | 928.47 | 9 | 103.16 | 4.27 | 0.00015 | *** |

On the other hand, variations in total phenolics were strongly influenced by accession and to a lower level by the stress treatments, while $H_2O_2$ content was more influenced by the stress treatment than accession. Both photosynthetic parameters, $PI_{abs}$ and $F_v/F_m$, were influenced by the accession and stress treatment, but not their interaction. Proline and root growth were significantly affected by treatment, accession and their interactions. Transcription factor NAC041 was also influenced by accession and stress treatment, while NAC084 was only affected by the kale accession.

Figure 10 shows the multivariate analysis of all parameters obtained for four selected accessions under abiotic stress treatments. As can be seen, high-tolerant (red dots) and low-tolerant (blue dots) accessions were grouped separately based on the measured parameters. The most reliable markers for distinguishing between tolerant and sensitive kale accessions are CAT activity and proline content. Other important parameters are: photosynthetic parameters $PI_{abs}$ and $F_v/F_m$, antioxidant activity measured by DPPH, $H_2O_2$, root growth, fructose and total phenolics (TPC). These variables were selected and presented in the figure according to the variable loading number (VIP) higher than the value 1 (Table S2). In addition, flavonoids, glucosinolates and transcript level of NAC041 also contributed significantly (VIP approximately 0.9; Table S2). DREB2A and HSFA7 transcript levels and MDA contributed to tolerance characterized by low VIP (below 0.5).

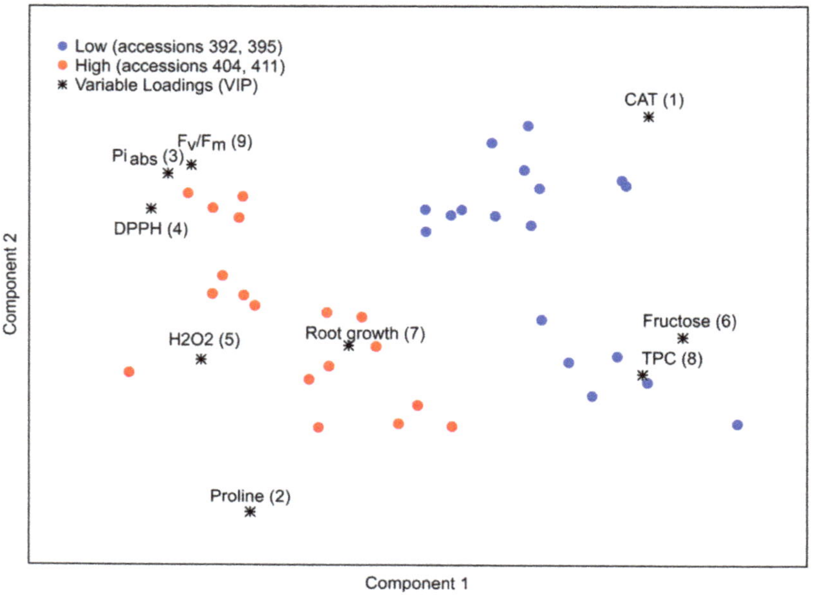


**Figure 10.** Multivariate analysis of parameters obtained for selected accessions 392, 395, 404, 411 under abiotic stress conditions (Table S2).

## 3. Discussion

Abiotic stress factors, such as high temperatures and drought associated with current climate changes, are affecting plant growth and crop yields for many agricultural crops, including Brassicaceae. Although many experimental designs examine individual stress factors, the situation in nature is more complex and plants are often exposed to complex stress factors in their habitat. To elucidate tolerance mechanisms in *Brassica*, we studied their responses to single elevated temperatures and osmotic stresses as well as to combined stress (elevated temperature + osmotic stress). Our previous work, including comparative studies of stress responses of *Brassica* among different species/varieties (Chinese cabbage (*Brassica rapa* ssp. *pekinensis*), white cabbage (*Brassica oleracea* var. *capitata*) and kale (*Brassica olerace* var. *acephala*)), showed that kale was the most tolerant white cabbage that was moderately tolerant and Chinese cabbage most sensitive to drought and salt stress [11,12,16]. We showed that bioactive molecules, such as phenolic compounds [16] and plant hormones [11,12,17], are highly involved in the mechanisms of abiotic stress tolerance in the studied *Brassica* plants. However, the mechanisms of Brassica tolerance to abiotic stresses involve complex networks of many players, such as reactive oxygen species (ROS), plant hormones, specialized metabolites, osmolytes, enzymatic and non-enzymatic antioxidants, etc., and are still far from being fully understood [18–20]. The stress response is highly dependent on the particular plant species/varieties. To better understand the mechanisms of stress response, we focused on kale (*Brassica oleracea* var. *acephala*) and screened 33 kale ecotypes or accessions for heat, osmotic and combined stress. Furthermore, we selected four accessions with different stress responses (392 and 395 as more sensitive and 404 and 411 as more tolerant) and used them for a more detailed analysis of stress response mechanisms. Among the stress treatments, the highest stress intensity, i.e., the most severe growth inhibition of seedlings, was associated with osmotic stress and combined stress, whereas temperature stress alone caused less noticeable changes compared with controls (Figures 1 and 2). The results are consistent with data published on *Brassica napus* exposed to drought, heat and combined stress [21]. Among the selected accessions (392, 395, 404 and 411) (Figure 2), it is evident that each of them responded differently to individual and combined stress. Thus, the 392 accession was the most sensitive, while 411 seems to be the most tolerant to all stress conditions. Accession 395 appears to be more tolerant to combined stress than the other accessions, while 404 is very tolerant to osmotic stress.

### 3.1. Stress Markers

Seedling biomass production, root growth and proline content were selected as strong and reliable stress markers, showing stress status of the ecotypes studied, and were used to screen 33 kale ecotypes for tolerance (Figures 1 and 2). A decrease in biomass accumulation, inhibition of root growth, decrease in water content and an increase in proline content were associated with higher stress status, particularly under osmotic and combined stress. This is in agreement with previous results obtained in *Brassica* crops treated by drought and salinity stress [11,12,17]. Plants under stress actively suppress their growth in order to survive under adverse conditions. Although beneficial for plant survival, active growth inhibition is often undesirable for crop productivity [22]. The effect of stress on plant growth can be measured as a decrease in plant growth rate or as a decrease in biomass accumulation. There is a strong positive correlation between biomass accumulation and root growth and stress tolerance under our experimental setup, showing that more tolerant accessions had lower suppression of biomass accumulation under stress conditions compared to the corresponding controls (Figures 1 and 2). It was particularly observable in 404 and 411 accessions that were selected as tolerant ones for further examination of tolerance mechanisms. The correlation between proline accumulation and stress tolerance is controversial [23]. Comparative studies between *Brassica* species/varieties under drought and salt stress indicate that a higher increase in proline content is associated with more sensitive species/varieties [11,13], which is in agreement with the results obtained here, while a positive correlation between proline accumulation and stress tolerance was found in a comparative study of different *Brassica juncea* cultivars [24] and in rapeseed (*B. napus*) [25] subjected to abiotic stress. Another interesting observation is that more tolerant accessions had higher basal proline content than sensitive accessions under the same control conditions, which is consistent with previously reported data in *Brassica* plants [11,16].

Additional stress markers analyzed in selected accessions (392, 395, 404, 411) that confirmed stress status were $H_2O_2$, CAT, lipid peroxidation (MDA) and GSH content (Figure 4). As can be seen, $H_2O_2$ level was strongly dependent on accession. It was generally elevated in all accessions under stress treatments, except for 395. Among the various ROS usually generated under stress conditions, freely diffusible and relatively long-lived $H_2O_2$ can cause certain damage to cell structures, but also act as signaling molecules that turn on stress-response mechanisms [26–28]. These mechanisms can then activate a network of processes that enhance tolerance to various abiotic stressors. Under the stress conditions prevailing in our experiment, the more tolerant accessions 404 and 411 accumulated higher $H_2O_2$ concentrations compared to the more sensitive ones (392 and especially 395) (Figure 4A). This was associated with lower CAT activity, lower proline content and lower concentration of MDA, as a marker of lipid peroxidation, suggesting lower stress intensity suffered by accessions 404 and 411. These observations are in agreement with publications showing that exogenous application of $H_2O_2$ to rapeseed (*B. napus*) attenuates the effects of drought stress by mediating hormonal status and oxidative response [29,30]. Mechanism of activity of $H_2O_2$ as a signaling molecule seems to be complex and it is still unclear. It was shown that $H_2O_2$ affects the expression of genes that are involved in plant responses to diverse environmental stresses. Stress-induced $H_2O_2$ acts as signal in complex cross-talk of plant hormones ABA and auxin in response to salinity and drought [27] by disrupting the interaction between tryptophan synthase β subunit 1 (TSB1) and β-glucosidase 1 (BG1) that are involved in ABA and auxin homeostasis. Furthermore, a recent study showed that $H_2O_2$ activates plant cold responses by sulfenylating cytosolic enolase2 (ENO2) and promoting its oligomerization, leading to enhanced nuclear translocation and transcriptional activation of C-repeat/DRE binding factor1 (CBF1) [28].

### 3.2. Photosynthetic Performance in Stress

Photosynthesis is a process that is highly influenced by abiotic stresses and impaired photosynthesis has negative effects on plant growth, biomass production and yield. Two commonly used photosynthesis parameters that serve as stress indicators in plants are the performance index ($PI_{abs}$) and the maximum quantum yield of PS II ($F_v/F_m$), both of which are determined by measuring chlorophyll a fluorescence that originates almost exclusively from PS II [31]. The values of $PI_{abs}$ parameter confirmed a much higher responsiveness compared to $F_v/F_m$ (Figure 3), which is consistent with published data obtained on different cultivars of white and red cabbages [32]. Our results showed that $PI_{abs}$ and $F_v/F_m$ were decreased significantly more in accessions 392 and 395, particularly under osmotic stress compared to the corresponding controls, than in more tolerant accessions 404 and 411 (Figure 3). This observation is in agreement with stress intensity level, measured as biomass accumulation, root growth and proline content. Moreover, our results correspond to data observed for *Brassica* species/varieties under drought stress [11]. Under our experimental conditions, temperature stress caused a significant decrease in photosynthetic performance only in the 395 accession, indicating a relative tolerance of this trait in kale to heat stress. Rodríguez et al. [33] compared white cabbage and kale under temperature stress and reported that low temperatures had a greater effect on *B. oleracea* physiology than high temperatures. A similar response, namely a greater decrease in chlorophyll content at low than at high temperatures, was observed in the study by Soengas et al. [34]. On the other hand, photosynthesis and respiration rates and the maximum quantum yield of photosystem II in developing seeds of *B. napus* were inhibited by heat stress that severely impairs yield and oil content [35]. Interestingly, our study showed that combined stress (high temperature + osmotic stress) caused fewer disturbances in photosynthetic performance than osmotic stress alone and did not cause a significant decrease in $PI_{abs}$ or $F_v/F_m$ parameters. In the case of *B. napus*, reductions in net photosynthetic assimilation rate were caused by combinations of heat and drought (heat + drought) treatments, by drought alone and, to a lesser extent, by heat alone [36]. Our results (Table 2) as well as the published data cited above suggest that the effect of abiotic stress on photosynthetic parameters depends on both species/varieties and stress treatment.

### 3.3. Specialized Metabolite Accumulation under Stress Conditions

Specialized metabolites, such as phenolics, are generally recognized as molecules involved in stress protection in plants [37]. Under our experimental conditions, total phenols did not change significantly compared with the control, except for accession 411 under osmotic and combined stress. Total flavonoids tended to decrease under all stress condition in all examined accessions, particularly under osmotic and combined stress (Figure 5). Among selected accessions, 404 contained the highest content of flavonoids under both control and stress conditions and, consequently, showed the highest antioxidant activity.

Glucosinolates are specialized metabolites specific to Brassicaceae species. Their content was higher in accessions 395, 404 and 411 than in 392 under control conditions. In all accessions, the total content of glucosinolate tended to decrease under stress conditions, most markedly under osmotic and combined stress. Elevated temperature was found to increase glucosinolate concentrations in *B. rapa* while decreasing their content in *B. napus* [38]. Thus, fluctuations in glusosinolates under stress conditions depend on plant species, plant developmental stage, stress intensity as well as stress duration. The authors speculated that down-regulation of glucosinolate pathways under stress conditions may serve as a protective measure to conserve energy to ensure survival under adverse environmental conditions. Extensive studies on the Brassicaceae family showed a positive correlation between abiotic stress tolerance and glucosinolate content in broccoli, canola, radish sprouts and pakchoi [39].

### 3.4. Soluble Sugar Fluctuations under Stress Conditions

Sugars are the major building blocks for carbohydrate storage, but also serve as signaling molecules and protective compounds during abiotic stress exposure. Among the metabolites accumulated as osmoprotectants, numerous carbohydrates, including fructose, sucrose, trehalose, raffinose and fructans that are of high solubility, have been shown to accumulate in response to abiotic stresses [40]. Our data showed that total soluble sugars increased significantly in accession 392 under temperature and combined stress (Figure 5A). The individual sugars (trehalose, sucrose, glucose and fructose) were analyzed by HPLC (Figure 5B). Fructose content was increased significantly under osmotic and temperature stress in sensitive accessions (392 and 395), while it decreased in 404 or was not changed in 411. Glucose content was decreased, particularly under osmotic and combined stress (392, 395, 404), or remained unchanged (411) compared to the controls. Trehalose+sucrose decreased significantly under osmotic stress and increased significantly under temperature stress in all accessions. Based on our results, there is no general conclusion about fluctuations in individual energy source sugars and stress tolerance. Fluctuations in individual sugars under abiotic stress in kale depend on plant ecotype, stress factors, intensity and duration of stress, which is in agreement with results obtained on canola (*B. napus*) [41].

It is interesting that accessions 392 and 395 contained higher levels of fructans compared to accessions 404 and 411 in the control conditions. Osmotic and combined stresses caused a significant decrease in fructans in 392 and 395 while heat did not change it. On the other hand, fructans content was significantly increased in 404 under all stress conditions. Fructans are the non-structural carbohydrates, possessing other physiological functions than carbon reserve [42]. Fructans have been reported to have a protecting role in plants against water deficit caused by drought and osmotic stress. Plants synthesize fructans in order to osmoregulate the cellular flux, therefore, reducing the membrane damage. In addition, fructans are recognized as excellent scavengers of ROS [43]. Our results suggest that one of the mechanisms of abiotic stress tolerance in kale is an increase in fructans content.

### 3.5. Stress-Related Transcription Factor Gene Expression and Heat Shock Protein Accumulation

Plant stress response to unfavorable environmental conditions includes developmental, physiological and biochemical changes, guided by stress-related transcription factors. Drought and heat stress are initially sensed by membrane-localized stress receptors, intracellularly relayed through secondary messengers, especially calcium ions that activate signaling pathways through mitogen-activated (MAPK) and calcium-dependent (CDPK) protein kinases. Several transcription factors, such as *HSFs*, *DREB* and *NAC* (NAM, ATAF and CUC domain), are upregulated by kinases (MAPKs and CDPKs) and induce synthesis of stress-responsive proteins that, in turn, ensure plant survival in stressful conditions. Tolerance or susceptibility to stress is, therefore, dependent on the ability of the plant to express a set of genes whose expression is often regulated by *HSF*, *DREB* and *NAC* transcription factors [44–46]. These transcription factors (TFs) regulate the expression of chaperones, such as heat shock proteins (HSPs), detoxification of reactive oxygen species (ROS), expression of antioxidant enzymes (ascorbate peroxidase, APX, catalase, CAT) and expression of other genes involved in signal transduction and regulatory pathways. They play critical roles, not only in heat stress response, but also in other abiotic stresses [19,32] and their transcriptional reprogramming or genetic manipulation is considered a valuable tool in engineering stress-tolerant varieties of crop plants [19,44,47]. Zhu et al. reported that the most *B. napus* HSFs were induced under heat as well as drought stress, suggesting their role in multiple abiotic stress responses in canola [46]. This is in agreement with our results, showing that transcript level of *HSFs* was significantly increased under heat stress, but also under osmotic stress in the most sensitive accession 392 and under combined stress in all, so the 411 accession was the most resistant (Figure 8). Our results suggest that more sensitive accessions increased expression of HSFs more prominently compared to the more tolerant accessions.

Unlike moderate increases in *HSFA2* and *HSFA7*, *DREB2*, a transcription factor that mediates high salinity, dehydration- and heat-stress-inducible transcription [9,48], had the highest induction under combined stress in all except the sensitive 392 accession, from which it can be concluded that exposure to drought- and heat-stress-tolerant cultivars has modest induction of *HSFA2* and *HSFA7* and more noticeable induction of DREB2A in comparison to more sensitive accessions.

There are multiple regulatory interactions between *HSFs* and HSPs [49]. Consequently, accumulation of HSPs was higher in sensitive accessions (392 and 395) compared to more tolerant (404 and 411) (Figure 9 and Figure S1). Earlier, HSPs were believed to produce under heat stress, as the name indicates, but now it is established that these biomolecules are produced in response to various biotic and abiotic stresses [50]. HSP90 works in association with HSP70 as a major part of chaperone complexes. While the major role of HSPs is protein folding, they also act as the key component in signal-transduction networks, cell-cycle control, protein trafficking, etc. A study on chickpea HSP70 reported that HSPs were first downregulated in the early stage of growth in drought-tolerant cultivars, which is in agreement with our results. In contrast, HSPs were abundant in drought-sensitive cultivars, which indicated that HSPs responded to drought not only in the specific genotypes but, also, during the developmental stage [51].

The role of *NAC* TFs in abiotic stress tolerance is well documented [19,52,53]. A number of *NAC* transcription factors has already been functionally characterized in model plants, demonstrating their role in abiotic stress. In response to heat and drought stress, many *NAC* transcription factors in *B. rapa* are repressed [52,54], which is consistent with our results, showing downregulation of *NAC041* under osmotic and combined stress in the kale accessions, except for 395, which had a significantly reduced basal *NAC041* expression level. On the other hand, *NAC084* showed a higher basal expression level in tolerant kale accessions and a tendency to overexpress under stress conditions, particularly in more tolerant accessions (Figures 7 and 8). Overall, our results suggest that the higher basal expression of *NAC04*, *NAC084* and *DREB2A* and the change in *HSFA2* and *HSFA7* can be used as valuable markers in screening heat- and drought-tolerant kale cultivars.

## 4. Materials and Methods

### 4.1. Plant Material

Seeds of 33 kale accessions (*Brassica oleracea* var. *acephala*) were collected from farmers along the Croatian coast and islands in 2018 and 2019, including four accessions from Bosnia and Herzegovina (385, 393, 395 and 413) (Table 1). None of the collected accessions are used for commercial purposes and they are reproduced and maintained by local farmers usually as a part of family inheritance. The collected seeds of all accessions were regenerated in the season 2019/2020 at the Institute of Agriculture and Tourism and used for this experiment.

### 4.2. Plant Growth and Abiotic Stress Treatments

The level of abiotic stress tolerance to osmotic stress, high-temperature stress and combined stress (osmotic + temperature stress) of 33 kale accessions was determined using the root-growth bioassay described earlier [17] with slight modifications. After sterilization, seeds were placed onto 1% agar plates and stratified at 4 °C for 5 days. Plates were then moved to a growth chamber (PHC Corporation, Tokyo, Japan), positioned vertically, at 22 °C, 16 h of light at 150 μmol $m^{-2}$ $s^{-1}$/8 h of darkness. One-day-old seedlings with approximately 1 cm long roots were subjected to stress as follows: osmotic (M)—seedlings transferred to 1% agar plates containing 0.3 M mannitol and incubated at 22 °C; elevated temperature (T)—seedlings transferred to 1% agar plates under the following temperature cycle: 24 °C for 10 h during the night period (10 p.m.–8 a.m.), 28 °C for 4 h (8:00 a.m.–12), 38 °C for 4.5 h (12–4:30 p.m.), 28 °C 5.5 h (4.30 p.m.–10 p.m.); combined osmotic + elevated temperature (TM)—seedlings transferred to 1% agar plates containing 0.3 M mannitol under the temperature cycle described above for T treatment. Seedlings transferred to 1% agar plates and incubated at 22 °C were used as control (C). Treatments were performed for 3 days, in growth chambers at a 16/8 photoperiod and a light intensity of 150 μmol $m^{-2}$ $s^{-1}$. Before harvesting the seedlings, the plates with seedlings were scanned for root analysis and the seedlings were weighed for biomass determination. The roots were analyzed using the ImageJ program. After photosynthetic measurements, the seedlings were quickly frozen in liquid nitrogen and stored at −80 °C. Plant material used for analyses of sugars, specialized metabolites, proline, malondialdehyde (MDA) and protein immunodetection was freeze dried before analyses.

### 4.3. Photosynthetic Measurements

Photosynthesis measurements were performed using six individual seedlings per treatment and control in vivo. Fast chlorophyll kinetics was measured on dark-adapted (30 min of light deprivation) cotyledons using Fluorpen (Photon Systems Instruments). The cotyledons were flattened on wet filter paper to make the surface area large enough for measurement. To induce chlorophyll fluorescence transients (OJIP), they were exposed to a saturation pulse (455 nm, max. 3000 μmol $m^{-2}$ $s^{-1}$) and fluorescence intensity was measured from 50 μs after the pulse ($F_0$) until 1 s ($F_m$). From these measurements the JIP test was performed to calculate related parameters [55] from which photosynthetic performance index ($PI_{abs}$) and the maximum quantum yield of PSII ($F_v/F_m$) were used.

### 4.4. Biochemical Stress Markers

$H_2O_2$ analysis was performed using plant material stored at −80 °C, while other parameters were determined using previously lyophilized plant material. Extracts for determination of $H_2O_2$ content were prepared by homogenization of frozen samples (−80 °C) with glass and ceramic beads in 70% ethanol. The solution was centrifuged at 5000× *g* for 20 min at 4 °C. $H_2O_2$ content was determined spectrophotometrically using FOX reagent (0.25 mM ammonium iron (II) sulfate, 124 μM xylenol orange, 99 mM sorbitol) [56]. Sample extracts were mixed with FOX reagent in the ratio of 1:10 (extract:reagent), incubated for 15 min at RT and absorbance was measured at 560 nm. $H_2O_2$ content was calculated using a standard curve of known $H_2O_2$ concentrations and expressed as μg per fresh weight (μg $g^{-1}$ FW).

For proline determination, lyophilized tissue was homogenized in 70% ethanol using glass beads and incubated for 30 min at 4 °C. The solution was centrifuged for 10 min at 16,000× *g* at 4 °C. Free proline was determined spectrophotometrically using acid ninhydrin (1% (*w*/*v*) ninhydrin, 60% (*v*/*v*) glacial acetic acid, 20% (*v*/*v*) 96% ethanol) as described [16]. The mixture was incubated at 95 °C for 20 min and absorbance was measured at 520 nm. Proline content was calculated using a standard curve of known L-proline concentrations and expressed as μmol L-proline per dry weight (μmol $g^{-1}$ DW).

Protein extracts were prepared by grinding lyophilized tissue in cold potassium phosphate buffer (0.1 M, pH 7.0) with the addition of polyvinylpolypyrrolidone. After centrifugation at 16,000× *g* for 20 min at 4 °C, protein concentration in the supernatant was measured using the Bradford assay [57]. The supernatants were stored at −20 °C and used for further analyses.

Catalase (CAT) activity was measured using protein extracts according to the method of Aebi [58]. Protein extracts were added to reaction buffer (0.1 M potassium phosphate buffer, 10 mM $H_2O_2$, pH 7.0) in a ratio of 1:40 (extract:reaction buffer) and the change in absorbance at 240 nm was measured every 10 s for 1 min. Catalase activity was calculated using the extinction coefficient of 40 $mM^{-1}$ $cm^{-1}$. Results were expressed as nmol of decomposed $H_2O_2$ per minute per milligram of total soluble proteins (nmol $min^{-1}$ $mg^{-1}$ proteins).

Lipid peroxidation was determined spectrophotometrically in protein extracts as described by Draper and Hadley [59]. Thawed protein extracts were added to 20% (*w*/*v*) trichloroacetic acid (TCA) and separately to TBA/TCA reagent (0.3% (*w*/*v*) thiobarbituric acid prepared in 20% TCA) in a ratio of 1:3 (extract:reagent). The mixtures were incubated for 30 min at 95 °C, centrifuged at 15,000× *g* for 10 min at 4 °C and the precipitate was discarded. Absorbance was measured at 532 and 600 nm and MDA content was calculated using the extinction coefficient of 155 $mM^{-1}$ $cm^{-1}$. Results were expressed as nmol MDA per protein (nmol $mg^{-1}$ proteins).

### 4.5. Protein Immmunodetection

Proteins were extracted in Tris-HCl buffer, pH 8.0 [60], quantified as described above and separated using vertical sodium dodecyl sulfate polyacrylamide gel electrophoresis (SDS-PAGE). A discontinuous Tris-Gly buffer (pH 8.3) system was used (4% stacking gel, 12% resolving gel). The proteins were denatured using Laemmli buffer [61] at 96 °C for 5 min and 6 μg of protein per sample was loaded onto the gel. The separated proteins were electrotransferred to nitrocellulose membrane in transfer buffer (0.335 % Tris, 1.44% Glycine, 10% methanol) at 60 V for 60 min. Before blocking, the efficiency of the transfer was checked by incubating the membrane in Ponceau rouge solution (0.05% Ponceau rouge). After washing in TBS buffer (0.24% Tris, 0.88% NaCl, pH 7.5), the membrane was blocked with 5% milk solution in TTBS buffer (TBS buffer, 0.05% Tween 20) for 1 h. The membrane was then incubated overnight at 4 °C in anti-HSP70 (Agrisera, AS08371) diluted 1:1000 in TTBS buffer containing 3% milk or anti-HSP90 (Agrisera, AS08346) diluted 1:3000 in TBS buffer. The membrane was washed in blocking solution, incubated with secondary antibody (anti-rabbit IgG-horseradish peroxidase (HRP) diluted 1:5000 in TTBS with 5% milk), 60 min at room temperature and washed in TTBS. Signals were detected using chemiluminescent HRP substrate and the intensity of the bands was quantified using ImageJ.

### 4.6. Specialized Metabolites, Antioxidant Capacity and GSH

Approximately 10 mg of freeze-dried plant tissue was homogenized with 500 μL of aqueous methanol (80:20, methanol:water, *v*/*v*). The homogenate was incubated in an ultrasonic bath (Sonorex, Bandelin) at 20 °C for 30 min and then centrifuged at 15,000× *g* for 5 min. The extracts were stored at −80 °C and used for the following analyses.

Total phenolics were assayed using the Folin–Ciocalteu method [62]. Briefly, 300 μL of 1.88 M $Na_2CO_3$ was added to 1580 μL of $dH_2O$, 20 μL of the extract and 100 μL of the Folin–Ciocalteu reagent. The reaction mixture was mixed and incubated in a dry-block heater at 45 °C for 1 h. The absorbance was measured at 765 nm and total phenols were expressed as equivalents of gallic acid per dry weight (mg GAE $g^{-1}$ DW).

Total flavonoids were determined by the $AlCl_3$ method [63], with minor modifications. The reaction mixture consisted of 100 μL of the extract, 20 μL 10% (*w*/*v*) $AlCl_3$, 500 μL 1 M potassium acetate and 380 μL $dH_2O$. The absorbance of the mixture was measured at 420 nm after 30 min of incubation at room temperature (25 °C). Results were expressed as quercetin equivalents per dry weight (mg Q $g^{-1}$ DW).

Total glucosinolates were measured as previously described [64], with some modifications. In short, 30 μL of the extract was mixed with 900 μL of 2 mM Na2PdCl4. After 30 min of incubation at room temperature (25 °C), the absorbance was measured at 425 nm and results were expressed as sinigrin equivalents per dry weight (mg SEQ $g^{-1}$ DW).

Total antioxidant capacity was evaluated using the DPPH radical scavenging capacity assay [65]. The extract (50 μL) was added to 950 μL of 0.1 mM DPPH reagent prepared in 96% ethanol, mixed and incubated for 30 min at room temperature (25 °C). The absorbance was read at 517 nm and results were expressed as Trolox equivalents per dry weight (μmol TE $g^{-1}$ DW).

GSH was measured according to Malar et al. [66]. Briefly, 20 μL of the extract was added to 940 μL of potassium phosphate buffer (0.1 M, pH 7.0) and 40 μL of 0.01 M

5,5 -dithio-bis-(2-nitrobenzoic acid) previously prepared in potassium phosphate buffer (0.1 M, pH 7.0). The mixture was incubated at room temperature for 5 min and absorbance was measured at 412 nm. GSH content was calculated using a standard curve of known GSH concentrations and expressed as mmol GSH per dry weight (mmol $g^{-1}$ DW).

### 4.7. Sugar Analysis

Total sugars were determined by the anthrone method [67]. To 1 mL of 2% (*w*/*v*) anthrone prepared in 71.33% (*v*/*v*) $H_2SO_4$, 200 µL of the extract was added. The reaction mixture was incubated at 90 °C for 5 min. After cooling to room temperature, absorbance was measured at 620 nm. Results were expressed as sucrose equivalents per dry weight (mg sucrose $g^{-1}$ DW).

For more detailed sugar analysis, the lyophilized plant tissue (approximately 150 mg) was homogenized with 2.4 mm metal beads (Omni kit 19-670, Kennesaw, GA, USA) for 1 min at 5 m s-1 in 3 mL of 80% methanol in water using a bead mill (Omni Bead Ruptor Elite, Kennesaw, GA, USA). The homogenates were left to macerate for 1 h on a rotator (Biosan RS-60, Riga, Latvia) and subsequently centrifuged for 5 min at 5000× *g*. The extracts were filtered through a 0.22 µm nylon filter prior to analysis. The analysis of fructans, trehalose, sucrose, glucose and fructose content was carried out using an HPLC system consisting of a system controller (Shimadzu CBM-40, Kyoto, Japan), a degassing unit (Shimadzu DGU-405, Kyoto, Japan), a solvent delivery unit (Shimadzu LC-20Ai, Kyoto, Japan), an autosampler (Shimadzu SIL-20AC, Kyoto, Japan), column oven (Shimadzu CTO-40S, Kyoto, Japan) and a refractive index detector (Shimadzu RID-20A, Kyoto, Japan). Chromatographic separation was achieved by injecting 10 µL of the sample on a 300 × 8 mm, 9 µm particle size, calcium cation exchange column (Dr. Maisch ReproGel Ca, Ammerbuch, Germany) held at 80 °C using deionized water as the mobile phase (0.6 mL min-1, isocratic elution). Retention times and peak areas of the investigated sugars were compared to analytical standards for identification and quantification, respectively. The retention time of sucrose and trehalose was identical and, therefore, the result was expressed as a sum of both sugars. Linear calibration curves were obtained with serial dilutions of 0.1, 0.5, 1.0, 5.0 and 10.0 g L-1 of inulin (standard for fructans) (y = 124.059x + 651.0289, coefficient of determination, R2 = 0.99999), sucrose (y = 135.759x − 1629.46, coefficient of determination, R2 = 0.99999), glucose (y = 138.942x − 1427.89, coefficient of determination, R2 = 0.9999) and fructose (y = 133.871x − 2662.79, coefficient of determination, R2 = 0.9999).

### 4.8. Quantitative Real-Time PCR (RT-PCR) Analysis

Total RNA was isolated from frozen 3-day old kale seedlings using the MagMAx Plant RNA Isolation Kit (Thermo Scientific, Waltham, MA, USA), according to the manufacturer's instructions. One biological replicate was composed of 10 seedlings grown either under control or stress conditions (mannitol, temperature treated or combined stress treated). After extraction RNA was quantified using NanoDropTM 1000 Spectrophotometer (Thermo Scientific). Isolated RNA (1 µg) was reverse transcribed in a total reaction volume of 20 µL using 200 U of RevertAid H Minus Reverse Transcriptase (Thermo Scientific), 1× Reaction Buffer (Thermo Scientific), 20 U of RiboLock RNase inhibitor (Thermo Scientific), 1 mM dNTPs (Sigma-Aldrich) and 2.5 µM Oligo(dT)18 primer (Thermo Scientific). For cDNA synthesis, the reactions were incubated at 65 °C for 5 min, at 42 °C for 1 h and at 70 °C for 15 min, followed by a five-fold dilution with water. Genomic DNA (gDNA) was extracted with CTAB [68]. All primers used (Table 3) were designed based on gene sequences of *Brassica rapa* ssp. *pekinensis* and *Brassica oleracea* ssp. *oleracea* and checked by standard PCR on gDNA and cDNA of kale cultivars. Standard PCR reactions contained 1× EmeraldAmp® GT PCR Master Mix (Takara Bio Inc., Kusatsu, Japan), 300 nM forward and reverse qB-OGIO primer and 2 µL (20 ng) of cDNA or genomic DNA in a total volume of 25 µL. PCR was performed in a thermocycler (Eppendorf Mastercycler, Hamburg, Germany) with the initial denaturation step set at 95 °C for 2 min, followed by 35 cycles of denaturation at 95 °C for 30 s, annealing at 58–60 °C for 30 s, extension at 72 °C for 1 min and a final extension step at 72 °C for 5 min. To check cDNA quality standard PCR reaction with qB-OGIO primers was performed. qB-OGIO forward and reverse primers are complementary to different exons and in case of undesirable gDNA presence in cDNA sample two fragments are synthesized indicating an unsuitable sample for qPCR. Quantitative RT-PCR was performed in duplicate on the MIC platform (Bio molecular Systems). A total reaction volume of 15 µL contained 1× GoTaq® qPCR Master Mix reagent (Promega, Madison, WI, USA), 133 nM of forward and reverse primers (Table 1) and 20 ng (2 µL) cDNA. The run profile of the PCR reaction was as follows: 95 °C for 5 min followed by 40 cycles of 95 °C for 5 s and 58 °C for 10 s. Melting curves were generated from 40 °C to 95 °C at a ramp speed 0.3 °C $s^{-1}$ to check for specific amplification. For normalization, reference genes OGIO (Bra028284) and PUX (Bra026205) were used as internal controls [69]. Relative gene expression was calculated according to Livak and Schmittgen [70].

**Table 3.** Primer sequences used for standard and quantitative real-time PCR analysis.

| Gene | Primer Name | Sequence 5′->3′ |
|---|---|---|
| *DREB2A*<br>Bra009112 | qB-DREB2A-Fw<br>qB-DREB2A-Rev | TTTGATGTTTCTGAGCTTCTTGG<br>CATTGTCTCCCAGGCATTGG |
| *HSFA2*<br>Bra000557 | qB-HSFA2-Fw<br>qB-HSFA2-Rev | ATGAATGTGATGATGGAAGATGGT<br>CTGCCCCAATCCAACGGTG |
| *HSFA7A*<br>Bra012828 | qB-HSFA7A-Fw<br>qB-HSFA7A-Rev | TCTGAGACAGCAGCAACAAAC<br>CTGGAGTAGCTGATACAGAAAC |
| *NAC041*<br>Bra021856 | qB-NAC041-Fw<br>qB-NAC041-Rev | CGAAGACGACAACAAGAGTGC<br>GAGTCACATTCAAATCGCAGC |
| *NAC084*<br>Bra006229 | qB-NAC084-Fw<br>qB-NAC084-Rev | AGGAAGAAGACAGAGGAAACC<br>GCTGAGGTAGGAGGAGATG |
| *OGIO*<br>Bra028284 | qB-OGIO-Fw<br>qB-OGIO-Rev | CAGTATCGTAGCTGAGGTAGC<br>AGAACGGAACACATACTTGACTC |
| *PUX*<br>Bra026205 | qB-PUX-Fw<br>qB-PUX-Rev | CAAACCCAAAGAGGTTGTTGC<br>TCATGTCGTTGTCTTCCAAGG |

### 4.9. Statistical Analysis

The experiments were performed with three biological replicates unless otherwise stated. Each plate represented one biological replicate consisted of 15 seedlings grown under the same conditions. To visualize similarities in the level of abiotic stress tolerance among 33 different kale accessions, heatmap and cluster analyses were performed on root length, biomass and water content and proline level data, using R 4.1.1. Software and "ComplexHeatmap" package [71]. The dendrogram was constructed using Euclidean distance.

Further analyses were performed on selected tolerant (404 and 411) and sensitive (392 and 395) kale accessions including all measured parameters except biomass and water content of seedlings as well as protein immunodetection. Two-way analysis was used to determine the influence of kale accession and stress treatment and one-way ANOVA followed by LSD test to find whether there were differences between the control and treatment groups of a particular kale accession. Results were considered significantly different at $p \leq 0.05$.

To further reveal the differences between accessions with lower or higher stress tolerance a partial least-squares–discriminant analysis (PLS-DA) model was developed based on the obtained results. To prevent overfitting of the developed model a cross-validation step was performed using the leave-one-out cross-validation method [72]. For each parameter a variable importance in projection (VIP) value was calculated and only those parameters with a value higher than one were deemed important. Statistical analyses were performed using TIBCO Statistica 13.5.0.17 software package (TIBCO Software Inc., New York, NY, USA).

## 5. Conclusions

To elucidate the mechanisms of abiotic stress tolerance, 33 kale accessions (*B. oleracea* var. *acephala*) were evaluated for individual (osmotic and elevated temperature stress) and combined stress (osmotic+temperature). Growth performance (root growth and biomass), water content and proline content, as reliable stress markers, were used to select accessions with different stress tolerance. The most sensitive accession to all stress factors was 392. Accession 395 was quite sensitive to heat and mannitol but showed better tolerance to combined stress. More tolerant accession, especially to osmotic stress, was 404 and the most tolerant accession to all stress conditions was 411. The more tolerant accessions had higher basal content of proline, total soluble sugars, glucosinolates, heat shock proteins and higher transcript level of *NAC* and *DREB* transcription factors. On the other hand, sensitive accessions were characterized by a high basal content of fructans. Among stress conditions, mannitol and combined stress caused more prominent changes in morphological and biochemical parameters compared to high-temperature stress. Under stress conditions, the most sensitive accession, 392, was characterized by a significant decrease in biomass accumulation, root growth, photosynthesis performance, fructans content, particularly under osmotic and combined stress, as well as a significant increase in *HSF* transcript expression and HSP accumulation under heat stress and a significant decrease in *NAC* expressions under all examined stress conditions. On the other hand, the most tolerant accession, 411, experienced the lowest changes in all analyzed parameters compared to other accessions. Based on multivariate analysis of all measured parameters for selected kale accessions under abiotic stress treatments, accessions 392 and 395 were distinguished as low tolerant from accessions 404 and 411, which appeared to be more tolerant. The most informative variables in tolerance definition were photosynthetic parameters $PI_{abs}$ and $F_v/F_m$, antioxidant activity, $H_2O_2$, proline, root growth, CAT, fructose and total phenolics.

**Supplementary Materials:** The following supporting information can be downloaded at: https://www.mdpi.com/article/10.3390/ijms231911494/s1.

**Author Contributions:** Conceptualization, B.S.-S. and S.G.B.; methodology, N.B., M.T. (Mirta Tkalec), N.M., A.T.V., S.V. and M.T. (Mirta Tokić); formal analysis, N.B., M.T. (Mirta Tkalec), N.M., A.T.V., M.T. (Mirta Tokić), S.V. and B.S.-S.; investigation, All; resources, S.G.B. and D.B.; writing—original draft preparation, B.S.-S.; writing—review and editing, All; visualization, N.B., M.T. (Mirta Tkalec), N.M., S.V. and B.S.-S.; supervision, B.S.-S., N.B. and M.T. (Mirta Tkalec); project administration, S.G.B.; funding acquisition, S.G.B. All authors have read and agreed to the published version of the manuscript.

**Funding:** This research was funded by the Operational Program Competitiveness and Cohesion 2014–2020 and the European regional fund under the specific Scheme "Scheme to strengthening applied research in proposing actions for climate change adaptation", grant number KK.05.1.1.02.0005.

**Institutional Review Board Statement:** Not applicable.

**Data Availability Statement:** Not applicable.

**Conflicts of Interest:** The authors declare no conflict of interest.

## References

1. Chen, S.; Nelson, M.N.; Chèvre, A.-M.; Jenczewski, E.; Li, Z.; Mason, A.S. Trigenomic bridges for Brassica improvement. *Crit. Rev. Plant Sci.* **2011**, *30*, 524–547. [CrossRef]
2. Šamec, D.; Pavlović, I.; Salopek-Sondi, B. White cabbage (*Brassica oleracea* var. *capitata* f. *alba*): Botanical, phytochemical and pharmacologicall overwiev. *Phytochem. Rev.* **2017**, *16*, 117–135. [CrossRef]
3. Šamec, D.; Urlić, B.; Salopek-Sondi, B. Kale (*Brassica oleracea* var. *acephala*) as a superfood: Review of the scientific evidence behind the statement. *Crit. Rev. Food Sci. Nutr.* **2019**, *59*, 2411–2422. [CrossRef] [PubMed]
4. Leisner, C.P. Review: Climate change impacts on food security- focus on perennial cropping systems and nutritional value. *Plant Sci.* **2020**, *293*, 110412. [CrossRef] [PubMed]

5. Hassan, M.U.; Chattha, M.U.; Khan, I.; Chattha, M.B.; Barbanti, L.; Aamer, M.; Iqbal, M.M.; Nawaz, M.; Mahmood, A.; Ali, A.; et al. Heat stress in cultivated plants: Nature, impact, mechanisms, and mitigation strategies—A review. *Plant Biosyst.* **2021**, *155*, 211–234. [CrossRef]
6. Corwin, D.L. Climate change impacts on soil salinity in agricultural areas. *Eur. J. Soil Sci.* **2021**, *72*, 842–862. [CrossRef]
7. Kaushal, N.; Bhandari, K.; Kadambot, H.M.; Siddique, K.H.M.; Nayyar, H. Food crops face rising temperatures: An overview of responses, adaptive mechanisms, and approaches to improve heat tolerance. *Cogent Food Agric.* **2016**, *2*, 1134380. [CrossRef]
8. Hirayama, T.; Shinozaki, K. Research on plant abiotic stress responses in the post-genome era: Past, present and future. *Plant J.* **2010**, *61*, 1041–1052. [CrossRef]
9. Yoon, Y.; Seo, D.H.; Shin, H.; Kim, H.J.; Kim, C.M.; Jang, G. The Role of Stress-Responsive Transcription Factors in Modulating Abiotic Stress Tolerance in Plants. *Agronomy* **2020**, *10*, 788. [CrossRef]
10. Nakashima, K.; Takasaki, H.; Mizoi, J.; Shinozaki, K.; Yamaguchi-Shinozaki, K. NAC transcription factors in plant abiotic stress responses. *Biochim. Biophys. Acta* **2012**, *1819*, 97–103. [CrossRef]
11. Pavlović, I.; Petřík, I.; Tarkowska, D.; Lepeduš, H.; Vujčić, V.; Radić Brkanac, S.; Novák, O.; Salopek-Sondi, B. Correlations between phytohormones and drought tolerance in selected Brassica crops: Chinese cabbage, white cabbage and kale. *Int. J. Mol. Sci.* **2018**, *19*, 2866. [CrossRef] [PubMed]
12. Pavlović, I.; Mlinarić, S.; Tarkowská, D.; Oklestková, J.; Novák, O.; Lepeduš, H.; Vujčić Bok, V.; Radić Brkanac, S.; Strnad, M.; Salopek-Sondi, B. Early Brassica crops responses to salinity stress: A Comparative Analysis between Chinese cabbage, White cabbage and Kale. *Front. Plant Sci.* **2019**, *10*, 450. [CrossRef] [PubMed]
13. USDA; National Agricultural Statistics Service. Census of Agriculture. 2012. Available online: https://www.agcensus.usda.gov/Publications/2012/Full_Report/Volume_1,_Chapter_1_US/st99_1_038_038.pdf (accessed on 18 June 2015).
14. Batelja, K.; Goreta Ban, S.; Žanić, K.; Miloš, B.; Dumičić, G.; Matotan, Z. Svojstva autohtonih populacija raštike (*Brassica oleracea* L. var. *acephala*) hrvatskog priobalja. *Poljoprivreda* **2009**, *2*, 1–7.
15. Agrimonti, C.; Lauro, M.; Visioli, G. Smart agriculture for food quality: Facing climate change in the 21st century. *Crit. Rev. Food Sci. Nutr.* **2021**, *61*, 971–981. [CrossRef] [PubMed]
16. Linić, I.; Šamec, D.; Grúz, J.; Vujčić Bok, V.; Strnad, M.; Salopek-Sondi, B. Involvement of phenolic acids in short-term adaptation to salinity stress is species specific among Brassicaceae. *Plants* **2019**, *8*, 155. [CrossRef] [PubMed]
17. Pavlović, I.; Pěnčík, A.; Novák, O.; Vujčić, V.; Radić Brkanac, S.; Lepeduš, H.; Strnad, M.; Salopek-Sondi, B. Short-term salinity stress response of Brassica rapa caused changes in auxin metabolism. *Plant Physiol. Biochem.* **2018**, *125*, 74–84. [CrossRef]
18. Beacham, A.M.; Hand, P.; Pink, D.A.C.; Monaghan, J.M. Analysis of Brassica oleracea early stage abiotic stress responses reveals tolerance in multiple crop types and for multiple sources of stress. *J. Sci. Food Agric.* **2017**, *97*, 5271–5277. [CrossRef]
19. Lohani, N.; Jain, D.; Singh, M.B.; Bhalla, P.L. Engineering multiple abiotic stress tolerance in Canola, *Brassica napus*. *Front. Plant Sci.* **2020**, *11*, 3. [CrossRef]
20. Singh, B.; Sran, A.S.; Sohi, G.S. Innovative strategies to develop abiotic and biotic stress tolerance in Mustard (Brassicaceae). In *Brassica Breeding and Biotechnology*; Islam, A.K.M.A., Hossain, M.A., Islam, A.K.M.M., Eds.; IntechOpen: London, UK, 2021. Available online: https://www.intechopen.com/chapters/75023 (accessed on 25 May 2022). [CrossRef]
21. Dikšaitytė, A.; Viršilė, A.; Žaltauskaitė, J.; Januškaitienė, I.; Juozapaitienė, G. Growth and photosynthetic responses in *Brassica napus* differ during stress and recovery periods when exposed to combined heat, drought and elevated $CO_2$. *Plant Physiol. Biochem.* **2019**, *142*, 59–72. [CrossRef]
22. Zhang, H.; Zhao, Y.; Zhu, J.-K. Thriving under stress: How plants balance growth and the stress response. *Dev. Cell* **2020**, *55*, 529–543. [CrossRef]
23. Petrić, I.; Šamec, D.; Karalija, E.; Salopek-Sondi, B. Beneficial microbes and molecules for mitigation of soil salinity in Brassica species: A Review. *Soil Syst.* **2022**, *6*, 18. [CrossRef]
24. Hayat, S.; Mir, B.A.; Wani, A.S.; Hasan, S.A.; Irfan, M.; Ahmad, A. Screening of salt-tolerant genotypes of Brassica juncea based on photosynthetic attributes. *J. Plant. Interact.* **2011**, *6*, 53–60. [CrossRef]
25. Mohamed, I.A.A.; Shalby, N.; Bai, C.; Qin, M.; Agami, R.A.; Jie, K.; Wang, B.; Zhou, G. Stomatal and photosynthetic traits are associated with investigating sodium chloride tolerance of *Brassica napus* L. cultivars. *Plants* **2020**, *9*, 62. [CrossRef] [PubMed]
26. Hossain, M.A.; Bhattacharjee, S.; Armin, S.-M.; Qian, P.; Xin, W.; Li, H.-Y.; Burritt, D.J.; Fujita, M.; Tran, L.-S.P. Hydrogen peroxide priming modulates abiotic oxidative stress tolerance: Insights from ROS detoxification and scavenging. *Front. Plant Sci.* **2015**, *6*, 420. [CrossRef]
27. Liu, W.-C.; Song, R.-F.; Zheng, S.-Q.; Zhang, B.-L.; Gao, X.; Lu, Y.-T. Coordination of plant growth and abiotic stress responses by tryptophan synthase β subunit 1 through modulation of tryptophan and ABA homeostasis in Arabidopsis. *Mol. Plant* **2022**, *15*, 973–990. [CrossRef]
28. Liu, W.-C.; Song, R.-F.; Qiu, Y.-M.; Zheng, S.-Q.; Li, T.-T.; Wu, Y.; Song, C.-P.; Lu, Y.-T.; Yuan, H.-M. Sulfenylation of ENOLASE2 facilitates $H_2O_2$-conferred freezing tolerance in Arabidopsis. *Dev. Cell* **2022**, *57*, 1883–1898. [CrossRef]
29. Khan, A.; Anwar, Y.; Hasan, M.M.; Iqbal, A.; Ali, M.; Alharby, H.F.; Hakeem, K.R.; Hasanuzzaman, M. Attenuation of drought stress in Brassica seedlings with exogenous application of $Ca^{2+}$ and $H_2O_2$. *Plants* **2017**, *6*, 20. [CrossRef]
30. Lee, B.-R.; La, V.H.; Park, S.-H.; Mamun, M.A.; Bae, D.-W.; Kim, T.-H. $H_2O_2$—Responsive hormonal status involves oxidative burst signaling and proline metabolism in rapeseed leaves. *Antioxidants* **2022**, *11*, 566. [CrossRef]

31. Kalaji, H.M.; Jajoo, A.; Oukarroum, A.; Brestic, M.; Zivcak, M.; Samborska, I.A.; Cetner, M.D.; Łukasik, I.; Goltsev, V.; Ladle, R.J. Chlorophyll a fluorescence as a tool to monitor physiological status of plants under abiotic stress conditions. *Acta Physiol. Plant.* **2016**, *38*, 3–11. [CrossRef]
32. Moradpour, M.; Abdullah, S.N.A.; Namasivayam, P. The impact of heat stress on morpho-physiological response and expression of specific genes in the heat stress-responsive transcriptional regulatory network in *Brassica oleracea*. *Plants* **2021**, *10*, 1064. [CrossRef]
33. Rodríguez, V.M.; Soengas, P.; Alonso-Villaverde, V.; Sotelo, T.; Cartea, M.E.; Velasco, P. Effect of temperature stress on the early vegetative development of *Brassica oleracea* L. *BMC Plant Biol.* **2015**, *15*, 145. [CrossRef] [PubMed]
34. Soengas, P.; Rodríguez, V.M.; Velasco, P.; Cartea, M.E. Effect of temperature stress on antioxidant defenses in Brassica oleracea. *ACS Omega* **2018**, *3*, 5237–5243. [CrossRef] [PubMed]
35. Huang, R.; Liu, Z.; Xing, M.; Yang, Y.; Wu, X.; Liu, H.; Liang, W. Heat stress suppresses *Brassica napus* seed oil accumulation by inhibition of photosynthesis and BnWRI1 pathway. *Plant Cell Physiol.* **2019**, *60*, 1457–1470. [CrossRef] [PubMed]
36. Elferjani, R.; Soolanayakanahally, R. Canola responses to drought, heat, and combined stress: Shared and specific effects on carbon assimilation, seed yield, and oil composition. *Front. Plant Sci.* **2018**, *9*, 1224. [CrossRef] [PubMed]
37. Šamec, D.; Karalija, E.; Šola, I.; Vujčić Bok, V.; Salopek-Sondi, B. The role of polyphenols in abiotic stress response: The influence of molecular structure. *Plants* **2021**, *10*, 118. [CrossRef]
38. Kourani, M.; Mohareb, F.; Rezwan, F.I.; Anastasiadi, M.; Hammond, J.P. Genetic and physiological responses to heat stress in *Brassica napus*. *Front. Plant Sci.* **2022**, *13*, 832147. [CrossRef]
39. Essoh, A.P.; Monteiro, F.; Pena, A.R.; Pais, M.S.; Moura, M.; Romeiras, M.M. Exploring glucosinolates diversity in Brassicaceae: A genomic and chemical assessment for deciphering abiotic stress tolerance. *Plant Physiol. Biochem.* **2020**, *150*, 151–161. [CrossRef]
40. Fareen, S.; Mohammad, Y.; Mohammad, F.; Ahmad, F.; Shamsul, H. Role of sugars under abiotic stress. *Plant Physiol. Biochem.* **2016**, *109*, 54–61. [CrossRef]
41. Viana, V.E.; Aranha, B.C.; Busanello, C.; Maltzahn, L.E.; Panozzo, L.E.; De Oliveira, A.C.; Rombaldi, C.V.; Pegoraro, C. Metabolic profile of canola (*Brassica napus* L.) seedlings under hydric, osmotic and temperature stresses. *Plant Stress* **2022**, *3*, 100059. [CrossRef]
42. Benkeblia, N. Insights on fructans and resistance of plants to drought stress. *Front. Sustain. Food Syst.* **2022**, *6*, 827758. [CrossRef]
43. Keller, K.; Rodrigues, C.M.; Neuhaus, H.E.; Pommerrenig, B. Improved resource allocation and stabilization of yield under abiotic stress. *J. Plant Physiol.* **2021**, *257*, 153336. [CrossRef] [PubMed]
44. Priya, M.; Dhanker, O.P.; Siddique, K.H.M.; Rao, B.H.; Nair, R.M.; Pandey, S.; Singh, S.; Varshney, R.K.; Prasad, P.V.V.; Nayyar, H. Drought and heat stress-related proteins: An update about their functional relevance in imparting stress tolerance in agricultural crops. *Theor. Appl. Genet.* **2019**, *132*, 1607–1638. [CrossRef] [PubMed]
45. Guo, M.; Liu, J.-H.; Ma, X.; Luo, D.-X.; Gong, Z.-H.; Lu, M.-H. The Plant Heat Stress Transcription Factors (HSFs): Structure, regulation, and function in response to abiotic stresses. *Front. Plant Sci.* **2016**, *7*, 114. [CrossRef]
46. Lohani, N.; Golicz, A.A.; Singh, M.B.; Prem, L.; Bhalla, P.L. Genome-wide analysis of the Hsf gene family in Brassica oleracea and a comparative analysis of the Hsf gene family in *B. oleracea*, *B. rapa* and *B. napus*. *Funct. Integr. Genomics* **2019**, *19*, 515–531. [CrossRef]
47. Zhu, X.; Wang, Y.; Liu, Y.; Zhou, W.; Yan, B.; Yang, J.; Shen, Y. Overexpression of BcHsfA1 transcription factor from *Brassica campestris* improved heat tolerance of transgenic tobacco. *PLoS ONE* **2018**, *13*, e0207277. [CrossRef]
48. Dong, X.; Yi, H.; Lee, J.; Nou, I.-S.; Han, C.-T.; Hur, Y. Global gene-expression analysis to identify differentially expressed genes critical for the heat stress response in *Brassica rapa*. *PLoS ONE* **2015**, *10*, e0130451. [CrossRef] [PubMed]
49. Hasanuzzaman, M.; Nahar, K.; Alam, M.M.; Roychowdhury, R.; Fujita, M. Physiological, biochemical, and molecular mechanisms of heat stress tolerance in plants. *Int. J. Mol. Sci.* **2013**, *14*, 9643–9684. [CrossRef] [PubMed]
50. Mishra, D.; Shekhar, S.; Singh, D.; Chakraborty, S.; Chakraborty, N. Heat shock proteins and abiotic stress tolerance in plants. In *Regulation of Heat Shock Protein Responses*; Asea, A.A.A., Kaur, P., Eds.; Springer: Dordrecht, The Netherlands, 2018. [CrossRef]
51. Haq, S.; Khan, A.; Ali, M.; Khattak, A.M.; Gai, W.-X.; Zhang, H.-X.; Wei, A.M.; Gong, Z.-H. Heat shock proteins: Dynamic biomolecules to counter plant biotic and abiotic stresses. *Int. J. Mol. Sci.* **2019**, *20*, 5321. [CrossRef]
52. Liu, T.; Song, X.; Duan, W.; Huang, Z.; Liu, G.; Li, Y.; Hou, X. Genome-wide analysis and expression patterns of NAC Transcription Factor Family under different developmental stages and abiotic stresses in Chinese Cabbage. *Plant Mol. Biol. Rep.* **2014**, *32*, 1041–1056. [CrossRef]
53. Zhong, H.; Guo, Q.-Q.; Chen, L.; Ren, F.; Wang, Q.-Q.; Zheng, Y.; Li, X.-B. Two *Brassica napus* genes encoding NAC transcription factors are involved in response to high-salinity stress. *Plant Cell Rep.* **2012**, *31*, 1991–2003. [CrossRef]
54. Ying, L.; Chen, H.; Cai, W. BnNAC485 is involved in abiotic stress responses andflowering time in *Brassica napus*. *Plant Physiol. Biochem.* **2014**, *79*, 77–87. [CrossRef] [PubMed]
55. Strasser, R.J.; Tsimilli-Michael, M.; Srivastava, A. Analysis of the chlorophyll a fluorescence transient. In *Chlorophyll a Fluorescence: A Signature of Photosynthesis*; Papageorgiou, G.C., Govinjee, Eds.; Springer: Dordrecht, The Netherlands, 2004; pp. 321–362. [CrossRef]
56. Mátai, A.; Hideg, É. A comparison of colorimetric assays detecting hydrogen peroxide in leaf extracts. *Anal. Methods* **2017**, *9*, 2357–2360. [CrossRef]

57. Bradford, M.M. A rapid and sensitive method for the quantitation of microgram quantities of protein utilizing the principle of protein-dye binding. *Anal. Biochem.* **1976**, *72*, 248–254. [CrossRef]
58. Aebi, H. Catalase in vitro. *Meth. Enzymol.* **1984**, *105*, 121–126. [CrossRef]
59. Draper, H.H.; Hadley, M. Malondialdehyde detemination as index of lipid peroxidation. *Meth. Enzymol.* **1990**, *186*, 421–431. [CrossRef]
60. Staples, R.C.; Stahmann, M.A. Changes in proteins and several enzymes in susceptible bean leaves after infection by the bean rust fungus. *Phytopathology* **1964**, *54*, 760–764.
61. Laemmli, U.K. Cleavage of structural proteins during the assembly of the head of bacteriophage T4. *Nature* **1970**, *227*, 680–685. [CrossRef]
62. Singleton, V.L.; Rossi, J.A. Colorimetry of total phenolics with phosphomolybdic-phosphotungstic acid reagents. *Am. J. Enol. Viticul.* **1965**, *16*, 144–158.
63. Woisky, R.G.; Salatino, A. Analysis of propolis: Some parameters and procedures for chemical quality control. *J. Apicul. Res.* **1998**, *37*, 99–105. [CrossRef]
64. Ishida, M.; Nagata, M.; Ohara, T.; Kakizaki, T.; Hatakeyama, K.; Nishio, T. Small variation of glucosinolate composition in Japanese cultivars of radish (Raphanus sativus L.) requires simple quantitative analysis for breeding of glucosinolate component. *Breed. Sci.* **2012**, *62*, 63–70. [CrossRef]
65. Brand-Williams, W.; Cuvelier, M.E.; Berset, C.L.W.T. Use of a free radical method to evaluate antioxidant activity. *LWT-Food Sci. Tech.* **1995**, *28*, 25–30. [CrossRef]
66. Malar, J.; Chairman, K.; Singh, A.R.J.; Vanmathi, J.S.; Balasubramanian, A.; Vasanthi, K. Antioxidative activity of different parts of the plant *Lepidium sativum* Linn. *Biotech. Rep.* **2014**, *3*, 95–98. [CrossRef]
67. Maness, N. Extraction and analysis of soluble carbohydrates. In *Plant Stress Tolerance*; Sunkar, R., Ed.; Methods in Molecular Biology; Humana Press: New York, NY, USA, 2010; Volume 639. [CrossRef]
68. Doyle, J.J.; Doyle, J.L. A rapid DNA isolation procedure for small quantities of fresh leaf tissue. *Phytochem. Bull.* **1987**, *19*, 11–15.
69. Škiljaica, A.; Jagić, M.; Vuk, T.; Leljak Levanić, D.; Bauer, N.; Markulin, L. Evaluation of reference genes for RT-qPCR gene expression analysis in Arabidopsis thaliana exposed to elevated temperatures. *Plant Biol.* **2022**, *24*, 367–379. [CrossRef]
70. Livak, K.J.; Schmittgen, T.D. Analysis of relative gene expression data using real-time quantitative PCR and the $2^{-\Delta\Delta CT}$ method. *Methods* **2001**, *25*, 402–408. [CrossRef]
71. Gu, Z.; Eils, R.; Schlesner, M. Complex heatmaps reveal patterns and correlations in multidimensional genomic data. *Bioinformatics* **2016**, *32*, 2847–2849. [CrossRef]
72. Sammut, C.; Webb, G.I. Leave-One-Out Cross-Validation. In *Encyclopedia of Machine Learning*; Sammut, C., Webb, G.I., Eds.; Springer: Boston, MA, USA, 2011; pp. 600–601. [CrossRef]

International Journal of
*Molecular Sciences*

MDPI

*Review*

# Current Studies of the Effects of Drought Stress on Root Exudates and Rhizosphere Microbiomes of Crop Plant Species

**Yalin Chen [1,2], Zongmu Yao [1,2], Yu Sun [2], Enze Wang [2], Chunjie Tian [1,2], Yang Sun [1], Juan Liu [3], Chunyu Sun [1,*] and Lei Tian [2,*]**

1 Key Laboratory of Straw Comprehensive Utilization and Black Soil Conservation, Ministry of Education, College of Life Science, Jilin Agricultural University, Changchun 130000, China; yalin12383@126.com (Y.C.); amuu0316@gmail.com (Z.Y.); tiancj@iga.ac.cn (C.T.); ysun@jlau.edu.cn (Y.S.)

2 Key Laboratory of Mollisols Agroecology, Northeast Institute of Geography and Agroecology, Chinese Academy of Sciences, Changchun 130102, China; sunyu08@iga.ac.cn (Y.S.); wangenze@iga.ac.cn (E.W.)

3 Sericultural Research Institute of Jilin Province, Jilin 132013, China; liujuan807@126.com

* Correspondence: cysun@jlau.edu.cn (C.S.); tianlei@iga.ac.cn (L.T.)

**Abstract:** With the warming global climate, drought stress is considered to be the most important abiotic factor limiting plant growth and yield in the world. Drought stress has serious impacts on crop production. Many researchers have studied the influences of drought stress on crop production and plant physiology; however, few researchers have combined root exudates with root-associated microbiomes for their mutual effects under drought conditions. In this review, we systematically illustrate the impact of drought stress on root exudates and root-associated microbiomes, and then we discuss the mutual regulation of root-associated microbiomes and the host plant in helping the plant adapt to drought. Finally, we construct a framework for the mutual connections between the plant, root exudates, and the microbiome. We hope this review can provide some significant guidelines to promote the study of drought resistance in plants in association with the rhizosphere microbiota.

**Keywords:** microbiomes; plant; root exudates; drought stress; rhizosphere

**Citation:** Chen, Y.; Yao, Z.; Sun, Y.; Wang, E.; Tian, C.; Sun, Y.; Liu, J.; Sun, C.; Tian, L. Current Studies of the Effects of Drought Stress on Root Exudates and Rhizosphere Microbiomes of Crop Plant Species. *Int. J. Mol. Sci.* **2022**, *23*, 2374. https://doi.org/10.3390/ijms23042374

Academic Editors: Isabel Marques, Ana I. Ribeiro-Barros and José C. Ramalho

Received: 27 December 2021
Accepted: 19 February 2022
Published: 21 February 2022



## 1. Introduction

Water is the most important resource for plants, and plant organs need to maintain 60–90% water content for sustainable activity. However, global climate change, caused by greenhouse gas emissions, has become more serious worldwide, leading to drought throughout the world [1]. Drought has been considered the most serious and recurrent abiotic factor limiting crop growth and yield in the world [2]. Drought stress decreases metabolism in plants, and serious drought stress will cause electrolyte disturbances in plant cells, which will lead to death of the plant. Thus, sufficient water in the soil is vital for plant growth and production.

Currently, the frequent changes in climate have increased the severity of drought events for plants [3]. The plant-associated and plant rhizosphere microbiomes, including plant growth-promoting bacteria (PGPB) and plant growth-promoting fungi (PGPF), are closely related to plant growth and environment change, including drought. Drought stress influences plant water uptake, which inversely affects the plant metabolism and root exudates, and the exudates affect the plant rhizosphere microbiome.

Furthermore, plant roots can help maintain the inherited microbial communities that may influence crop growth, nutrition, and health [4]. A previous study also showed that drought may produce legacy effects on soil microbial communities [5]. In contrast, the root-associated microbiome can help plants resist and adapt to drought.

Although the composition of the root-associated microbiome has been determined in many crop species, the influence of abiotic stresses, especially drought stress, on the root-associated microbiota has been less studied [6]. In this paper, we review how drought stress affects plant root exudates and the root-associated microbiome, and then illustrate how the root-associated microbiome affects plants in resisting drought. We hope that this review can provide some significant guidelines for understating the mutual connections between plant root exudates and microbiomes under drought stress and can provide practice for developing beneficial microbiota in promoting plant resistance to drought stress.

## 2. Drought Stress Influences Root Exudates

Root exudates are plant metabolites and are secreted in the rhizosphere, including various compounds secreted or released from different parts of the root system into the rhizosphere environment during the growth of plants [7]. Root exudates can account for more than 10% of the plant photosynthate [8], including low molecular weight primary metabolites (especially sugars, amino acids, and organic acids) and secondary metabolites (phenols, flavonoids, and terpenoids) [7].

Root exudates, responding to the changing environment, can serve as positive or negative factors in plant growth [7,9]. A previous study demonstrated that drought promoted the secretion of organic acids in corn root, and malic acid was the main organic acid secreted [10], which may help to solubilize phosphate for plant utilization and promote drought resistance in plants. However, excessive root exudates will decrease the carbon storage in crop plants [11].

Root exudates serve as important carrier materials for material exchange and information transmission between plants and soil [12–15]. Root exudates are also the key factor to maintaining the vitality and function of the rhizosphere micro-ecosystem and are also an important part of the rhizosphere material circulation [12–15]. Root exudates can improve the bioavailability of soil nutrients and improve plant growth by changing the physical, chemical, or biological properties of the rhizosphere [12–15].

With the rapid development of modern instrumental analysis methods and the wide application of root exudates in the fields of plant nutrition, agroecological and environment studies, etc., the research on root exudates has entered a new upsurge. Especially in the last decade, with the construction and improvement of rhizosphere microecology, root exudates have been considered an important part of plant nutrition and rhizosphere microecology research.

Unlike animals, plants cannot escape various adverse environmental stresses and can only grow and develop in their germinating place where they finally complete their life cycle. Root secretion activity helps plants adapt to and gradually change the soil environment contacted by the roots, which is an adaptive response mechanism for plants to cope with environmental stress [12–14]. Therefore, the composition and amount of root exudates not only depend on the species and genotype of plants [16] but are also influenced and controlled by the external environment, such as drought. The composition and amount of plant root exudates under drought stress will also change to adapt to the stress [17].

In general, the content of root exudates of plants under drought stress will be higher than that under normal conditions. Many studies showed that the total content of root exudates (i.e., soluble sugars, amino acids, and organic acids) of the plants increased with the level of drought stress intensity [18,19]. The content of organic acids (malic acid, lactic acid, acetic acid, succinic acid, citric acid, and maleic acid) increased significantly in root exudates of maize under drought stress [18]. The increase in plant root exudates, especially organic acids, may help plants resist osmotic stress under drought stress.

Root exudates can improve the drought tolerance in plants. Drought stress will not only change the amount of plant root exudates but also significantly change their composition [8]. Gargallo-Garriga and others also studied the root exudates of *Quercus ilex* under gradient drought stress and subsequent recovery and found that the increased drought stress strongly affected the secondary metabolites (accounting for 71% of total metabolites), with the increased synthesis of alkaloids and terpenoids [20]. However, in the drought recovery stage, the composition of root exudates became dominated by primary metabolites (accounting for 81% of the total metabolites), and under extreme drought conditions, the changes in root exudates were irreversible, and plants could not recover [20].

Another study also showed that potassium (K+) and organic acid were the most important contributors to resisting drought stress in plants [21]. K+ is extracted from the soil by roots, where the roots can exchange it with organic acids [21]. Thus, with the frequent occurrence of drought now, we should not only strengthen research on the induction mechanism of specific root exudates under drought stress but also focus on the synthesis, transfer, and secretion pathways as well as the mechanisms of root exudates under drought stress. These studies may provide a theoretical basis for the development and utilization of plant resources in acclimating to drought stress.

## 3. Drought Affects Plant Hormone Production and Regulates the Metabolism

Plant hormones, which are produced in plants and transported from one place to another, play a regulatory role in the life activities of plants. The levels of endogenous hormones, abscisic acid (ABA), auxin (IAA), gibberellin (GA), cytokinin (CTK), and ethylene (ETH) are decreased or increased under drought (Figure 1) [22–24]. Many researchers have shown that, under drought conditions, the concentrations of ABA and ETH increase significantly [24,25], while it is less clear what happens to the concentrations of CTK, IAA, and GA. Studies showed that the concentrations of CTK, IAA, and GA decreased to a certain extent, and the endogenous hormones of varieties with strong drought resistance also changed greatly, which indicated that drought stress was negatively correlated with these hormone concentrations [26,27].

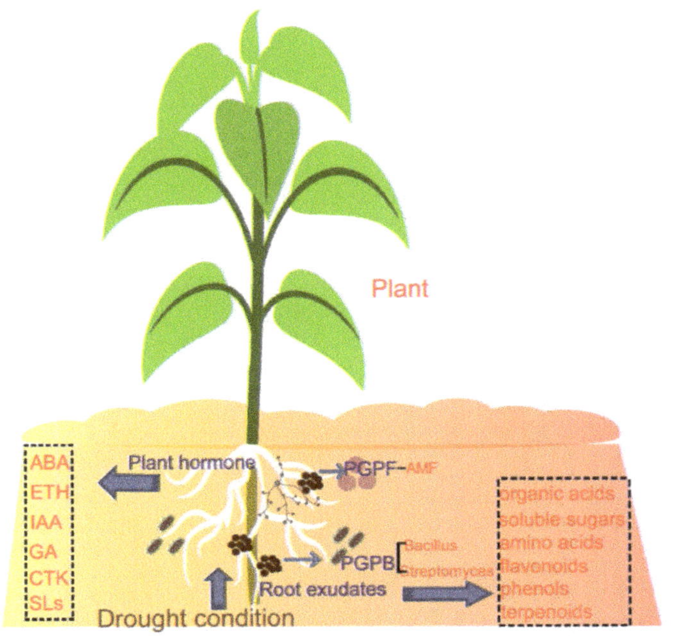


**Figure 1.** Drought stress influences root exudates and associated microbiotas. ABA, abscisic acid; ETH, ethylene; IAA, auxin; GA, gibberellin; CTK, cytokinin; SLs, strigolactones; PGPB, plant growth-promoting bacteria; and PGPF, plant growth-promoting fungi.

ABA is the most studied plant endogenous hormone under drought stress and is considered the most important hormone in plants for regulating signal transduction under drought stress [28,29]. Under drought stress, the plant roots receive the stress signal first, which stimulates the activity of ABA synthetase. Then, ABA, as a drought signal substance, is rapidly synthesized in the root cap and actively transported to the growth part, reducing stomatal conductance and water loss, thus improving the drought resistance [30]. Furthermore, plants can induce the expression of many genes under stress, and ABA can serve as the signaling molecule that induces the activation of these genes, which will lead to protein expression and other plant phenotypes [31,32].

In the potato budding stage, with the increase under drought stress, the ABA concentration and expression of the 9-cis-epoxycarotenoid dioxygenase (StNCED) gene (a key enzyme for ABA synthesis) in tubers showed an increasing trend. Under severe stress, the ABA concentration in tubers increased by 33%, and the transcription levels of the StNCED1 and StNCED2 genes increased by four and nine times, respectively [33]. Ding et al. (2016) studied the changes of endogenous hormones in rice seedlings under mild drought stress and concluded that the ABA concentration in rice root increased significantly with decreased precipitation [34].

It was shown that the concentration of strigolactones (SLs), one of the plant hormones, also decreases in plants under drought stress. However, evidence showed that SL levels increase in plants when faced with drought stress with the presence of arbuscular mycorrhizal fungi (AMF) spores adhering to the plant roots (Figure 1) [35,36]. SLs are also a signal substance for helping plants connect with AMF, and inoculated AMF helps plants resist biotic or abiotic stress.

The dialogue between a plant and fungus comes from plant photosynthesis and root exudation, molecular cues (SLs), secreted into the rhizosphere [37]. It was shown that ABA can promote SL production under drought stress [38]. However, crosstalk analysis between cytokinins and SLs demonstrated that they have opposite effects on plant root drought adaptation, which indicated that CTKs act as negative regulators, while SLs act as positive regulators in responding to drought [36].

## 4. Drought Stress Influences Root-Associated Microbiomes Directly or Indirectly

Soil microbes can respond accurately and quickly to changes in soil environment, and small changes in soil environment can lead to changes in soil microbes, which can sensitively warn of small changes in terrestrial ecosystems. The soil water content is closely related to soil microbes, and drought will affect the rhizosphere microbial communities. First, drought will cause osmotic stress on soil microbes, which will directly affect them, resulting in their death and cell lysis. Second, drought will change the quality and quantity of carbon sources available to rhizosphere microbes by affecting plant photosynthesis and then indirectly affect the roots.

Fuchslueger et al. (2014) demonstrated that water deficit caused a change in the microbial community composition; an increase of Gram-positive bacteria was driven by the drought stress, and drought decreased the C allocation below ground but did not influence the transfer of recently plant-assimilated C to fungi [39]. Comparative analysis of the effect of drought stress on the root and rhizosphere microbiomes of 18 species of monocot plants showed that a prominent enrichment in Actinobacteria during water deficit is common among the hosts [40]. Actinobacteria were more enriched in the root endosphere than in the surrounding soils [40], which indicated that Actinobacteria may be connected more closely with plant roots when facing water stress.

Furthermore, the enrichment of Streptomyces (belonging to the Actinobacteria) under drought was shown to play a subsequent role in improving plant drought tolerance in the study [41]. Through the study of the interaction between soil components and wheat root microbiota under drought stress, it was found that the soil pH value and microbial biomass could significantly explain the changes in the microbial community, wheat genotype, and soil sodium and iron levels [42]. Under drought stress, the biodiversity in soil decreased significantly but increased in the rhizosphere community, in which specific soil parameters seemed to determine the enrichment of bacterial flora.

Drought stress leads to specific recombination of the rice root-related microbial community [6]. It was shown that enrichment of Actinobacteria and Chloroflexi, as well as depletion of several Acidobacteria and Deltaproteobacteria, was common for rice root-associated microbiota under drought stress. The study also showed that the plant genotype, as well as drought stress, influenced the root-associated fungal community in rice and that some fungi potentially improved plant drought tolerance [43].

In the soil environment, Gram-positive bacteria are more tolerant to drought than Gram-negative bacteria because their cell walls are thicker and some Gram-positive bacteria have the ability to form spores [41]. However, it has been observed that the enrichment degree of the Gram-positive bacteria in the relevant parts of plants is the highest, which indicates that the enriched Gram-positive bacteria described here are at least partly driven by the interaction within the plant host, not only by the ability of the Gram-positive bacteria to resist the dry environment [41].

Carbohydrates and amino acids produced by plants are powerful candidates among compounds that can promote the growth of a single germ layer [41]. It is well-known that root exudates have been demonstrated to selectively induce the growth of rhizosphere bacteria [44]. In addition, the recent work on the plant-related microbiome has confirmed that the correlation between the abundance of the Gram-positive bacteria (Actinobacteria, *Verticillium virens*, and Sclerotinia) and specific amino sugars and sugar alcohols, indicated that the Gram-positive bacteria are related to a group of different compounds, including polysaccharides and glycoside hydrolases.

In addition, the metabolomics of roots under drought stress showed that the yields of many carbohydrates (including xylose and glucose) and amino acids (such as proline, threonine, and asparagine) increased significantly [45]. Malic acid, a well-known root exudate, is also an effective chemoattractant for *Bacillus subtilis* (e.g., in the soybean rhizosphere) when plants are under drought stress [10]. Santana et al. (2020) demonstrated that root exudates promoted the enrichment of *Bacillus* sp. ESA 402, one of the plant growth promoting bacteria, in the rhizosphere of sorghum (*Sorghum bicolor*), which benefited the growth of the sorghum under drought stress [46]. However, how the connection between the Gram-positive bacteria and the root exudates helps plants resist drought stress has not been clearly illustrated.

As for rhizosphere microbiotas, some scholars have used the dilution plate counting method to study the number of bacteria [40]. They found that fungi and Actinobacteria in the rhizosphere soil of ginger (*Zingiber officinale*), sorghum (*S. bicolor*), corn (*Zea maize*), wheat (*Triticum monococcum*), and others decreased with an increased level of drought stress intensity, while the number of bacteria and Actinobacteria in rhizosphere soil of cherry increased with the initial stress intensity and the decreased with a high level of drought stress [40].

Santos-Medellín et al. (2021) also showed that Streptomyces were the most enriched Actinobacteria in the rice root microbiome after drought stress [5]. Zhang et al. (2021) also showed that drought stress increased the diversity and abundance of the rhizosphere bacteria in both wild (*Glycine soja*) and cultivated (*Glycine max*) soybean [47]. However, the number of fungi decreased under drought stress in both studies [5,47]. Therefore, drought has a significant impact on soil microorganisms, and proper (mild or moderate) drought is beneficial for maintaining microbial diversity in rhizosphere soil [5,47]. Drought stress changes the composition of bacteria and fungi in rhizosphere soil, making bacteria dominant and then leading to the decline of the quality of organic substances.

The ratio of fungi to bacteria is an important index for measuring the microbial community structure, which also reflects the nutrient status of the substrate [48,49]. The fungal pathway mainly occurs in soil with low nutrition, difficult decomposition, and a high carbon–nitrogen ratio, and the substrate circulation time is relatively long. This indicates that drought will have a great impact on the microbial community structure diversity of rhizosphere soil, and a certain intensity of drought can improve the microbial community structure diversity of rhizosphere soil, while excessive drought will lead to a decrease in the microbial community structure diversity of rhizosphere soil.

## 5. Root-Associated Microbiomes Regulate Plants in Resisting Drought Stress

Studies showed that not only did drought influence root-associated microbiomes but, conversely, root-associated microbiomes regulated plants in resisting drought stress. Plant growth-promoting rhizobacteria (PGPR) can not only help promote plant growth but also help plants resist biotic and abiotic stress [50]. It was also shown that PGPR can alleviate plant damage by inducing drought-responsive genes, such as the aquaporin (TaTIP1;1) and helicase genes [51]. Furthermore, plants contain various other resistance strategies to adapt to drought stress.

In order to effectively absorb water and reduce water evaporation under drought conditions, the plant will decrease the leaf size and extend the root system into deeper soil [51,52]. Plants also benefit from the synthesis of different compounds for osmotic regulation (such as proline, glycine, betaine, and potassium), plant hormones (such as abscisic acid, salicylic acid, auxin, and gibberellin), and antioxidation (such as polyamine) when facing drought stress [52,53].

Although these strategies are promoted by plants, the contribution of bacteria also affects this protection. For example, although proline serves as the most vital osmotic agent in the plant cytoplasm, under environmental stress, proline accumulation occurs not only in higher plants but also in bacteria, which may also help the bacteria for drought tolerance and then in association with plant in resisting drought [54,55].

Rolli et al. (2016) showed that improved plant resistance to drought is promoted by the root-associated microbiome as a water stress-dependent trait [56]. The study also showed that the rice genotype, as well as drought stress, affected the root-associated fungal community, and conversely, some special fungi were shown to help improve the drought tolerance of rice [43]. The rhizobacteria can also produce plant hormones (such as IAA and CTK) or inhibit plant hormone production, which may directly affect plant growth [57,58].

The mechanisms of the plant in responding to drought stress by the rhizobacteria have been shown to include the production of abscisic acid, cytokinin, indoleacetic acid, trehalose, 1-aminocyclopropane-1-carboxylate (ACC) deaminase, volatile organic compounds, and exopolysaccharides [59,60]. The most studied beneficial fungi, arbuscular mycorrhiza fungi (AMF), are one of the most widely distributed fungi in the ecosystem, and they play an important role in the material circulation and energy flow in nature [61–63]. AMF can form mycorrhiza with host plants, and more than 80% of terrestrial plants have mycorrhiza [64,65].

They can also promote the plant absorption of nitrogen, phosphorus, and other elements in the soil, while plants can provide AMF with carbon sources generated by photosynthesis [65–67]. Rice, corn, wheat, soybean, and so on, which are the major food crops for human survival, are all mycorrhizal plants [68–70]. AMF not only forms a symbiotic relationship with rice in loose and breathable soil but also forms a symbiotic relationship with rice under flooded conditions [71–73]. Pavithra et al. (2018) confirmed that AMF inoculation can improve soybean biomass and proline content under drought conditions [74]. Drought may also improve the vesicle formation and hyphae development of AMF root colonization [75].

It is also shown that the extraradical hyphae of AMF colonized in roots extend into the soil. The extended and dense hyphae in the soil can contact and extract water from the soil pores, which is not accessible to plant roots and the root-hair zone [76]. As a strictly symbiotic fungus, AMF often forms nutrition complementation with plants while forming a symbiosis with them, and AMF plays important roles in promoting plant resistance to biotic and abiotic stresses [77–79]. AMF can improve trehalose in rice at low temperatures and help rice cope with low temperatures [73].

Trehalose, as an important non-reducing sugar, can not only be used in plant energy storage but can also be of great significance in assisting plants to cope with drought and low temperature [80]. Furthermore, when AMF hyphae infect plants or occupy favorable ecological positions, AMF can often inhibit the infection of other fungal hyphae to plant roots [81]. It can be speculated that AMF has a better effect on improving drought stress resistance of wild rice, and AMF infection can significantly improve the expression of plant-related genes and the ABA hormone level of wild rice [82]. Researchers demonstrated that the root exudates, strigolactones, and flavonoid regulated plant–AMF interactions can alleviate drought stress in plants [2,83].

## 6. Framework of the Mutual Connection among Plant, Root Exudates, and Associated Microbiota

Based on the review and illustration of the studied and current understanding of the relationship among the plant, root exudates, and microbiome, we constructed a framework for mutual connections among them under wild and moderate drought conditions (Figure 2). Drought stress induces ABA and ethylene production, and the increased level of plant hormones regulates the plant phenotype and gene expression, which will decrease the plant metabolism and photosynthesis.

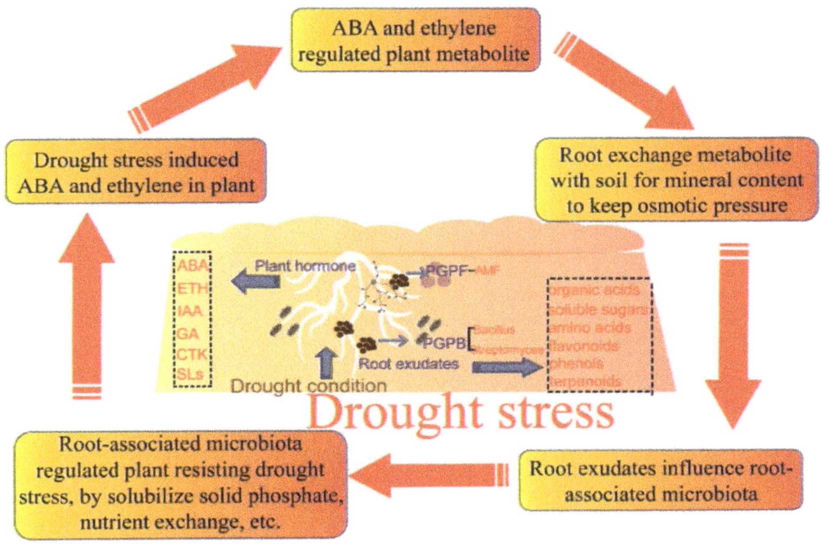


**Figure 2.** Framework of the mutual connection among plant, root exudates and associated microbiota. ABA, abscisic acid; ETH, ethylene; IAA, auxin; GA, gibberellin; CTK, cytokinin; SLs, strigolactones; PGPB, plant growth-promoting bacteria; and PGPF, plant growth-promoting fungi.

The root cell extracts will contain more mineral material to keep the osmotic pressure by exchanging cell metabolism outside. Then the root exudates will change the root-associated microbial community structure, such as by increasing the abundance of Streptomyces. Then, the enriched PGPB or PGPF can stimulate the mineral extraction, which will help to regulate drought resistance in plants.

However, there are also questions about the beneficial exchange of the root exudates for mineral between the plant root and soil microbes as well as the balance of the framework under the drought condition. Thus, in future studies, several aspects in connecting the rhizosphere microbiomes and root exudates of crop plants under drought conditions should be focused on: (1) the drought condition influence on root exudates, including primary and secondary metabolites; (2) the plant hormones secreted from the roots that regulate root-associated microbiota directly or indirectly; and (3) the crop plant exchange through the root exudates with the soil for mineral elements with the help of PGPB and PGPF.

## 7. Conclusions

In summary, drought stress has been shown to seriously influence root exudates and root-associated microbiomes in previous studies. Drought stress influences root exudates, including sugars, amino acids, flavonoids, hormones, etc. This stress not only influences soil microbiomes but also seriously affects rhizosphere communities, which may help plants resist drought stress. However, the crosstalk of the root-associated microbiomes with the root exudates has not been clearly illustrated. Thus, further studies should be focused on this topic in the future as this will have theoretical and practical significance for crop production.

**Author Contributions:** Conceptualization, L.T. and C.S.; writing—original draft preparation, Y.C.; writing—review and editing, L.T. and C.S.; visualization, Y.C., Z.Y., Y.S. (Yu Sun), E.W., C.T., Y.S. (Yang Sun), and J.L.; supervision L.T. and C.S.; project administration, L.T. and C.S.; funding acquisition, L.T. All authors have read and agreed to the published version of the manuscript.

**Funding:** This work is financially supported by the National Natural Science Foundation of China (42007043, 41920104008), Science and Technology Project for Black Soil Granary (XDA28080200, XDA28020400), Special Foundation for Basic Research Program in Wild China of CAS (XDA23070501), Key Laboratory Foundation of Mollisols Agroecology (2020ZKHT-02), and Key Technology Research and Development Project in Jilin Province of China (20180201067SF).

**Institutional Review Board Statement:** Not applicable.

**Informed Consent Statement:** Not applicable.

**Data Availability Statement:** Not applicable.

**Conflicts of Interest:** The authors declare no conflict of interest.

## References

1. Shen, M.; Huang, W.; Chen, M.; Song, B.; Zeng, G.; Zhang, Y. (Micro) plastic crisis: Un-ignorable contribution to global greenhouse gas emissions and climate change. *J. Clean. Prod.* **2020**, *254*, 120138. [CrossRef]
2. Ruiz-Lozano, J.M.; Aroca, R.; Zamarreño, Á.M.; Molina, S.; Andreo-Jiménez, B.; Porcel, R.; García-Mina, J.M.; Ruyter-Spira, C.; López-Ráez, J.A. Arbuscular mycorrhizal symbiosis induces strigolactone biosynthesis under drought and improves drought tolerance in lettuce and tomato. *Plant Cell Environ.* **2016**, *39*, 441–452. [CrossRef] [PubMed]
3. Canarini, A.; Schmidt, H.; Fuchslueger, L.; Martin, V.; Herbold, C.W.; Zezula, D.; Gundler, P.; Hasibeder, R.; Jecmenica, M.; Bahn, M.; et al. Ecological memory of recurrent drought modifies soil processes via changes in soil microbial community. *Nat. Commun.* **2021**, *12*, 5308. [CrossRef] [PubMed]
4. García de Salamone, I.E.; Funes, J.M.; Di Salvo, L.P.; Escobar-Ortega, J.S.; D'Auria, F.; Ferrando, L.; Fernandez-Scavino, A. Inoculation of paddy rice with Azospirillum brasilense and Pseudomonas fluorescens: Impact of plant genotypes on rhizosphere microbial communities and field crop production. *Appl. Soil Ecol.* **2012**, *61*, 196–204. [CrossRef]
5. Santos-Medellín, C.; Liechty, Z.; Edwards, J.; Nguyen, B.; Huang, B.; Weimer, B.C.; Sundaresan, V. Prolonged drought imparts lasting compositional changes to the rice root microbiome. *Nat. Plants* **2021**, *7*, 1065–1077. [CrossRef]
6. Santos-Medellin, C.; Edwards, J.; Liechty, Z.; Nguyen, B.; Sundaresan, V. Drought stress results in a compartment-specific restructuring of the rice root-associated microbiomes. *mBio* **2017**, *8*, e00764-17. [CrossRef]
7. Williams, A.; de Vries, F.T. Plant root exudation under drought: Implications for ecosystem functioning. *New Phytol.* **2020**, *225*, 1899–1905. [CrossRef]
8. de Vries, F.T.; Williams, A.; Stringer, F.; Willcocks, R.; McEwing, R.; Langridge, H.; Straathof, A.L. Changes in root-exudate-induced respiration reveal a novel mechanism through which drought affects ecosystem carbon cycling. *New Phytol.* **2019**, *224*, 132–145. [CrossRef]
9. Karlowsky, S.; Augusti, A.; Ingrisch, J.; Akanda, M.K.U.; Bahn, M.; Gleixner, G. Drought-induced accumulation of root exudates supports post-drought recovery of microbes in mountain grassland. *Front. Plant Sci.* **2018**, *9*, 1593. [CrossRef]
10. Henry, A.; Doucette, W.; Norton, J.; Bugbee, B. Changes in crested wheatgrass root exudation caused by flood, drought, and nutrient stress. *J. Environ. Qual.* **2007**, *36*, 904–912. [CrossRef]
11. Kumar, R.; Pandey, S.; Pandey, A. Plant roots and carbon sequestration. *Curr. Sci. India* **2006**, *91*, 885–890.
12. Montiel-Rozas, M.M.; Madejón, E.; Madejón, P. Effect of heavy metals and organic matter on root exudates (low molecular weight organic acids) of herbaceous species: An assessment in sand and soil conditions under different levels of contamination. *Environ. Pollut.* **2016**, *216*, 273–281. [CrossRef] [PubMed]
13. Zhao, M.; Zhao, J.; Yuan, J.; Hale, L.; Wen, T.; Huang, Q.; Vivanco, J.M.; Zhou, J.; Kowalchuk, G.A.; Shen, Q. Root exudates drive soil-microbe-nutrient feedbacks in response to plant growth. *Plant Cell Environ.* **2021**, *44*, 613–628. [CrossRef] [PubMed]

14. Marschener, H. Role of root growth, arbuscular mycorrhiza, and root exudates for the efficiency in nutrient acquisition. *Field Crop. Res.* **1998**, *56*, 203–207. [CrossRef]
15. Neumann, G. Root Exudates and Nutrient Cycling. In *Nutrient Cycling in Terrestrial Ecosystems*; Marschner, P., Rengel, Z., Eds.; Springer: Berlin/Heidelberg, Germany, 2007; pp. 123–157. [CrossRef]
16. Sun, Y.; Tian, L.; Chang, J.; Shi, S.; Zhang, J.; Xie, H.; Cai, Y.; Chen, D.; Kuramae, E.E.; van Veen, J.A.; et al. Rice domestication influences the composition and function of the rhizosphere bacterial chemotaxis systems. *Plant Soil* **2021**, *466*, 81–99. [CrossRef]
17. Canarini, A.; Merchant, A.; Dijkstra, F.A. Drought effects on Helianthus annuus and Glycine max metabolites: From phloem to root exudates. *Rhizosphere* **2016**, *2*, 85–97. [CrossRef]
18. Song, F.; Han, X.; Zhu, X.; Herbert, S.J. Response to water stress of soil enzymes and root exudates from drought and non-drought tolerant corn hybrids at different growth stages. *Can. J. Soil Sci.* **2012**, *92*, 501–507. [CrossRef]
19. Ulrich, D.E.M.; Sevanto, S.; Ryan, M.; Albright, M.B.N.; Johansen, R.B.; Dunbar, J.M. Plant-microbe interactions before drought influence plant physiological responses to subsequent severe drought. *Sci. Rep.* **2019**, *9*, 249. [CrossRef]
20. Gargallo-Garriga, A.; Preece, C.; Sardans, J.; Oravec, M.; Urban, O.; Peñuelas, J. Root exudate metabolomes change under drought and show limited capacity for recovery. *Sci. Rep.* **2018**, *8*, 12696. [CrossRef]
21. Hummel, I.; Pantin, F.; Sulpice, R.; Piques, M.; Rolland, G.; Dauzat, M.; Christophe, A.; Pervent, M.; Bouteillé, M.; Stitt, M.; et al. Arabidopsis plants acclimate to water deficit at low cost through changes of carbon usage: An integrated perspective using growth, metabolite, enzyme, and gene expression analysis. *Plant Physiol.* **2010**, *154*, 357–372. [CrossRef]
22. Singh, R.; Jwa, N.-S. The rice MAPKK–MAPK interactome: The biological significance of MAPK components in hormone signal transduction. *Plant Cell Rep.* **2013**, *32*, 923–931. [CrossRef] [PubMed]
23. Campalans, A.; Messeguer, R.; Goday, A.; Pagès, M. Plant responses to drought, from ABA signal transduction events to the action of the induced proteins. *Plant Physiol. Bioch.* **1999**, *37*, 327–340. [CrossRef]
24. Andi, B.; Jensen, C.R.; Folkard, A.; Mogensen, V.O. Drought-induced changes in xylem pH, ionic composition, and ABA concentration act as early signals in field-grown maize (*Zea mays* L.) *J. Exp. Bot.* **2002**, *53*, 251–263.
25. Liu, F.; Jensen, C.R.; Andersen, M.N. A review of drought adaptation in crop plants: Changes in vegetative and reproductive physiology induced by ABA-based chemical signals. *Aust. J. Agric. Res.* **2005**, *56*, 1245–1252. [CrossRef]
26. Barnawal, D.; Bharti, N.; Pandey, S.S.; Pandey, A.; Chanotiya, C.S.; Kalra, A. Plant growth-promoting rhizobacteria enhance wheat salt and drought stress tolerance by altering endogenous phytohormone levels and TaCTR1/TaDREB2 expression. *Physiol. Plant.* **2017**, *161*, 502–514. [CrossRef]
27. Bano, A.; Yasmeen, S. Role of phytohormones under induced drought stress in wheat. *Pak. J. Bot.* **2010**, *42*, 2579–2587.
28. Chen, Y.; Chen, Y.; Shi, Z.; Jin, Y.; Sun, H.; Xie, F.; Zhang, L. Biosynthesis and signal transduction of ABA, JA, and BRs in response to drought stress of Kentucky bluegrass. *Int. J. Mol. Sci.* **2019**, *20*, 1289. [CrossRef]
29. La, V.H.; Lee, B.-R.; Islam, M.T.; Park, S.-H.; Jung, H.-i.; Bae, D.-W.; Kim, T.-H. Characterization of salicylic acid-mediated modulation of the drought stress responses: Reactive oxygen species, proline, and redox state in Brassica napus. *Environ. Exp. Bot.* **2019**, *157*, 1–10. [CrossRef]
30. Blum, A. Genomics for drought resistance–getting down to earth. *Funct. Plant Biol.* **2014**, *41*, 1191–1198. [CrossRef]
31. Zhao, Y.; Chan, Z.; Gao, J.; Xing, L.; Cao, M.; Yu, C.; Hu, Y.; You, J.; Shi, H.; Zhu, Y.; et al. ABA receptor PYL9 promotes drought resistance and leaf senescence. *Proc. Natl. Acad. Sci. USA* **2016**, *113*, 1949–1954. [CrossRef]
32. Raghavendra, A.S.; Gonugunta, V.K.; Christmann, A.; Grill, E. ABA perception and signalling. *Trends Plant Sci.* **2010**, *15*, 395–401. [CrossRef] [PubMed]
33. Destefano-Beltrán, L.; Knauber, D.; Huckle, L.; Suttle, J.C. Effects of postharvest storage and dormancy status on ABA content, metabolism, and expression of genes involved in ABA biosynthesis and metabolism in potato tuber tissues. *Plant Mol. Biol.* **2006**, *61*, 687–697. [CrossRef] [PubMed]
34. Ding, L.; Li, Y.; Wang, Y.; Gao, L.; Wang, M.; Chaumont, F.; Shen, Q.; Guo, S. Root ABA accumulation enhances rice seedling drought tolerance under ammonium supply: Interaction with aquaporins. *Front. Plant Sci.* **2016**, *7*, 1206. [CrossRef] [PubMed]
35. Zheng, X.; Li, Y.; Xi, X.; Ma, C.; Sun, Z.; Yang, X.; Li, X.; Tian, Y.; Wang, C. Exogenous Strigolactones alleviate KCl stress by regulating photosynthesis, ROS migration and ion transport in Malus hupehensis Rehd. *Plant Physiol. Bioch.* **2021**, *159*, 113–122. [CrossRef]
36. Li, W.; Herrera-Estrella, L.; Tran, L.-S.P. Do Cytokinins and strigolactones crosstalk during drought adaptation? *Trends Plant Sci.* **2019**, *24*, 669–672. [CrossRef]
37. López-Ráez, J.A. How drought and salinity affect arbuscular mycorrhizal symbiosis and strigolactone biosynthesis? *Planta* **2016**, *243*, 1375–1385. [CrossRef]
38. López-Ráez, J.A.; Kohlen, W.; Charnikhova, T.; Mulder, P.; Undas, A.K.; Sergeant, M.J.; Verstappen, F.; Bugg, T.D.H.; Thompson, A.J.; Ruyter-Spira, C.; et al. Does abscisic acid affect strigolactone biosynthesis? *New Phytol.* **2010**, *187*, 343–354. [CrossRef]
39. Fuchslueger, L.; Bahn, M.; Fritz, K.; Hasibeder, R.; Richter, A. Experimental drought reduces the transfer of recently fixed plant carbon to soil microbes and alters the bacterial community composition in a mountain meadow. *New Phytol.* **2014**, *201*, 916–927. [CrossRef]
40. Naylor, D.; DeGraaf, S.; Purdom, E.; Coleman-Derr, D. Drought and host selection influence bacterial community dynamics in the grass root microbiome. *ISME J.* **2017**, *11*, 2691–2704. [CrossRef]

41. Xu, L.; Coleman-Derr, D. Causes and consequences of a conserved bacterial root microbiome response to drought stress. *Curr. Opin. Microbiol.* **2019**, *49*, 1–6. [CrossRef]
42. Si, J.; Froussart, E.; Viaene, T.; Vázquez-Castellanos, J.F.; Hamonts, K.; Tang, L.; Beirinckx, S.; De Keyser, A.; Deckers, T.; Amery, F.; et al. Interactions between soil compositions and the wheat root microbiome under drought stress: From an in silico to in planta perspective. *Comput. Struct. Biotec.* **2021**, *19*, 4235–4247. [CrossRef] [PubMed]
43. Andreo-Jimenez, B.; Vandenkoornhuyse, P.; Lê Van, A.; Heutinck, A.; Duhamel, M.; Kadam, N.; Jagadish, K.; Ruyter-Spira, C.; Bouwmeester, H. Plant host and drought shape the root associated fungal microbiota in rice. *PeerJ* **2019**, *7*, e7463. [CrossRef] [PubMed]
44. Xiong, Y.-W.; Li, X.-W.; Wang, T.-T.; Gong, Y.; Zhang, C.-M.; Xing, K.; Qin, S. Root exudates-driven rhizosphere recruitment of the plant growth-promoting rhizobacterium Bacillus flexus KLBMP 4941 and its growth-promoting effect on the coastal halophyte Limonium sinense under salt stress. *Ecotoxicol. Environ. Saf.* **2020**, *194*, 110374. [CrossRef] [PubMed]
45. Zhang, Y.; Li, P.; Cheng, L. Developmental changes of carbohydrates, organic acids, amino acids, and phenolic compounds in 'Honeycrisp' apple flesh. *Food Chem.* **2010**, *123*, 1013–1018. [CrossRef]
46. Santana, S.R.A.; Voltolini, T.V.; Antunes, G.d.R.; da Silva, V.M.; Simões, W.L.; Morgante, C.V.; de Freitas, A.D.S.; Chaves, A.R.d.M.; Aidar, S.d.T.; Fernandes-Júnior, P.I. Inoculation of plant growth-promoting bacteria attenuates the negative effects of drought on sorghum. *Arch. Microbiol.* **2020**, *202*, 1015–1024. [CrossRef]
47. Zhang, J.; Nasir, F.; Kong, Y.; Tian, L.; Batool, A.; Tian, C. Drought stress shapes the rootassociated bacterial and fungal community structure in soybean genotypes. *Pak. J. Bot.* **2017**, *49*, 1933–1942.
48. Bailey, V.L.; Smith, J.L.; Bolton, H. Fungal-to-bacterial ratios in soils investigated for enhanced C sequestration. *Soil Biol. Biochem.* **2002**, *34*, 997–1007. [CrossRef]
49. Soares, M.; Rousk, J. Microbial growth and carbon use efficiency in soil: Links to fungal-bacterial dominance, SOC-quality and stoichiometry. *Soil Biol. Biochem.* **2019**, *131*, 195–205. [CrossRef]
50. Nadeem, S.M.; Ahmad, M.; Zahir, Z.A.; Javaid, A.; Ashraf, M. The role of mycorrhizae and plant growth promoting rhizobacteria (PGPR) in improving crop productivity under stressful environments. *Biotechnol. Adv.* **2014**, *32*, 429–448. [CrossRef]
51. Wang, R.; Wang, M.; Chen, K.; Wang, S.; Mur, L.A.J.; Guo, S. Exploring the roles of aquaporins in plant–microbe interactions. *Cells* **2018**, *7*, 267. [CrossRef]
52. Shahzad, R.; Khan, A.L.; Bilal, S.; Waqas, M.; Kang, S.-M.; Lee, I.-J. Inoculation of abscisic acid-producing endophytic bacteria enhances salinity stress tolerance in Oryza sativa. *Environ. Exp. Bot.* **2017**, *136*, 68–77. [CrossRef]
53. Sofy, A.R.; Dawoud, R.A.; Sofy, M.R.; Mohamed, H.I.; Hmed, A.A.; El-Dougdoug, N.K. Improving regulation of enzymatic and non-enzymatic antioxidants and stress-related gene stimulation in cucumber mosaic cucumovirus-infected cucumber plants treated with glycine betaine, chitosan and combination. *Molecules* **2020**, *25*, 2341. [CrossRef] [PubMed]
54. Park, M.-H.; Park, C.-H.; Sim, Y.B.; Hwang, S.-J. Response of Scenedesmus quadricauda (Chlorophyceae) to salt stress considering nutrient enrichment and intracellular proline accumulation. *Int. J. Environ. Res. Public Health* **2020**, *17*, 3624. [CrossRef]
55. Brewer, T.E.; Wagner, A. Translation stalling proline motifs are enriched in slow-growing, thermophilic, and multicellular bacteria. *ISME J.* **2021**. [CrossRef] [PubMed]
56. Rolli, E.; Marasco, R.; Vigani, G.; Ettoumi, B.; Mapelli, F.; Deangelis, M.L.; Gandolfi, C.; Casati, E.; Previtali, F.; Gerbino, R.; et al. Improved plant resistance to drought is promoted by the root-associated microbiome as a water stress-dependent trait. *Environ. Microbiol.* **2015**, *17*, 316–331. [CrossRef]
57. Arkhipova, T.N.; Prinsen, E.; Veselov, S.U.; Martinenko, E.V.; Melentiev, A.I.; Kudoyarova, G.R. Cytokinin producing bacteria enhance plant growth in drying soil. *Plant Soil* **2007**, *292*, 305–315. [CrossRef]
58. Glick, B.R. Bacteria with ACC deaminase can promote plant growth and help to feed the world. *Microbiol. Res.* **2014**, *169*, 30–39. [CrossRef]
59. Forni, C.; Duca, D.; Glick, B.R. Mechanisms of plant response to salt and drought stress and their alteration by rhizobacteria. *Plant Soil* **2017**, *410*, 335–356. [CrossRef]
60. Liu, Z.; Li, Y.; Ma, L.; Wei, H.; Zhang, J.; He, X.; Tian, C. Coordinated regulation of arbuscular mycorrhizal fungi and soybean MAPK pathway genes improved mycorrhizal soybean drought tolerance. *Mol. Plant-Microbe Interact.* **2015**, *28*, 408–419. [CrossRef]
61. Tian, C.; Kasiborski, B.; Koul, R.; Lammers, P.J.; Bücking, H.; Shachar-Hill, Y. Regulation of the nitrogen transfer pathway in the arbuscular mycorrhizal symbiosis: Gene characterization and the coordination of expression with nitrogen flux. *Plant Physiol.* **2010**, *153*, 1175–1187. [CrossRef]
62. Riaz, M.; Kamran, M.; Fang, Y.; Wang, Q.; Cao, H.; Yang, G.; Deng, L.; Wang, Y.; Zhou, Y.; Anastopoulos, I.; et al. Arbuscular mycorrhizal fungi-induced mitigation of heavy metal phytotoxicity in metal contaminated soils: A critical review. *J. Hazard. Mater.* **2021**, *402*, 123919. [CrossRef] [PubMed]
63. Konvalinková, T.; Püschel, D.; Řezáčová, V.; Gryndlerová, H.; Jansa, J. Carbon flow from plant to arbuscular mycorrhizal fungi is reduced under phosphorus fertilization. *Plant Soil* **2017**, *419*, 319–333. [CrossRef]
64. Lee, E.-H.; Eo, J.-K.; Ka, K.-H.; Eom, A.-H. Diversity of arbuscular mycorrhizal fungi and their roles in ecosystems. *Mycobiology* **2013**, *41*, 121–125. [CrossRef] [PubMed]
65. Wang, L.; Liu, Y.; Zhu, X.; Zhang, Y.; Yang, H.; Dobbie, S.; Zhang, X.; Deng, A.; Qian, H.; Zhang, W. Effects of arbuscular mycorrhizal fungi on crop growth and soil N2O emissions in the legume system. *Agric. Ecosyst. Environ.* **2021**, *322*, 107641. [CrossRef]

66. El-Sawah, A.M.; El-Keblawy, A.; Ali, D.F.I.; Ibrahim, H.M.; El-Sheikh, M.A.; Sharma, A.; Alhaj Hamoud, Y.; Shaghaleh, H.; Brestic, M.; Skalicky, M.; et al. Arbuscular mycorrhizal fungi and plant growth-promoting rhizobacteria enhance soil key enzymes, plant growth, seed yield, and qualitative attributes of guar. *Agriculture* **2021**, *11*, 194. [CrossRef]
67. Diagne, N.; Ngom, M.; Djighaly, P.I.; Fall, D.; Hocher, V.; Svistoonoff, S. Roles of arbuscular mycorrhizal fungi on plant growth and performance: Importance in biotic and abiotic stressed regulation. *Diversity* **2020**, *12*, 370. [CrossRef]
68. Tian, L.; Chang, C.; Ma, L.; Nasir, F.; Zhang, J.; Li, W.; Tran, L.-S.P.; Tian, C. Comparative study of the mycorrhizal root transcriptomes of wild and cultivated rice in response to the pathogen Magnaporthe oryzae. *Rice* **2019**, *12*, 35. [CrossRef]
69. Sawers, R.J.H.; Ramirez-Flores, M.R.; Olalde-Portugal, V.; Paszkowski, U. The impact of domestication and crop improvement on arbuscular mycorrhizal symbiosis in cereals: Insights from genetics and genomics. *New Phytol.* **2018**, *220*, 1135–1140. [CrossRef]
70. Li, S.; Yang, W.; Guo, J.; Li, X.; Lin, J.; Zhu, X. Changes in photosynthesis and respiratory metabolism of maize seedlings growing under low temperature stress may be regulated by arbuscular mycorrhizal fungi. *Plant Physiol. Bioch.* **2020**, *154*, 1–10. [CrossRef]
71. Mitra, D.; BE, G.S.; Khoshru, B.; De Los Santos Villalobos, S.; Belz, C.; Chaudhary, P.; Shahri, F.N.; Djebaili, R.; Adeyemi, N.O.; El-Ballat, E.M.; et al. Impacts of arbuscular mycorrhizal fungi on rice growth, development, and stress management with a particular emphasis on strigolactone effects on root development. *Commun. Soil Sci. Plant* **2021**, *52*, 1591–1621. [CrossRef]
72. Panneerselvam, P.; Kumar, U.; Senapati, A.; Parameswaran, C.; Anandan, A.; Kumar, A.; Jahan, A.; Padhy, S.R.; Nayak, A.K. Influence of elevated CO2 on arbuscular mycorrhizal fungal community elucidated using Illumina MiSeq platform in sub-humid tropical paddy soil. *Appl. Soil Ecol.* **2020**, *145*, 103344. [CrossRef]
73. Liu, Z.; Ma, L.; He, X.; Tian, C. Water strategy of mycorrhizal rice at low temperature through the regulation of PIP aquaporins with the involvement of trehalose. *Appl. Soil Ecol.* **2014**, *84*, 185–191. [CrossRef]
74. Pavithra, D.; Yapa, N. Arbuscular mycorrhizal fungi inoculation enhances drought stress tolerance of plants. *Groundw. Sustain. Dev.* **2018**, *7*, 490–494. [CrossRef]
75. Davies, F.T.; Olalde-Portugal, V.; Aguilera-Gomez, L.; Alvarado, M.J.; Ferrera-Cerrato, R.C.; Boutton, T.W. Alleviation of drought stress of Chile ancho pepper (*Capsicum annuum* L. cv. San Luis) with arbuscular mycorrhiza indigenous to Mexico. *Sci. Hortic.* **2002**, *92*, 347–359. [CrossRef]
76. Begum, N.; Qin, C.; Ahanger, M.A.; Raza, S.; Khan, M.I.; Ashraf, M.; Ahmed, N.; Zhang, L. Role of arbuscular mycorrhizal fungi in plant growth regulation: Implications in abiotic stress tolerance. *Front. Plant Sci.* **2019**, *10*, 1068. [CrossRef] [PubMed]
77. Li, J.; Meng, B.; Chai, H.; Yang, X.; Song, W.; Li, S.; Lu, A.; Zhang, T.; Sun, W. Arbuscular mycorrhizal fungi alleviate drought stress in C3 (*Leymus chinensis*) and C4 (*Hemarthria altissima*) grasses via altering antioxidant enzyme activities and photosynthesis. *Front. Plant Sci.* **2019**, *10*, 499. [CrossRef]
78. Tian, L.; Lin, X.; Tian, J.; Ji, L.; Chen, Y.; Tran, L.-S.P.; Tian, C. Research advances of beneficial microbiota associated with crop plants. *Int. J. Mol. Sci.* **2020**, *21*, 1792. [CrossRef]
79. Rahman, M.A.; Parvin, M.; Das, U.; Ela, E.J.; Lee, S.-H.; Lee, K.-W.; Kabir, A.H. Arbuscular mycorrhizal symbiosis mitigates iron (Fe)-deficiency retardation in alfalfa (*Medicago sativa* L. ) through the enhancement of Fe accumulation and sulfur-assisted antioxidant defense. *Int. J. Mol. Sci.* **2020**, *21*, 2219.
80. Redillas, M.C.F.R.; Park, S.-H.; Lee, J.W.; Kim, Y.S.; Jeong, J.S.; Jung, H.; Bang, S.W.; Hahn, T.-R.; Kim, J.-K. Accumulation of trehalose increases soluble sugar contents in rice plants conferring tolerance to drought and salt stress. *Plant Biotechnol. Rep.* **2012**, *6*, 89–96. [CrossRef]
81. Cruz, A.F.; Ishii, T. Arbuscular mycorrhizal fungal spores host bacteria that affect nutrient biodynamics and biocontrol of soil-borne plant pathogens. *Biol. Open* **2012**, *1*, 52–57. [CrossRef]
82. Martín-Robles, N.; Lehmann, A.; Seco, E.; Aroca, R.; Rillig, M.C.; Milla, R. Impacts of domestication on the arbuscular mycorrhizal symbiosis of 27 crop species. *New Phytol.* **2018**, *218*, 322–334. [CrossRef] [PubMed]
83. Sugiyama, A.; Yazaki, K. Flavonoids in plant rhizospheres: Secretion, fate and their effects on biological communication. *Plant Biotechnol.* **2014**, *31*, 431–443. [CrossRef]

International Journal of
*Molecular Sciences*

MDPI

*Article*

# Interplay between $Ca^{2+}$/Calmodulin-Mediated Signaling and AtSR1/CAMTA3 during Increased Temperature Resulting in Compromised Immune Response in Plants

**Peiguo Yuan and B. W. Poovaiah ***

Department of Horticulture, Washington State University, Pullman, WA 99164-6414, USA; pomology2010@gmail.com
* Correspondence: poovaiah@wsu.edu

**Abstract:** Changing temperatures are known to affect plant–microbe interactions; however, the molecular mechanism involved in plant disease resistance is not well understood. Here, we report the effects of a moderate change in temperature on plant immune response through $Ca^{2+}$/calmodulin-mediated signaling. At 30 °C, *Pst* DC3000 triggered significantly weak and relatively slow $Ca^{2+}$ influx in plant cells, as compared to that at 18 °C. Increased temperature contributed to an enhanced disease susceptibility in plants; the enhanced disease susceptibility is the result of the compromised stomatal closure induced by pathogens at high temperature. A $Ca^{2+}$ receptor, AtSR1, contributes to the decreased plant immunity at high temperatures and the calmodulin-binding domain (CaMBD) is required for its function. Furthermore, both salicylic acid biosynthesis (ICS) and salicylic acid receptor (NPR1) are involved in this process. In addition to stomatal control, AtSR1 is involved in high temperature-compromised apoplastic immune response through the salicylic acid signaling pathway. The qRT-PCR data revealed that AtSR1 contributed to increased temperatures-mediated susceptible immune response by regulating SA-related genes in *atsr1*, such as *PR1*, *ICS1*, *NPR1*, as well as *EDS1*. Our results indicate that $Ca^{2+}$ signaling has broad effects on the molecular interplay between changing temperatures as well as plant defense during plant–pathogen interactions.

**Keywords:** AtSR1/CAMTA3; calcium signaling; stomatal immunity; SA signaling; temperature

**Citation:** Yuan, P.; Poovaiah, B.W. Interplay between $Ca^{2+}$/Calmodulin-Mediated Signaling and AtSR1/CAMTA3 during Increased Temperature Resulting in Compromised Immune Response in Plants. *Int. J. Mol. Sci.* **2022**, *23*, 2175. https://doi.org/10.3390/ijms23042175

Academic Editors: Isabel Marques, Ana I Ribeiro-Barros and José C. Ramalho

Received: 27 January 2022
Accepted: 12 February 2022
Published: 16 February 2022



## 1. Introduction

Extreme temperatures cause adverse impacts on plant growth and development, which can lead to significant crop losses all over the world [1]. In recent years, global warming has resulted in more frequent extreme temperature events [2]. Plant disease is another major cause of agricultural loss [3–5]. Environmental temperature changes antagonistically interact with the plant immune response, for example, plant immunity is usually repressed at high temperatures, as compared to low temperatures [3,6–8]. However, the molecular mechanisms involved in plant responses at varying temperatures have remained elusive.

Plants use hormones to adapt to various environmental stimuli, including pathogen infections [9]. Salicylic acid (SA) acts as one of the main defense phytohormones against biotrophic and hemi-biotrophic microbes in both local and systemic resistance [10]. Facing the pathogen challenges, plants establish immune resistance through reprograming SA-related genes, (such as Enhanced Disease Susceptibility 1 (EDS1), Phytoalexin Deficient 4 (PAD4), Isochorismate Synthase 1 (ICS1) and *Nonexpresser of Pathogenesis-Related Genes 1 (NPR1)*) and activating the biosynthesis of SA [10,11]. SA also plays a role in the association between plant immune response and temperature. For instance, the accumulation of SA induced by pathogen was compromised at 30 °C, and the gene expressions of *PAD4* and *EDS1* were reduced at higher temperatures [12]. Unlike SA signals, temperature regulated jasmonate (JA) and/or ethylene (ET) act in an opposing manner [13]. JA and/or

ET signaling was suppressed at low temperature, while high temperature promoted JA and/or ET-mediated plant immunity [14–16].

In addition to phytohormone-regulated defense, high temperature also reduced resistance (R) protein (which are nucleotide binding-leucine rich repeat, NB-LRR, protein) mediated immune response in plants [17,18]. Pathogens secrete effectors into plant cells to suppress the innate immune response and improve their virulence, however, the microbe-derived effector is recognized by R protein [19]. The recognition and activation of R protein results in a rapid and strong resistance (also known as effector-triggered immunity, ETI), and sometimes associated with programmed cell death at the infected site, termed hypersensitive response (HR) [19]. At high temperature, ETI or HR was inhibited [20]. *Pseudomonas syringae* pv. Tomato (*Pst*) DC3000 carrying AvrRpt2, AvrRmp1 and AvrRps4 induced HR in *Arabidopsis* at 22 °C, however, the ETI-HR was inhibited at 28 °C [6]. Several mutants are also reported to display temperature-sensitive autoimmune phenotypes, such as enhanced SA accumulation, retarded growth, and constitutively activated defense pathways. Some mutants, such as *Bonzai 1* (*bon1*) [12], *Suppressor of Npr1 Constitutive 1* (*snc1*) [12], and *Mapk/Erk Kinase Kinase 1* (*mekk1*) [21] display the autoimmune phenotype at normal temperature (22 °C), which was compromised at a higher temperature (28 °C); whereas, other mutants, such as *Suppressor of Salicylic acid Insensitive 4* (*ssi4*) [22], *chilling-sensitive 2* (*chs2*) [23], *Ler/Kashmir 2* (*kas-2*) [24], and *Uk1* [25] display the autoimmune phenotype only at low temperatures (14–16 °C), which was compromised at normal temperature (22 °C). Additionally, the increased temperature repressed the temperature-dependent autoimmunity phenotype in some mutated plants, such as *atsr1* or *rps4-OE* [26,27].

Guard cells represent one of the most significant cell types in terrestrial plants, which forms the microscopic pores in the epidermis to ensure gas ($CO_2$ and $O_2$) exchange and transpiration [28]. Plants regulate the movement of guard cells by ABA in response to drought and/or salt stress [29]. Some studies have revealed that the stomatal opening and closure are essential for plant immune response, since the stomates were found to act as the invasion entry into the leaf interior for a large number of bacterial pathogens [30–32]. Plants detect the pathogen-associated molecular patterns (PAMPs) to induce stomatal closure to prevent pathogen entry [30,32]. The PAMPs preceptors (such as FLS2) and the pathogen-triggered accumulation of SA are required for the pathogen-induced stomatal closure [32]. Meanwhile, virulent pathogens generate coronatine (COR), which is structurally similar to jasmonoyl-isoleucine (JA-Ile), to suppress SA-regulated stomatal immunity through activated JA signaling [32]. Bacterial pathogens have evolved with type III secretion systems to deliver the effector into the plant cell to suppress PTI, which is also known as effector-triggered susceptibility (ETS) [19]. The expression of HopM1 (an effector protein from *Pst* DC3000) in *Arabidopsis*-compromised flg22 triggered production of reactive oxygen species (ROS) and stomatal closure [33]. In addition to stomatal closure, a novel mechanism of stomatal immunity was revealed where the stomal cell death was triggered to prevent the pathogen invasion; a recent study revealed that *Arabidopsis* plants can sense fungal chitin to induce guard cell death to evade fungal infection [30]. Chitin also induced guard cell movement through $Ca^{2+}$ and Calcium-dependent Protein Kinases (CPKs), such as CPK6 [30].

Calcium ion ($Ca^{2+}$) acts as a second messenger in plant cells and is responsible for a large number of environmental stimuli and developmental cues [34–36]. Calcium signaling plays a vital role in sensing environmental stimuli and establishing a proper response to maintain optimal growth and development [37–39]. The role of $Ca^{2+}$ signaling in plant defense has been well documented. Plants employ Pattern Recognition Receptors (PRRs) to sense the Microbe-Associated Molecular Patterns (MAMP)s and trigger $Ca^{2+}$ influx into plant cells [40–42]. Effectors from pathogens can induce an increase in $Ca^{2+}$ influx as compared to MAMPs [11]. CPKs/CDPKs activate TFs to induce transcriptional expression of defense-related genes [43,44]. Furthermore, CaM-binding transcription factors are known to regulate SA signaling [45,46] as well as temperature-modulated $Ca^{2+}$ signaling [47]. Low temperature is known to induce $Ca^{2+}$ influx in roots [48]. In recent years, it has become clear that a change in temperature induces a rise in cytosolic $Ca^{2+}$. Cold stress is known to induce a rise in $Ca^{2+}$ in plant cells, whereas heat stress increases the free $Ca^{2+}$ concentration in the chloroplast, but not in the cytoplast [49,50]. In addition, a chloroplast-specific $Ca^{2+}$ rise was also found in the light to dark transition [51].

Although $Ca^{2+}$ signaling mediates temperature stress and immune response, the role of $Ca^{2+}$ signaling in temperature-mediated plant immune response needs further research to better understand the underlying mechanisms involved in this process. Here, we report that the inoculation with Pst DC3000 induced a strong $Ca^{2+}$ spike in plant cells, while the pathogen-triggered $Ca^{2+}$ spike was compromised to some extent at high temperatures. High temperatures repressed plant immunity, and AtSR1 is involved in this process. Furthermore, through the SA signaling pathway AtSR1 also plays a role in increased plant disease susceptibility at high temperatures.

## 2. Results

### 2.1. *$Ca^{2+}$ Influx by Pathogen Was Repressed at High Temperature*

Our previous study revealed that temperature can affect plant immune response [26]. To investigate whether the inoculation with pathogens triggered different $Ca^{2+}$ influxes at different temperatures, we used WT plants carrying *aequorin (AEQ)* grown at 18 or 30 °C. At the lower temperature, pathogens induced an increase in $Ca^{2+}$ concentration at 4 min post inoculation and reached the peak at 5 min; whereas at 30 °C, pathogens induced a rise of $Ca^{2+}$ at 6 min post inoculation and reached the peak at 7 min (Figure 1A,B). Further analysis revealed that pathogens triggered nearly a two-fold increase in $Ca^{2+}$ influx at low temperature, as compared to that at high temperature (Figure 1C). These observations suggest that the increased temperature delayed and decreased the pathogen-induced $Ca^{2+}$ spikes.

### 2.2. *AtSR1 Contributed to the Suppression of High Temperature-Mediated Changes in Stomatal Aperture*

$Ca^{2+}$ changes induced by pathogens lead to stomatal closure, which prevents pathogen entry into plants [52]. Thus, we used dip inoculation to assess the effect of temperature on stomatal immunity. At 3 days post inoculation (3 d.p.i), the propagation of pathogens was greatly increased at high temperature, as compared to low temperature (Figure 2A). To further examine whether temperature regulated stomatal changes by pathogen infection, we inoculated peeled plant leaf sections with the pathogen. To rule out the possibility that the different temperatures lead to the different stomatal apertures, we exposed the peeled leaf samples at strong light for 3 h at 18 and 30 °C to ensure that the stomates were completely opened before inoculation (Figures S1 and S2). Consistent with disease resistance data, the width of the stomatal aperture was decreased at low temperature 1 h after inoculation, as compared to high temperature (Figure 2B,C).

In our previous studies, AtSR1 was reported to act as a $Ca^{2+}$/CaM-mediated transcription factor to regulate plant defense signaling [26]. Therefore, we hypothesized that temperature-regulated stomatal movement involves AtSR1. *Atsr1* mutant displayed more resistance to pathogen attack at 18 °C, as compared to WT (Figure 2A); while the enhanced resistance in *atsr1* was partially maintained at 30 °C (Figure 2A), suggesting that AtSR1 modulates high temperature-promoted plant disease susceptibility. Furthermore, the average width of stomatal aperture was significantly decreased in *atsr1*, as compared to WT, at both 18 and 30 °C (Figure 2B,C). Our results suggest that AtSR1 contributes to high temperature-mediated compromised stomatal immunity.

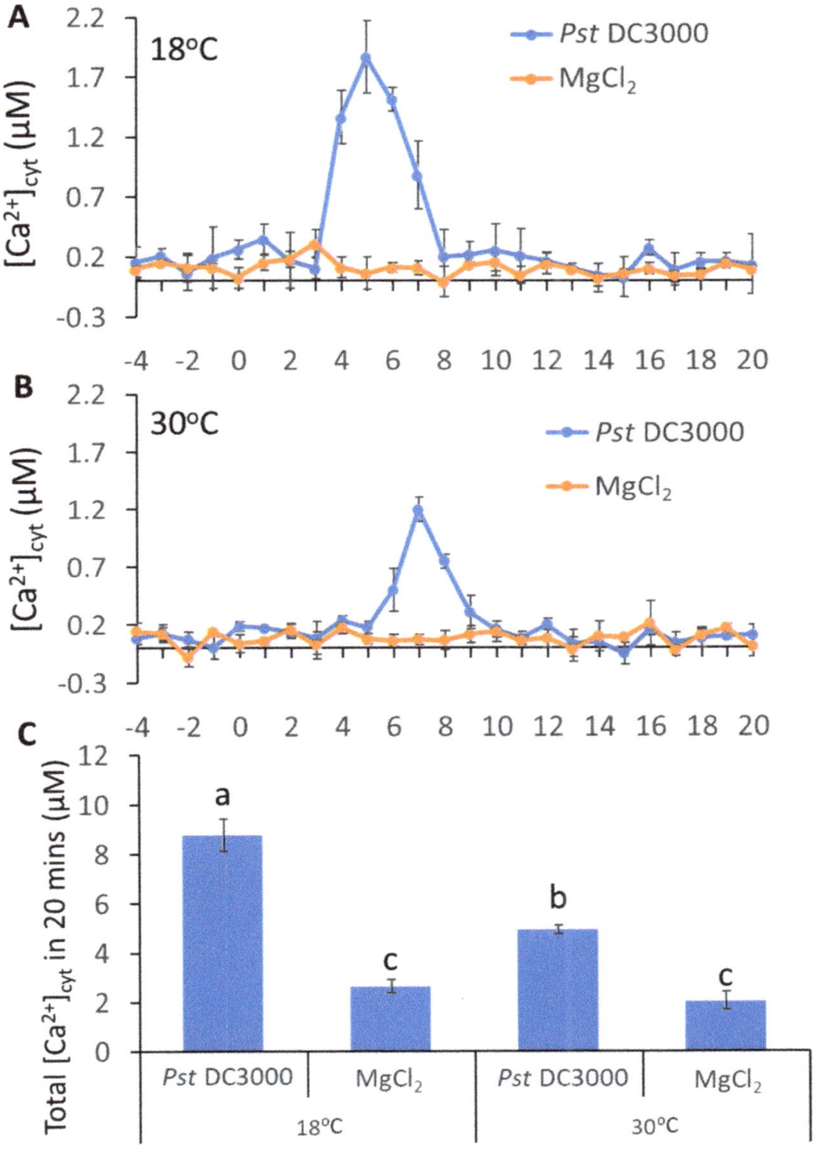


**Figure 1. Pathogen-induced different $Ca^{2+}$ spikes in plants grown at different temperatures.** (**A**,**B**), pathogen-triggered cytosolic $Ca^{2+}$ elevation in leaves of *aequorin*-expressing *Arabidopsis* plants. The dip inoculation of *Pst* DC3000 (OD = 0.01) for leaf discs at 18 °C (**A**) and 30 °C (**B**). Results shown are mean values ± SD ($n = 4$). (**C**) The histogram shows total $[Ca^{2+}]_{cyt}$ 20 min after pathogen addition. The data were analyzed by a two-factor ANOVA with all pairs Tukey's HSD post hoc analysis ($p < 0.05$) for statistical tests: different letters indicate statistical significance; samples sharing letters are not significantly different at 18 or 30 °C.

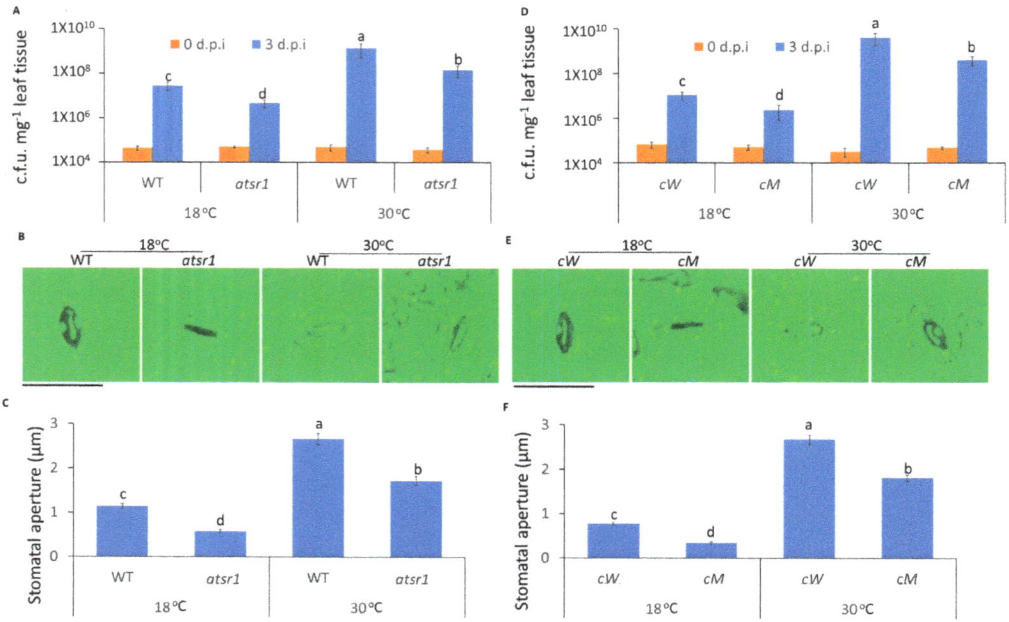


**Figure 2. The stomatal immunity was compromised at high temperature and AtSR1 contributed to high temperature-mediated increased plant disease susceptibility.** (**A**) *Arabidopsis* WT and *atsr1* mutant plants were dip inoculated with *Pst* DC3000 (OD600 = 0.1). The colony forming units (c.f.u.) were calculated at 0 and 3 d.p.i. The growth of *Pst* DC3000 in WT and *atsr1* at 18 °C and 30 °C, respectively, are shown. Error bars represent standard error of six biological repeats. (**B**) Photograph shows the stomatal movement induced by the dip inoculation of *Pst* DC3000 (OD600 = 0.1) in WT and *atsr1* at different temperatures. The scale bar represents 20 μM. (**C**) Stomatal apertures were determined in *Arabidopsis* WT and *atsr1* mutant plants 1 h after incubation of *Pst* DC3000 (OD600 = 0.1). (**D**) *Arabidopsis cW* and *cM* mutant plants were dip inoculated with *Pst* DC3000 (OD600 = 0.1). The colony forming units (c.f.u.) were calculated at 0 and 3 d.p.i. The growth of *Pst* DC3000 in *cW* and *cM* at 18 and 30°C, respectively, is shown. Error bars represent standard error of six biological repeats. (**E**) Photo shows the stomatal movement induced by the incubation of *Pst* DC3000 (OD600 = 0.1) in *cW* and *cM* at different temperatures. The scale bar represents 20 μM. (**F**) Stomatal aperture in *Arabidopsis cW* and *cM* mutant plants 1 h after incubation of *Pst* DC3000 (OD600 = 0.1). All data are representative as means s.e.m. from four independent experiments. The data were analyzed by a two-factor ANOVA with all pairs Tukey's HSD post hoc analysis ($p < 0.05$) for statistical tests: different letters indicate statistical significance; samples sharing letters are not significantly different at 18 or 30 °C.

It was shown that the calmodulin-binding domain (CaMBD) of AtSR1 is required for the suppression of AtSR1 during plant immunity [11]. To further test whether CaMBD is required for AtSR1-mediated stomatal immunity at different temperatures, two complemented lines were used: *cW* (expressing WT AtSR1 in *atsr1* mutant) and *cM* (expressing mutant AtSR1 in *atsr1* mutant K907E at CaMBD, which lacks the CaM-binding ability). As shown in Figure 1D–F, the complemented line, *cW*, was restored to WT *Arabidopsis* with regards to high temperature-induced disease susceptibility and wider stomata aperture. In contrast, *cM* plants displayed the phenotype of the loss-of-function of AtSR1 (Figure 1D–F). These results indicate that the disruption of CaMBD in AtSR1 compromises its mediation in temperature-regulated stomatal defense.

### 2.3. AtSR1 Contributes to Decreased Stomatal Immunity at 30 °C in a SA-Dependent Manner

We next tested the requirement of SA signaling and its role in the AtSR1-mediated stomatal immunity at different temperatures. ICS1 was reported to be a key enzyme for SA biosynthesis in *Arabidopsis*. The growth of pathogen was repressed in *atsr1 ics1* double mutant at low temperature, as compared to *ics1* mutant (Figure 3A); whereas, the decreased pathogen growth in *atsr1 ics1* double mutant was compromised at high temperature (Figure 3A). Similarly, the average width of the stomatal aperture was decreased in the double mutant at low temperature (Figure 3B,C). However, at 30 °C, the average width of the stomatal aperture was not significantly decreased in the double mutant as compared to *ics1* single mutant (Figure 3B,C). This suggests that the biosynthesis of SA is required for AtSR1-mediated reduced stomatal immunity at high temperature.

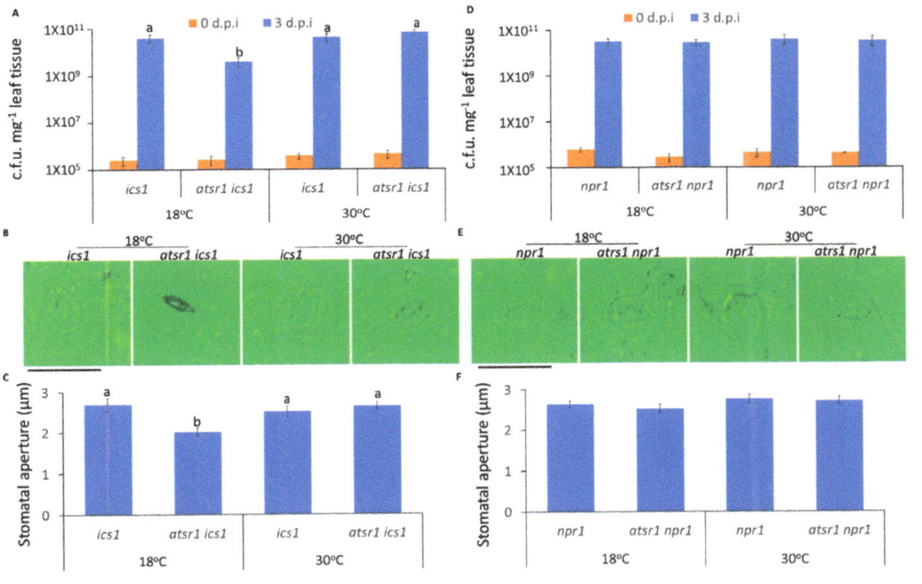


**Figure 3. AtSR1 contributed to high temperature-mediated repressed stomatal immunity in a SA-dependent manner.** (**A**) *Arabidopsis ics1* and *atsr1 ics1* mutant plants were dip inoculated with *Pst* DC3000 (OD600 = 0.1). The colony forming units (c.f.u.) were calculated at 0 and 3 d.p.i. The growth of *Pst* DC3000 in *ics1* and *atsr1 ics1* at 18 and 30 °C, respectively, is shown. Error bars represent standard error of six biological repeats. (**B**) Photograph shows the stomatal movement induced by the incubation of *Pst* DC3000 (OD600 = 0.1) in *ics1* and *atsr1 ics1* at different temperatures. The scale bar represents 20 μM. (**C**) Stomatal aperture in *Arabidopsis ics1* and *atsr1 ics1* mutant plants 1 h after incubation of *Pst* DC3000 (OD600 = 0.1). (**D**) The *Arabidopsis npr1* and *atsr1 npr1* mutant plants were dip inoculated with *Pst* DC3000 (OD600 = 0.1). The colony forming units (c.f.u.) were calculated at 0 and 3 d.p.i. The growth of *Pst* DC3000 in *npr1* and *atsr1 npr1* at 18 and 30 °C, respectively, is shown. Error bars represent standard error of six biological repeats. (**E**) Photograph shows the stomatal movement induced by the incubation of *Pst* DC3000 (OD600 = 0.1) in *npr1* and *atsr1 npr1* at different temperatures. The scale bar represents 20 μM. (**F**) Stomatal aperture in *Arabidopsis npr1* and *atsr1 npr1* mutant plants 1 h after incubation of *Pst* DC3000 (OD600 = 0.1). All data is representative as means s.e.m. from four independent experiments. The data were analyzed by a two-factor ANOVA with all pairs Tukey's HSD post hoc analysis ($p < 0.05$) for statistical tests: different letters indicate statistical significance; samples sharing letters are not significantly different at 18 or 30 °C.

The SA receptor, NPR1, is also an important component of SA-regulated plant immunity [53,54]. Unlike *ics1*, the enhanced resistance in *atsr1* was compromised in the *npr1* mutant background (Figure 3A–C), indicating that the SA receptor is necessary for AtSR1-involved compromised stomatal immunity at 30 °C, as compared to 18 °C.

### 2.4. The Suppression of Plant Apoplastic Immunity at Increased Temperature Is AtSR1-Dependent

In addition to the control of pathogen entry, we further determined the temperature-mediated plant immunity to restrict the propagation of pathogens. To test the role of AtSR1 in plant apoplastic immune response, we used the infiltrating inoculation of the pathogen to rule out the effect of AtSR1 on stomatal closure. As shown in Figure 4A, *atsr1* was more resistant to pathogen infection at low temperature, as compared to WT; there is decreased bacterial growth in *atsr1*, as compared to WT (Figure 4B). Interestingly, the enhanced disease resistant phenotype in *atsr1* was partially retained at 30 °C (Figure 4A,B), which indicates that AtSR1 contributes to high temperature-promoted disease susceptibility.

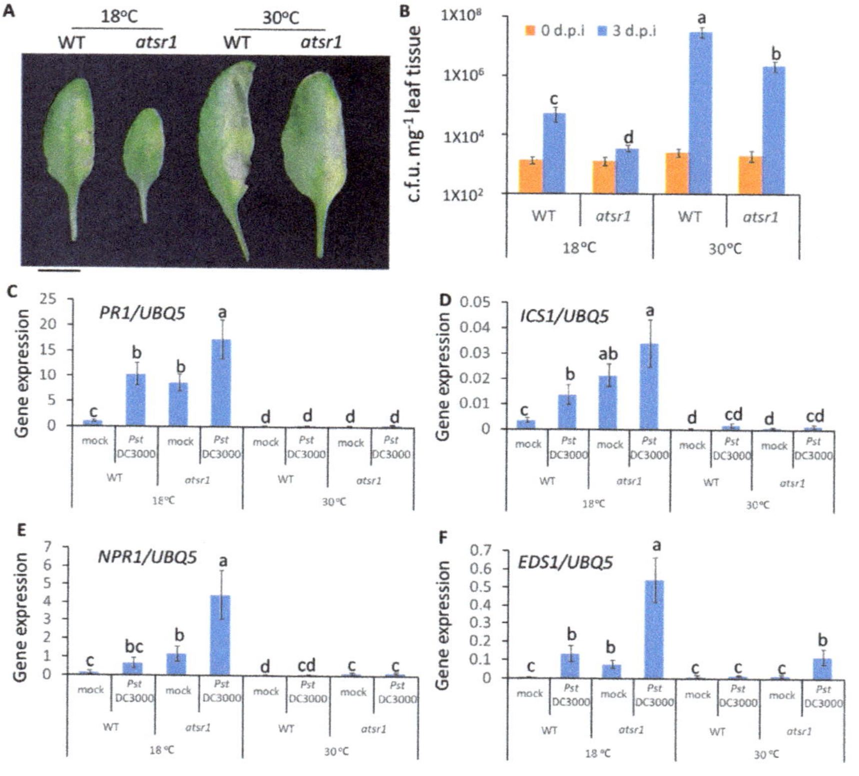


**Figure 4. AtSR1 mediates increased temperature-promoted susceptibility in plant apoplast immunity.** (**A**) Photograph shows the disease symptoms of rosette leaves in WT and *atsr1* mutant plants at 3 days post inoculation (d.p.i.) with the infiltrating inoculation of *Pst* DC3000 (OD600 = 0.001). The scale bar represents 1cM. (**B**) The colony forming units (c.f.u.) were calculated at 0 and 3 d.p.i. The growth of *Pst* DC3000 in WT and *atsr1* at 18 and 30 °C, respectively, is shown. Error bars represent standard error of six biological repeats. (**C–F**) Pathogen induced the expression of defense genes in WT and *atsr1* at both 18 and 30 °C. *PR1* (**C**), *ICS1* (**D**), *NPR1* (**E**), and ESD1 (**F**) at 1 day post inoculation of *Pst* DC3000 (OD600 = 0.001) *AtUBQ5* was used as an internal control. All data were representative as means s.e.m. from four independent experiments. The data were analyzed by a two-factor ANOVA with Tukey's HSD post hoc analysis ($p < 0.05$) for statistical tests: different letters indicate statistical significance; samples sharing letters are not significantly different at 18 or 30 °C.

Previous studies have revealed that AtSR1 suppresses the transcriptional expression of SA-related genes which are involved in plant immune response at 20 °C [11,40]. Hence, we determined the transcriptional expressions of SA-related genes at different temperatures. At 18 °C, pathogens significantly induced *PR1* expression in *atsr1*, as compared to WT, and the constitutively induced PR1 expression in *atsr1* is pathogen-independent (Figure 4C). The enhanced *PR1* expression in *atsr1* was compromised at 30 °C. Similar results were observed in the *ICS1* and *NPR1* expressions (Figure 4D,E). Earlier studies have revealed that the nucleotide-binding domain leucine-rich repeat (NLR)-signaling was also involved in temperature-mediated plant immunity [8]. Hence, we measured the transcriptional expression of *EDS1*, regulated by NLR signaling, which was reported to be regulated by AtSR1 [26]. The expression of *EDS1* induced by pathogens was regulated by AtSR1 at both 18 and 30 °C (Figure 4F).

*2.5. CaMBD Is Required for the Regulation of AtSR1 for Enhanced Apoplastic Disease Susceptibility at 30 °C*

We further tested the role of CaMBD in the AtSR1-regulated temperature-dependent apoplastic immune response. As shown in Figure 5, the complemented line, *cW*, was restored to WT *Arabidopsis* with regards to high temperature-induced susceptibility and the transcriptional expression of defense-related genes, such as *PR1*, *ICS1*, *NPR1*, and *EDS1*. In contrast, *cM* plants displayed the phenotype of the loss-of-function of AtSR1 (Figure 5). These observations suggest that the CaMBD in AtSR1 is essential for its involvement in high temperature-mediated compromised apoplastic immune response in plants.

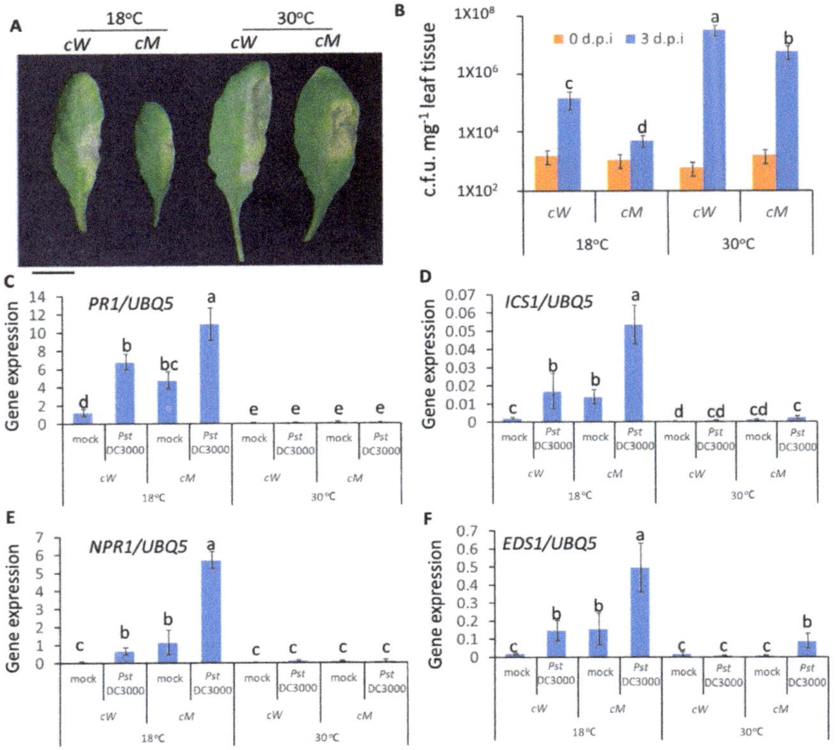


**Figure 5. CaMBD in AtSR1 is required for AtSR1-mediated decreased plant apoplast immune response at high temperature.** (**A**) Photographs show the disease symptoms of rosette leaves in *cW*

and *cM* mutant plant at 3 days post inoculation (d.p.i.) with the infiltrating inoculation of *Pst* DC3000 (OD600 = 0.001). The scale bar represents 1cM. (**B**) The colony forming units (c.f.u.) were calculated at 0 and 3 d.p.i. The growth of *Pst* DC3000 in *cW* and *cM* at 18 and 30 °C, respectively, is shown. Error bars represent standard error of six biological repeats. (**C–F**) Pathogen induced the expression of defense genes in WT and *atsr1* at both 18 and 30 °C. *PR1* (**C**), *ICS1* (**D**), *NPR1* (**E**), and *ESD1* (**F**) at 1 day post inoculation of *Pst* DC3000 (OD600 = 0.001). *AtUBQ5* was used as an internal control. All data were representative as means s.e.m. from four independent experiments. The data were analyzed by a two-factor ANOVA with Tukey's HSD post hoc analysis ($p < 0.05$) for statistical tests: different letters indicate statistical significance; samples sharing letters are not significantly different at 18 or 30 °C.

### *2.6. Involvement of AtSR1 in Enhanced Apoplastic Susceptibility at Increased Temperature Is Dependent on SA Signaling*

To test whether SA signaling is required for AtSR1-regulated plant apoplastic defense at different temperatures, we carried out disease resistance assays using infiltration inoculation in *atsr1 ics1* and *atsr1 npr1* double mutants. We observed that *atsr1 ics1* displayed improved resistance as compared to *ics1* single mutants at 18 °C, however, the improved resistant phenotype was compromised at high temperature (Figure 6A,B). Unlike the mutants of *ics1*, *atsr1 npr1* double mutant plants displayed similar resistance as compared to *npr1* single mutant plants at both 18 and 30 °C (Figure 6C,D), indicating that AtSR1 regulates increased apoplastic susceptibility at high temperature in a SA-dependent manner.

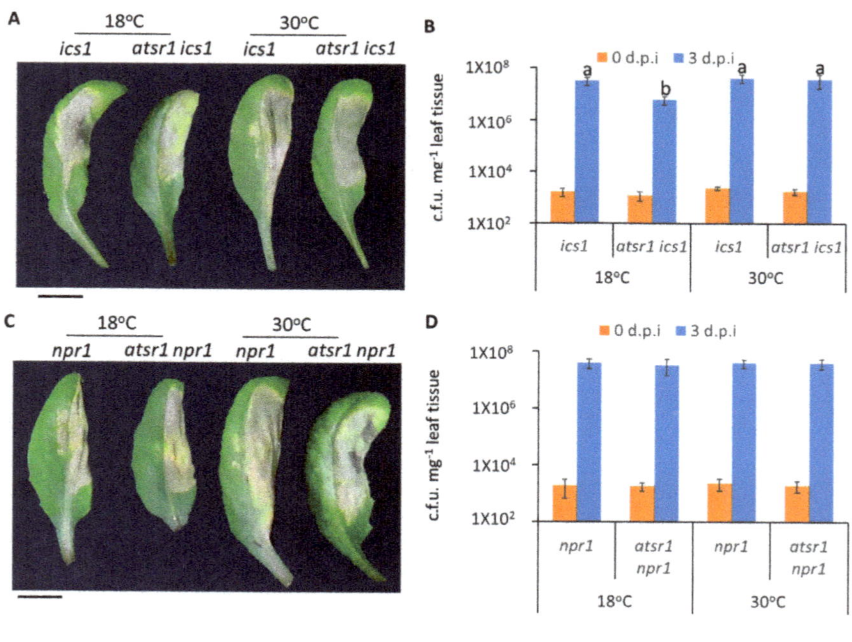


**Figure 6. AtSR1 is involved in high temperature-regulated compromised apoplastic immunity in plants through the SA-signaling pathway.** (**A**) Photograph shows the disease symptoms of rosette leaves in *ics1* and *astsr1 ics1* mutant plant at 3 days post inoculation (d.p.i.) with the infiltrating inoculation of *Pst* DC3000 (OD600 = 0.001). The scale bar represents 1cM. (**B**) The colony forming units (c.f.u.) were calculated at 0 and 3 d.p.i. The growth of *Pst* DC3000 in *ics1* and *astsr1 ics1* at 18 and 30 °C, respectively, is shown. Error bars represent standard error of six biological repeats. (**C**) The photograph shows the disease symptoms of rosette leaves in *npr1* and *astsr1 npr1* mutant plant at

3 days post inoculation (d.p.i.) with the infiltrating inoculation of *Pst* DC3000 (OD600 = 0.001). The scale bar represents 1cM. (**D**) The colony forming units (c.f.u.) were calculated at 0 and 3 d.p.i. The growth of *Pst* DC3000 in *npr1* and *astsr1 npr1* at 18 and 30 °C, respectively, is shown. Error bars represent standard error of six biological repeats. The data were analyzed by a two-factor ANOVA with Tukey's HSD post hoc analysis ($p < 0.05$) for statistical tests: different letters indicate statistical significance; samples sharing letters or no letter are not significantly different at 18 or 30 °C.

## 3. Discussion

Based on the 50-year-old concept of the "disease triangle", successful survival of plants facing pathogen attack requires established effective immune response, suppressed pathogen virulence, and favorable environmental conditions [55]. Hence, our goal was to further investigate the plant–microbe interactions at varying temperatures to extend our knowledge of plant immune response.

Temperature is one of the most important environmental factors which impacts plant growth and development [47,56]. In addition, ambient temperature is a major contributor to plant immunity and growth regulation [57]. Temperature also influences plant and microbe interactions [58]. Plants sense pathogen attack, which triggers $Ca^{2+}$ influx in the plant cell [59,60]. Plants regulate a complicated network of signaling pathways to establish immune responses to pathogen invasion; $Ca^{2+}$ signaling cascade is a key determinant for plants to integrate the various environmental stimuli to prevent invading pathogens [42,61]. However, whether pathogens at different temperatures induce different $Ca^{2+}$ influxes are not clearly understood. In this study, we observed that the rise of $Ca^{2+}$ in plant cells triggered by pathogen is greatly reduced at high temperature, and the time to reach the highest $Ca^{2+}$ flux was slightly delayed as compared to low temperature (Figure 1). This indicates a possible connection between the increased susceptibility at high temperature and compromised $Ca^{2+}$ signaling, although further studies are needed in this area. Previous studies have supported the observation that both of the increased basal and increased flg22-induced $Ca^{2+}$ concentration in the *aca4/11* double mutant, as compared to WT, were suppressed at high temperature [62]. In addition, AtACA4 and AtACA11 were identified as two tonoplast-localized $Ca^{2+}$ pumps. The reasonable explanation is that the altered temperature affects the $Ca^{2+}$ pumps or $Ca^{2+}$ channel to regulate $Ca^{2+}$ influx during plant–microbe interaction. Another possible explanation is that the compromised rise of $Ca^{2+}$ at high temperature is possibly due to the increase in the free $Ca^{2+}$ concentration within the stroma of chloroplasts [50]. However, a reduced $Ca^{2+}$ spike in the chloroplast caused by a decrease in SA accumulation subsequently reduced SA-related gene expressions [63], although the underlying mechanisms of these interactions are still not clear.

It is becoming clear that stomates are the natural entry point for bacterial pathogens into plants and that $Ca^{2+}$ signaling plays a critical role in regulating the stomatal immunity. A recent study reported that reduced hyperosmolality-induced $[Ca^{2+}]i$ increase 1.3 (OSCA1.3), acted as a $Ca^{2+}$-permeable channel which regulates stomatal closure during pathogen infection [64]. In addition, two-pore channel 1 (TPC1), known as $Ca^{2+}$-dependent $Ca^{2+}$-release channel localized in the vacuole, regulates stomatal movement [65]. The stomatal closure was impaired in *tpc1*, suggesting $Ca^{2+}$ influx is necessary for stomatal closure to stop the entry of pathogens into plants [65]. Our results indicate that the increased temperature repressed the stomatal immunity. The average width of stomatal apertures was increased at 30 °C, as compared to 18 °C (Figure 2). Moreover, the growth of pathogens in plants was greatly increased at 30 °C when we used dip inoculation to mimic plants facing pathogen attack in nature. Previous studies have revealed that AtSR1 is a suppressor of plant defense signaling and is involved in abiotic stress, especially cold stress [66,67]. Our results further confirm that AtSR1 is also a suppressor of stomatal immunity. The reduced growth of pathogens was tested in *atsr1*; moreover, reduced width of stomatal aperture was observed in the mutants (Figure 2). Hence, these results suggest that AtSR1 contributes to high temperature-mediated enhanced stomatal susceptibility.

Typically, the intact and functional CaMBD plays a critical role in CaM-binding proteins [68,69]. The mutated CaMBD in CaM-binding proteins leads to a loss-of-function protein. Previous studies have revealed that in mutated AtSR1 (K907E), a single amino acid mutated at the CaMBD failed to bind to calmodulin and the *atsr1* mutants complemented with mutated AtSR1 (K907E, *cM*) resembled the *atsr1* plants, but not WT *Arabidopsis*. Similarly, the mutated complemented line, *cM*, displayed increased resistance and reduced width of stomatal aperture (Figure 3D–F). In addition, the temperature-regulated stomatal response was partially retained in *cM* mutated plants. This observation indicates that $Ca^{2+}$/CaM-binding is required for the function of AtSR1 to suppress stomatal immunity at high temperatures.

The defense phytohormone, SA, plays a key role in stomatal immunity [70]. The stomatal closure induced by pathogens was impaired in WT carrying NahG, encoding salicylate hydroxylase that converts SA to catechol [32]. In this study, we observed that SA signaling is required for AtSR1-regulated stomatal immunity (Figure 3). Unlike *atsr1*, *atsr1 ics1* double mutant displayed a decreased plant immune response in stomates, and increased width of stomatal aperture triggered by pathogen (Figure 3). Our results are consistent with previous studies that indicate that high temperature promoted pathogen susceptibility in plants through the suppression of SA accumulation [71,72]. In addition to SA biosynthesis, we also tested the SA receptor, NPR1. We observed a similar disease symptom in *atsr1 npr1* double mutant. These results suggest that the production of SA is required for AtSR1-mediated stomatal immunity, and the SA receptor, NPR1, is also required.

In plant innate immune response, stomatal immunity contributed to limiting the pathogen invasion into the plant and apoplastic immunity repressed the growth and propagation of the pathogens after the entry into the plants [73]. In addition to temperature-mediated stomatal immunity, we determined that the temperature-mediated apoplastic immune response is required for AtSR1. As is the case for apoplastic immunity, *atsr1* mutant is more resistant to pathogen attack at 30 °C, as compared to WT; the pathogen propagation was also repressed in *atsr1* at 30 °C (Figure 4). These results suggest that AtSR1 regulates enhanced temperature-promoted plant susceptibility. Moreover, increased temperatures decreased the induction of SA-biosynthesis genes, such as *EDS1*, *PAD4*, and *ICS1*, resulting in reduced plant defense against pathogen as compared to that in low temperature [71,72]. Our results support the previous study that the enhanced expression of SA-related genes (*PR1*, *ICS1*, and *NPR1*) at 18 °C were compromised at 30 °C in *atsr1*, indicating that AtSR1 regulates enhanced susceptibility at high temperature in a SA-dependent way (Figure 4). $Ca^{2+}$ and calmodulin play a key role in AtSR1-regulated temperature-depended plant immune response, as shown by similar results observed in *cM* as compared to *cW* (Figure 5). However, apoplastic immunity is not always consistent with stomatal immunity. A previous study identified some mutants, such as *scord2* and *scord4*, which displayed normal stomatal defense, but reduced apoplastic immune response [74].

Previous studies have revealed that the accumulation of SA is required for high temperature to confer the suppression of plant immune response. Our observation further confirms that AtSR1 protein represses plant immune response at 30 °C in a SA-dependent manner, resulting in increased plant immunity at high temperature in *atsr1*, which was compromised in *ics1* background mutants (Figure 6). Our studies also indicate that SA biosynthesis as well as the SA receptor, NPR1, are required for AtSR1-mediated temperature-dependent plant immune response (Figure 6). Both *atsr1* and the *suppressor of npr1, constitutive 1* (*snc1*) mutant plants displayed a temperature-sensitive autoimmunity phenotype. The temperature sensitive autoimmunity phenotype in *snc1* was compromised in *Phytochrome Interacting Factor 4* (*pif4*) or *Sap and Miz1 domain-containing Ligase 1* (*siz1*) mutants, respectively [75,76], which raises a question about whether PIF4- and/or SIZ1 mediated the autoimmunity phenotype in *atsr1*.

## 4. Methods and Materials

### 4.1. Plant Materials and Growth Conditions

The genetic resources for this study are wild-type (WT) Columbia (Col-0), loss-of-function *atsr1* mutant (Salk_001152C), loss-of-function *ics1* mutant (Salk_088254), and loss-of-function *npr1* line (Salk_204100C), which were ordered from ABRC; as well as complimentary AtSR1 lines in *atsr1*, i.e., *cW* and *cM* (K907E) which were generated in a previous study [26]. The homozygous knock-out mutants were verified by PCR and RT-PCR.

The seeds were surface sterilized with 10% diluted bleach for 5 min and then 70% ethanol for another 5 min. The sterilized seeds were washed 5 times with sterilized water and placed on half-strength MS medium (Caisson Laboratories Inc., Smithfield, UT, USA) containing 0.05% MES and 1% sucrose, adjusting pH to 5.7 with KOH at 4 °C dark for 3 days and germinated in a growth chamber (humidity 60–70%) under 12 h light (light condition: 100–150 $\mu E \cdot m^{-2} \cdot s^{-1}$) and dark periods at 18 or 30 °C, respectively. One-week-old seedlings were transferred to pots containing soil mix (Metro Mix 360 Rsi, Sun Gro Horticulture, Agawam, MA, USA). Plants were maintained in a growth chamber under a 12 h photoperiod at 18 °C for low temperature or 30 °C for high temperature (humidity: 60%; light condition: 100–150 $\mu E \cdot m^{-2} \cdot s^{-1}$).

### 4.2. Calcium Measurement

The calcium spikes in leaves were measured with AEQ-based calcium assay [77–79]. The *Arabidopsis* Col-0 plants carrying *AEQ* were grown in soil at 18 or 30 °C. The leaf discs (5 mm diameter) obtained from 4-week-old plants were immersed into 1 mL of 5 μM coelenterazine solution (NanoLight Technologies, Aurora, CO, USA) in 24-well microplates. The plate was left under vacuum for 10 min twice, and then further incubated overnight in the dark at 18 or 30 °C, respectively. The AEQ-based bioluminescence was quantified in a microplate reader for 5 min as baseline. An equal volume of double-strength pathogen was added (the final concentration of *Pst* DC3000 is OD600 = 0.01) and quantified for 20 min, as L (luminescence intensity per second). The total remaining $Ca^{2+}$ in each microplate well was discharged by treatment with equal volume of 2 M $CaCl_2$ in 20% ethanol to release remaining AEQ, as $L_{max}$. $Ca^{2+}$ concentrations in plant cells were calculated as described previously (Tanaka et al., 2013). The equation is: $[Ca^{2+}]_{cyt}$ (nM) = $[X + (X*55) - 1]/(1 - X)/0.02$, where $X = (L/L_{max})1/3$.

### 4.3. Disease Resistance Assay

*Pst* DC3000 was cultured in King's B medium (20 g/L peptone, 1.5 g/L $MgSO_4$, 1.5 g/L $K_2HPO_4$, pH = 7), containing 50 ug/mL rifamycin and 25 ug/mL kanamycin, overnight at 28 °C. The cells were harvested (until an OD600 = 0.5 was reached) by centrifugation (10,000× *g*, for 10 min), washed twice by autoclaved water, and diluted into the desired density as described below.

For dip inoculation: the cells were diluted to OD600 of 0.1, containing 0.05% Silwet L-77 in 10 mM $MgCl_2$. The 4- or 5-week-old WT and mutated *Arabidopsis* plants were dipped in bacterial suspension with gentle shaking for 5 min and kept under high humidity in the dark overnight. At 1 h after inoculation (as day 0) and 3 days after inoculation (as 3 d.p.i, at 18 or 30 °C, respectively), the leaf samples were harvested for disease resistance test. The leaf samples were weighed and ground in 1 mL autoclaved water; serial dilutions were dropped on King's B medium containing antibiotic as described above. Then, 48 h after being grown at 28 °C, the bacterial colony forming units (c.f.u.) were calculated.

For infiltrated inoculation: the leaf inoculation was performed as previously described [34]. Briefly, the pathogens were prepared as described above. Leaves of 4- to 5-week-old plants were infiltrated with *Pst* DC3000 at OD600 of 0.001 in 10 mM $MgCl_2$, using 1 mL needleless syringe for time course induction (for testing the gene expression of defense-related genes) and disease resistance test. At 1 h after inoculation (as day 0) and 3 days after inoculation (as 3 d.p.i., at 18 or 30 °C, respectively), the leaf samples were harvested for disease resistance test. The leaf sample was weighed and ground in 1 mL autoclaved water; serial dilutions were dropped on King's B medium containing antibiotic as descripted above. Then, 48 h after growth at 28 °C, the bacterial colony forming units (c.f.u.) were calculated. Data were shown as an average of six biological replicates; the results are presented as mean ± S.D.

### 4.4. Measurements of Stomatal Aperture

Stomatal aperture was measured in epidermal peels excised from the abaxial side of leaves of 4- or 5-week-old WT or mutated *Arabidopsis* plants described previously [80]. In order to ensure that all genotypes of the plants at different temperature had almost fully open stomata, the peeled leaf samples were incubated at 10mL stomatal opening solution (30 mM KCl, 1 mM $CaCl_2$, 10 mM Tris, pH 5.8) to expose to white light (150–200 $\mu E \cdot m^{-2} \cdot s^{-1}$) for at least 3 h, at 18 or 30 °C, respectively. An equal volume (10 mL) of pathogen (OD600 = 0.2, in 10 mM $MgCl_2$) was added into the peeled leaf samples with stomatal opening solution, at 18 or 30 °C, respectively. The stomatal movement was imaged and counted using light microscopy (200×). The stomatal aperture was measured using ImageJ software. Data are calculated as the average of 10 stomatal aperture for the leaf sample from 4 individual seedlings.

### 4.5. RNA Extraction and Transcriptional Expression Analysis

Four-week-old WT and mutated *Arabidopsis* seedlings were used to measure gene expressions. At one day post inoculation, 100 mg leaf tissues were harvested from control and infected leaf samples at different temperatures of different genotypes and immediately frozen in liquid nitrogen. The frozen tissues were ground to powder in 1.5 mL Eppendorf tubes. Total RNA was prepared using TRIzol Reagent (Invitrogen) based on the manufacturer's protocol, followed by DNase-I (Roche) treatment. In total, 2 µg total RNA were used to synthesize cDNA with an oligo (dT) primer and random hexadeoxynucleotides primer (Thermo Fisher Scientific, Waltham, MA, USA). The cDNA was diluted 10 times and 1 µL/reaction (10 µL) was used as a template. Real-time PCR was performed on a MyiQTM single-color real-time PCR detection system using SYBR Green Supermix (Bio-Rad). Target gene expression levels were normalized to that of *AtUBQ5* (AT3G62250). A minimum of two technical replicates and four biological replicates were used for each sample.

### 4.6. Data Analysis

Results were analyzed using Microsoft Excel. Error bars in all of the figures represent standard error [81]. Number of replicates is described in the figure legends. For multiple group samples, statistical analyses were performed by two-way ANOVA with Tukey's HSD (honest significant difference) test. The different letters (a, b, c) indicate samples with statistically significant differences ($p < 0.05$), while the same letter indicates no statistically significant difference.

## 5. Conclusions

Our results suggest that $Ca^{2+}$ transients triggered by pathogen were compromised at high temperature. High temperature contributed to enhanced plant susceptibility in both stomatal defense and apoplastic immune response. AtSR1/CAMTA3, as a $Ca^{2+}$/CaM receptor, is involved in increased temperature-mediated stomatal defense and apoplastic immunity. In addition, the contribution of AtSR1 to the temperature-modulated plant immune response requires functional CaMBD in AtSR1. This and other studies indicate that $Ca^{2+}$ signaling acts as a general defense response to pathogen infection in the context of temperature.

**Supplementary Materials:** Supplementary materials can be found at https://www.mdpi.com/article/10.3390/ijms23042175/s1.

**Author Contributions:** P.Y. carried out the experiments. B.W.P. directed this study and provided the necessary support. P.Y. and B.W.P. prepared the manuscript. All authors have read and agreed to the published version of the manuscript.

**Funding:** This work was supported by the National Science Foundation grants 1021344 and 1557813 as well as USDA NIFA Hatch project no. 1015621.

**Acknowledgments:** The authors thank Ralph Panstruga of the Max Planck Institute, Cologne, Germany and Marc Knight of Durham University, Durham, UK for providing the AEQ-expressing *Arabidopsis* seeds. The help of Lorie Mochel in manuscript preparation and Kanthi Poovaiah for her help in the laboratory are acknowledged.

**Conflicts of Interest:** The authors declare no conflict of interest.

## References

1. Schauberger, B.; Archontoulis, S.; Arneth, A.; Balkovic, J.; Ciais, P.; Deryng, D.; Elliott, J.; Folberth, C.; Khabarov, N.; Müller, C.; et al. Consistent negative response of US crops to high temperatures in observations and crop models. *Nat. Commun.* **2017**, *8*, 13931. [CrossRef] [PubMed]
2. Zhao, C.; Liu, B.; Piao, S.; Wang, X.; Lobell, D.B.; Huang, Y.; Huang, M.; Yao, Y.; Bassu, S.; Ciais, P.; et al. Temperature increase reduces global yields of major crops in four independent estimates. *Proc. Natl. Acad. Sci. USA* **2017**, *114*, 9326. [CrossRef] [PubMed]
3. Huot, B.; Castroverde, C.D.M.; Velásquez, A.C.; Hubbard, E.; Pulman, J.A.; Yao, J.; Childs, K.L.; Tsuda, K.; Montgomery, B.L.; He, S.Y. Dual impact of elevated temperature on plant defence and bacterial virulence in Arabidopsis. *Nat. Commun.* **2017**, *8*, 1808. [CrossRef] [PubMed]
4. Savary, S.; Willocquet, L.; Pethybridge, S.J.; Esker, P.; McRoberts, N.; Nelson, A. The global burden of pathogens and pests on major food crops. *Nat. Ecol. Evol.* **2019**, *3*, 430–439. [CrossRef] [PubMed]
5. Deutsch, C.A.; Tewksbury, J.J.; Tigchelaar, M.; Battisti, D.S.; Merrill, S.C.; Huey, R.B.; Naylor, R.L. Increase in crop losses to insect pests in a warming climate. *Science* **2018**, *361*, 916–919. [CrossRef]
6. Menna, A.; Nguyen, D.; Guttman, D.S.; Desveaux, D. Elevated Temperature Differentially Influences Effector-Triggered Immunity Outputs in Arabidopsis. *Front. Plant Sci.* **2015**, *6*, 995. [CrossRef]
7. Wang, Y.; Bao, Z.; Zhu, Y.; Hua, J. Analysis of temperature modulation of plant defense against biotrophic microbes. *Mol. Plant Microbe Interact.* **2009**, *22*, 498–506. [CrossRef]
8. Zhu, Y.; Qian, W.; Hua, J. Temperature Modulates Plant Defense Responses through NB-LRR Proteins. *PLoS Pathog.* **2010**, *6*, e1000844. [CrossRef]
9. Verma, V.; Ravindran, P.; Kumar, P.P. Plant hormone-mediated regulation of stress responses. *BMC Plant Biol.* **2016**, *16*, 86. [CrossRef]
10. Zhang, Y.; Li, X. Salicylic acid: Biosynthesis, perception, and contributions to plant immunity. *Curr. Opin. Plant Biol.* **2019**, *50*, 29–36. [CrossRef]
11. Yuan, P.; Tanaka, K.; Poovaiah, B.W. Calmodulin-binding transcription activator AtSR1/CAMTA3 fine-tunes plant immune response by transcriptional regulation of the salicylate receptor NPR1. *Plant Cell Environ.* **2021**, *44*, 3140–3154. [CrossRef]
12. Yang, S.; Hua, J. A Haplotype-Specific Resistance Gene Regulated by BONZAI1 Mediates Temperature-Dependent Growth Control in Arabidopsis. *Plant Cell* **2004**, *16*, 1060–1071. [CrossRef] [PubMed]
13. Dong, X. SA, JA, ethylene, and disease resistance in plants. *Curr. Opin. Plant Biol.* **1998**, *1*, 316–323. [CrossRef]
14. Li, Z.; Liu, H.; Ding, Z.; Yan, J.; Yu, H.; Pan, R.; Hu, J.; Guan, Y.; Hua, J. Low Temperature Enhances Plant Immunity via Salicylic Acid Pathway Genes That Are Repressed by Ethylene. *Plant Physiol.* **2020**, *182*, 626–639. [CrossRef] [PubMed]
15. Huang, J.; Zhao, X.; Bürger, M.; Wang, Y.; Chory, J. Two interacting ethylene response factors regulate heat stress response. *Plant Cell* **2020**, *33*, 338–357. [CrossRef] [PubMed]
16. Havko, N.E.; Das, M.R.; McClain, A.M.; Kapali, G.; Sharkey, T.D.; Howe, G.A. Insect herbivory antagonizes leaf cooling responses to elevated temperature in tomato. *Proc. Natl. Acad. Sci. USA* **2020**, *117*, 2211–2217. [CrossRef]
17. Jacob, P.; Kim, N.H.; Wu, F.; El-Kasmi, F.; Chi, Y.; Walton, W.G.; Furzer, O.J.; Lietzan, A.D.; Sunil, S.; Kempthorn, K.; et al. Plant "helper" immune receptors are $Ca^{2+}$-permeable nonselective cation channels. *Science* **2021**, *373*, 420–425.
18. Liu, J.; Huang, Y.; Kong, L.; Yu, X.; Feng, B.; Liu, D.; Zhao, B.; Mendes, G.C.; Yuan, P.; Ge, D.; et al. The malectin-like receptor-like kinase LETUM1 modulates NLR protein SUMM2 activation via MEKK2 scaffolding. *Nat. Plants* **2020**, *6*, 1106–1115. [CrossRef]
19. Jones, J.D.G.; Dangl, J.L. The plant immune system. *Nature* **2006**, *444*, 323–329. [CrossRef]
20. Cheng, C.; Gao, X.; Feng, B.; Sheen, J.; Shan, L.; He, P. Plant immune response to pathogens differs with changing temperatures. *Nat. Commun.* **2013**, *4*, 2530. [CrossRef]
21. Ichimura, K.; Casais, C.; Peck, S.C.; Shinozaki, K.; Shirasu, K. MEKK1 Is Required for MPK4 Activation and Regulates Tissue-specific and Temperature-dependent Cell Death in *Arabidopsis*. *J. Biol. Chem.* **2006**, *281*, 36969–36976. [CrossRef] [PubMed]

22. Shirano, Y.; Kachroo, P.; Shah, J.; Klessig, D.F. A Gain-of-Function Mutation in an Arabidopsis Toll Interleukin1 Receptor–Nucleotide Binding Site–Leucine-Rich Repeat Type R Gene Triggers Defense Responses and Results in Enhanced Disease Resistance. *Plant Cell* **2002**, *14*, 3149–3162. [CrossRef] [PubMed]
23. Huang, X.; Li, J.; Bao, F.; Zhang, X.; Yang, S. A gain-of-function mutation in the Arabidopsis disease resistance gene RPP4 confers sensitivity to low temperature. *Plant Physiol.* **2010**, *154*, 796–809. [CrossRef] [PubMed]
24. Alcázar, R.; García, A.V.; Parker, J.E.; Reymond, M. Incremental steps toward incompatibility revealed by Arabidopsis epistatic interactions modulating salicylic acid pathway activation. *Proc. Natl. Acad. Sci. USA* **2009**, *106*, 334–339. [CrossRef]
25. Bomblies, K.; Lempe, J.; Epple, P.; Warthmann, N.; Lanz, C.; Dangl, J.L.; Weigel, D. Autoimmune Response as a Mechanism for a Dobzhansky-Muller-Type Incompatibility Syndrome in Plants. *PLoS Biol.* **2007**, *5*, e236. [CrossRef]
26. Du, L.; Ali, G.S.; Simons, K.A.; Hou, J.; Yang, T.; Reddy, A.S.N.; Poovaiah, B.W. $Ca^{2+}$/calmodulin regulates salicylic-acid-mediated plant immunity. *Nature* **2009**, *457*, 1154–1158. [CrossRef]
27. Heidrich, K.; Tsuda, K.; Blanvillain-Baufumé, S.; Wirthmueller, L.; Bautor, J.; Parker, J. Arabidopsis TNL-WRKY domain receptor RRS1 contributes to temperature-conditioned RPS4 auto-immunity. *Front. Plant Sci.* **2013**, *4*, 403. [CrossRef]
28. Schulze-Lefert, P.; Robatzek, S. Plant Pathogens Trick Guard Cells into Opening the Gates. *Cell* **2006**, *126*, 831–834. [CrossRef]
29. Sah, S.K.; Reddy, K.R.; Li, J. Abscisic Acid and Abiotic Stress Tolerance in Crop Plants. *Front. Plant Sci.* **2016**, *7*, 571. [CrossRef]
30. Ye, W.; Munemasa, S.; Shinya, T.; Wu, W.; Ma, T.; Lu, J.; Kinoshita, T.; Kaku, H.; Shibuya, N.; Murata, Y. Stomatal immunity against fungal invasion comprises not only chitin-induced stomatal closure but also chitosan-induced guard cell death. *Proc. Natl. Acad. Sci. USA* **2020**, *117*, 20932–20942. [CrossRef]
31. Yan, J.; Yu, H.; Li, B.; Fan, A.; Melkonian, J.; Wang, X.; Zhou, T.; Hua, J. Cell autonomous and non-autonomous functions of plant intracellular immune receptors in stomatal defense and apoplastic defense. *PLoS Pathog.* **2019**, *15*, e1008094. [CrossRef] [PubMed]
32. Melotto, M.; Underwood, W.; Koczan, J.; Nomura, K.; He, S.Y. Plant Stomata Function in Innate Immunity against Bacterial Invasion. *Cell* **2006**, *126*, 969–980. [CrossRef] [PubMed]
33. Lozano-Durán, R.; Bourdais, G.; He, S.Y.; Robatzek, S. The bacterial effector HopM1 suppresses PAMP-triggered oxidative burst and stomatal immunity. *New Phytol.* **2014**, *202*, 259–269. [CrossRef] [PubMed]
34. Yuan, P.; Du, L.; Poovaiah, B. $Ca^{2+}$/Calmodulin-Dependent AtSR1/CAMTA3 Plays Critical Roles in Balancing Plant Growth and Immunity. *Int. J. Mol. Sci.* **2018**, *19*, 1764. [CrossRef] [PubMed]
35. Tuteja, N.; Mahajan, S. Calcium Signaling Network in Plants: An Overview. *Plant Signal. Behav.* **2007**, *2*, 79–85. [CrossRef] [PubMed]
36. Prasad, K.V.S.K.; Abdel-Hameed, A.A.E.; Xing, D.; Reddy, A.S.N. Global gene expression analysis using RNA-seq uncovered a new role for SR1/CAMTA3 transcription factor in salt stress. *Sci. Rep.* **2016**, *6*, 27021. [CrossRef]
37. Thor, K. Calcium—Nutrient and Messenger. *Front. Plant Sci.* **2019**, *10*, 440. [CrossRef]
38. Chin, K.; DeFalco, T.A.; Moeder, W.; Yoshioka, K. The Arabidopsis Cyclic Nucleotide-Gated Ion Channels AtCNGC2 and AtCNGC4 Work in the Same Signaling Pathway to Regulate Pathogen Defense and Floral Transition. *Plant Physiol.* **2013**, *163*, 611–624. [CrossRef]
39. Abdel-Hameed, A.A.E.; Prasad, K.V.S.K.; Jiang, Q.; Reddy, A.S.N. Salt-Induced Stability of SR1/CAMTA3 mRNA Is Mediated by Reactive Oxygen Species and Requires the 3′ End of Its Open Reading Frame. *Plant Cell Physiol.* **2020**, *61*, 748–760. [CrossRef]
40. Yuan, P.; Jewell, J.B.; Behera, S.; Tanaka, K.; Poovaiah, B.W. Distinct Molecular Pattern-Induced Calcium Signatures Lead to Different Downstream Transcriptional Regulations via AtSR1/CAMTA3. *Int. J. Mol. Sci.* **2020**, *21*, 8163. [CrossRef]
41. Tian, W.; Hou, C.; Ren, Z.; Wang, C.; Zhao, F.; Dahlbeck, D.; Hu, S.; Zhang, L.; Niu, Q.; Li, L.; et al. A calmodulin-gated calcium channel links pathogen patterns to plant immunity. *Nature* **2019**, *572*, 131–135. [CrossRef] [PubMed]
42. DeFalco, T.A.; Moeder, W.; Yoshioka, K. Opening the Gates: Insights into Cyclic Nucleotide-Gated Channel-Mediated Signaling. *Trends Plant Sci.* **2016**, *21*, 903–906. [CrossRef] [PubMed]
43. Ranty, B.; Aldon, D.; Cotelle, V.; Galaud, J.-P.; Thuleau, P.; Mazars, C. Calcium Sensors as Key Hubs in Plant Responses to Biotic and Abiotic Stresses. *Front. Plant Sci.* **2016**, *7*, 327. [CrossRef] [PubMed]
44. Yuan, P.; Jauregui, E.; Du, L.; Tanaka, K.; Poovaiah, B.W. Calcium signatures and signaling events orchestrate plant–microbe interactions. *Curr. Opin. Plant Biol.* **2017**, *38*, 173–183. [CrossRef] [PubMed]
45. Yuan, P.; Tanaka, K.; Du, L.; Poovaiah, B.W. Calcium Signaling in Plant Autoimmunity: A Guard Model for AtSR1/CAMTA3-Mediated Immune Response. *Mol. Plant* **2018**, *11*, 637–639. [CrossRef]
46. Yuan, P.; Tanaka, K.; Poovaiah, B.W. Calcium/Calmodulin-Mediated Defense Signaling: What Is Looming on the Horizon for AtSR1/CAMTA3-Mediated Signaling in Plant Immunity. *Front. Plant Sci.* **2022**, *11*, 795353. [CrossRef]
47. Yuan, P.; Yang, T.; Poovaiah, B.W. Calcium Signaling-Mediated Plant Response to Cold Stress. *Int. J. Mol. Sci.* **2018**, *19*, 3896. [CrossRef]
48. Wilkins, K.A.; Matthus, E.; Swarbreck, S.M.; Davies, J.M. Calcium-Mediated Abiotic Stress Signaling in Roots. *Front. Plant Sci.* **2016**, *7*, 1296. [CrossRef]
49. Mori, K.; Renhu, N.; Naito, M.; Nakamura, A.; Shiba, H.; Yamamoto, T.; Suzaki, T.; Iida, H.; Miura, K. $Ca^{2+}$-permeable mechanosensitive channels MCA1 and MCA2 mediate cold-induced cytosolic $Ca^{2+}$ increase and cold tolerance in Arabidopsis. *Sci. Rep.* **2018**, *8*, 550. [CrossRef]
50. Lenzoni, G.; Knight, M.R. Increases in Absolute Temperature Stimulate Free Calcium Concentration Elevations in the Chloroplast. *Plant Cell Physiol.* **2019**, *60*, 538–548. [CrossRef]

51. Knight, H.; Knight, M.R. Imaging spatial and cellular characteristics of low temperature calcium signature after cold acclimation in Arabidopsis. *J. Exp. Bot.* **2000**, *51*, 1679–1686. [CrossRef]
52. Melotto, M.; Underwood, W.; He, S.Y. Role of Stomata in Plant Innate Immunity and Foliar Bacterial Diseases. *Annu. Rev. Phytopathol.* **2008**, *46*, 101–122. [CrossRef]
53. Ding, Y.; Sun, T.; Ao, K.; Peng, Y.; Zhang, Y.; Li, X.; Zhang, Y. Opposite Roles of Salicylic Acid Receptors NPR1 and NPR3/NPR4 in Transcriptional Regulation of Plant Immunity. *Cell* **2018**, *173*, 1454–1467.e15. [CrossRef] [PubMed]
54. Cao, H.; Glazebrook, J.; Clarke, J.D.; Volko, S.; Dong, X. The Arabidopsis NPR1 Gene That Controls Systemic Acquired Resistance Encodes a Novel Protein Containing Ankyrin Repeats. *Cell* **1997**, *88*, 57–63. [CrossRef]
55. Scholthof, K.-B.G. The disease triangle: Pathogens, the environment and society. *Nat. Rev. Microbiol.* **2007**, *5*, 152–156. [CrossRef] [PubMed]
56. Went, F.W. The Effect of Temperature on Plant Growth. *Annu. Rev. Plant Physiol.* **1953**, *4*, 347–362. [CrossRef]
57. Martins, S.; Montiel-Jorda, A.; Cayrel, A.; Huguet, S.; Roux, C.P.-L.; Ljung, K.; Vert, G. Brassinosteroid signaling-dependent root responses to prolonged elevated ambient temperature. *Nat. Commun.* **2017**, *8*, 309. [CrossRef]
58. Velásquez, A.C.; Castroverde, C.D.M.; He, S.Y. Plant–Pathogen Warfare under Changing Climate Conditions. *Curr. Biol.* **2018**, *28*, R619–R634. [CrossRef]
59. Bose, J.; Pottosin, I.I.; Shabala, S.S.; Palmgren, M.G.; Shabala, S. Calcium Efflux Systems in Stress Signaling and Adaptation in Plants. *Front. Plant Sci.* **2011**, *2*, 85. [CrossRef]
60. DeFalco, T.A.; Toyota, M.; Phan, V.; Karia, P.; Moeder, W.; Gilroy, S.; Yoshioka, K. Using GCaMP3 to Study $Ca^{2+}$ Signaling in Nicotiana Species. *Plant Cell Physiol.* **2017**, *58*, 1173–1184. [CrossRef]
61. Zipfel, C.; Oldroyd, G.E.D. Plant signalling in symbiosis and immunity. *Nature* **2017**, *543*, 328–336. [CrossRef] [PubMed]
62. Hilleary, R.; Paez-Valencia, J.; Vens, C.; Toyota, M.; Palmgren, M.; Gilroy, S. Tonoplast-localized $Ca^{2+}$ pumps regulate $Ca^{2+}$ signals during pattern-triggered immunity in *Arabidopsis thaliana*. *Proc. Natl. Acad. Sci. USA* **2020**, *117*, 18849–18857. [CrossRef] [PubMed]
63. Nomura, H.; Komori, T.; Uemura, S.; Kanda, Y.; Shimotani, K.; Nakai, K.; Furuichi, T.; Takebayashi, K.; Sugimoto, T.; Sano, S.; et al. Chloroplast-mediated activation of plant immune signalling in Arabidopsis. *Nat Commun* **2012**, *3*, 926. [CrossRef] [PubMed]
64. Thor, K.; Jiang, S.; Michard, E.; George, J.; Scherzer, S.; Huang, S.; Dindas, J.; Derbyshire, P.; Leitão, N.; DeFalco, T.A.; et al. The calcium-permeable channel OSCA1.3 regulates plant stomatal immunity. *Nature* **2020**, *585*, 569–573. [CrossRef]
65. Peiter, E.; Maathuis, F.J.M.; Mills, L.N.; Knight, H.; Pelloux, J.; Hetherington, A.M.; Sanders, D. The vacuolar $Ca^{2+}$-activated channel TPC1 regulates germination and stomatal movement. *Nature* **2005**, *434*, 404–408. [CrossRef] [PubMed]
66. Kim, Y.S.; An, C.; Park, S.; Gilmour, S.J.; Wang, L.; Renna, L.; Brandizzi, F.; Grumet, R.; Thomashow, M. CAMTA-Mediated Regulation of Salicylic Acid Immunity Pathway Genes in Arabidopsis Exposed to Low Temperature and Pathogen Infection. *Plant Cell* **2017**, *29*, 2465–2477. [CrossRef]
67. Kidokoro, S.; Yoneda, K.; Takasaki, H.; Takahashi, F.; Shinozaki, K.; Yamaguchi-Shinozaki, K. Different Cold-Signaling Pathways Function in the Responses to Rapid and Gradual Decreases in Temperature. *Plant Cell* **2017**, *29*, 760–774. [CrossRef] [PubMed]
68. DeFalco, T.A.; Marshall, C.B.; Munro, K.; Kang, H.-G.; Moeder, W.; Ikura, M.; Snedden, W.A.; Yoshioka, K. Multiple Calmodulin-binding Sites Positively and Negatively Regulate Arabidopsis CYCLIC NUCLEOTIDE-GATED CHANNEL12. *Plant Cell* **2016**, *28*, 1738–1751. [CrossRef]
69. Reddy, A.S.N.; Ali, G.S.; Celesnik, H.; Day, I.S. Coping with Stresses: Roles of Calcium- and Calcium/Calmodulin-Regulated Gene Expression. *Plant Cell* **2011**, *23*, 2010–2032. [CrossRef]
70. Lefevere, H.; Bauters, L.; Gheysen, G. Salicylic Acid Biosynthesis in Plants. *Front. Plant Sci.* **2020**, *11*, 338. [CrossRef]
71. Li, N.; Euring, D.; Cha, J.Y.; Lin, Z.; Lu, M.; Huang, L.-J.; Kim, W.Y. Plant Hormone-Mediated Regulation of Heat Tolerance in Response to Global Climate Change. *Front. Plant Sci.* **2021**, *11*, 2318. [CrossRef] [PubMed]
72. Castroverde, C.D.M.; Dina, D. Temperature regulation of plant hormone signaling during stress and development. *J. Exp. Bot.* **2021**, *72*, 7436–7458. [CrossRef] [PubMed]
73. Panchal, S.; Melotto, M. Stomate-based defense and environmental cues. *Plant Signal. Behav.* **2017**, *12*, e1362517. [CrossRef]
74. Zeng, W.; Brutus, A.; Kremer, J.M.; Withers, J.C.; Gao, X.; Jones, A.D.; He, S.Y. A Genetic Screen Reveals Arabidopsis Stomatal and/or Apoplastic Defenses against Pseudomonas syringae pv. tomato DC3000. *PLoS Pathog.* **2011**, *7*, e1002291. [CrossRef] [PubMed]
75. Hammoudi, V.; Fokkens, L.; Beerens, B.; Vlachakis, G.; Chatterjee, S.; Arroyo-Mateos, M.; Wackers, P.F.K.; Jonker, M.J.; van den Burg, H.A. The Arabidopsis SUMO E3 ligase SIZ1 mediates the temperature dependent trade-off between plant immunity and growth. *PLoS Genet.* **2018**, *14*, e1007157. [CrossRef] [PubMed]
76. Gangappa, S.N.; Berriri, S.; Kumar, S.V. PIF4 Coordinates Thermosensory Growth and Immunity in *Arabidopsis*. *Curr. Biol.* **2017**, *27*, 243–249. [CrossRef] [PubMed]
77. Tanaka, K.; Choi, J.; Stacey, G. Aequorin Luminescence-Based Functional Calcium Assay for Heterotrimeric G-Proteins in Arabidopsis. In *G Protein-Coupled Receptor Signaling in Plants: Methods and Protocols*; Running, M.P., Ed.; Humana Press: Totowa, NJ, USA, 2013; pp. 45–54.
78. Knight, H.; Trewavas, A.J.; Knight, M.R. Cold calcium signaling in Arabidopsis involves two cellular pools and a change in calcium signature after acclimation. *Plant Cell* **1996**, *8*, 489–503.
79. Maintz, J.; Cavdar, M.; Tamborski, J.; Kwaaitaal, M.; Huisman, R.; Meesters, C.; Kombrink, E.; Panstruga, R. Comparative Analysis of MAMP-induced Calcium Influx in Arabidopsis Seedlings and Protoplasts. *Plant Cell Physiol.* **2014**, *55*, 1813–1825. [CrossRef]

80. Su, J.; Zhang, M.; Zhang, L.; Sun, T.; Liu, Y.; Lukowitz, W.; Xu, J.; Zhang, S. Regulation of Stomatal Immunity by Interdependent Functions of a Pathogen-Responsive MPK3/MPK6 Cascade and Abscisic Acid. *Plant Cell* **2017**, *29*, 526–542. [CrossRef]
81. Yuan, P.; Pan, H.; Boak, E.N.; Pierson, L.S.; Pierson, E.A. Phenazine-Producing Rhizobacteria Promote Plant Growth and Reduce Redox and Osmotic Stress in Wheat Seedlings Under Saline Conditions. *Front. Plant Sci.* **2020**, *11*, 1442. [CrossRef]

International Journal of
*Molecular Sciences*

MDPI

*Article*

# Genome-Wide Identification and Functions against Tomato Spotted Wilt Tospovirus of PR-10 in *Solanum lycopersicum*

**Md. Monirul Islam [1,2], Shiming Qi [1,2], Shijie Zhang [1,2], Bakht Amin [1,2], Vivek Yadav [1,2], Ahmed H. El-Sappah [1,2,3], Fei Zhang [1,2] and Yan Liang [1,2,*]**

1 College of Horticulture, Northwest A&F University, Xianyang 712100, China; monirul@nwafu.edu.cn (M.M.I.); qishiming2008@nwafu.edu.cn (S.Q.); 13289316916zsj@nwafu.edu.cn (S.Z.); bakhtamin96@nwafu.edu.cn (B.A.); vivekyadav@nwafu.edu.cn (V.Y.); ahmed_elsappah2006@yahoo.com (A.H.E.-S.); feizhang@nwsuaf.edu.cn (F.Z.)

2 State Agriculture Ministry Laboratory of Northwest Horticultural Plant Germplasm Resources & Genetic Improvement, Northwest A&F University, Xianyang 712100, China

3 Genetics Department, Faculty of Agriculture, Zagazig University, Zagazig 44511, Egypt

* Correspondence: liangyan@nwsuaf.edu.cn; Tel.: +86-29-8708-2613

**Abstract:** Tomato spotted wilt virus impacts negatively on a wide range of economically important plants, especially tomatoes. When plants facing any pathogen attack or infection, increase the transcription level of plant genes that are produced pathogenesis-related (PR) proteins. The aim of this study is a genome-wide identification of PR-10 superfamily and comparative analysis of *PR-10* and *Sw-5b* gene functions against tomato responses to biotic stress (TSWV) to systemic resistance in tomato. Forty-five candidate genes were identified, with a length of 64–210 amino acid residues and a molecular weight of 7.6–24.4 kDa. The *PR-10* gene was found on ten of the twelve chromosomes, and it was determined through a genetic ontology that they were involved in six biological processes and molecular activities, and nine cellular components. Analysis of the transcription level of PR-10 family members showed that the *PR-10* gene (Solyc09g090980) has high expression levels in some parts of the tomato plant. *PR-10* and *Sw-5b* gene transcription and activity in tomato leaves were strongly induced by TSWV infection, whereas H8 plants having the highest significantly upregulated expression of *PR-10* and *Sw-5b* gene after the inoculation of TSWV, and TSWV inoculated in M82 plants showed significantly upregulated expression of *PR-10* gene comparatively lower than H8 plants. There was no significant expression of *Sw-5b* gene of TSWV inoculated in M82 plants and then showed highly significant correlations between *PR-10* and *Sw-5b* genes at different time points in H8 plants showed significant correlations compared to M82 plants after the inoculation of TSWV; a heat map showed that these two genes may also participate in regulating the defense response after the inoculation of TSWV in tomato.

**Keywords:** PR-10 protein; resistance; *Sw-5b*; tomato; TSWV

**Citation:** Islam, M.M.; Qi, S.; Zhang, S.; Amin, B.; Yadav, V.; El-Sappah, A.H.; Zhang, F.; Liang, Y. Genome-Wide Identification and Functions against Tomato Spotted Wilt Tospovirus of PR-10 in *Solanum lycopersicum*. *Int. J. Mol. Sci.* **2022**, *23*, 1502. https://doi.org/10.3390/ijms23031502

Academic Editors: Isabel Marques, Ana I Ribeiro-Barros and José C. Ramalho

Received: 2 January 2022
Accepted: 20 January 2022
Published: 28 January 2022



## 1. Introduction

When pathogens infect a plant, pathogenesis-related proteins are produced. They are now categorized into 17 groups based on structural and functional features, and they play a key role in plant growth and development under a variety of stressors [1]. Through their enzymatic activity, PR proteins can directly impact pathogen integrity and/or create signal molecules, which act as a trigger to activate other plant defense processes [2]. PR-10 is a naturally occurring acidic protein with a molecular weight of 16–19 kDa and three-dimensional β-sheet topologies covered by a compact, bipartite framework held together by hydrophobic interactions. PR-10 proteins have antimicrobial action and are involved in a variety of biological processes, including ribonuclease and other secondary metabolism enzymatic functions, along with plant defense against biotic and abiotic stressors [3]. The majority of PR-10 proteins are made up of two domains: Bet_ v _1 and P-loop. Bet_ v

_1 is a conserved domain of the PR-10 protein family that serves as a defense against pathogen infection. The RNase activity of these proteins contains a P-loop motif, such as GxGGxGxxK, which serves as a nucleotide-binding site [4].

Tomatoes (*Solanum lycopersicum*) are the most widely grown commercial vegetable crop in the world. The tomato plant, on the other hand, is susceptible to a variety of diseases. More than 136 viral species that are detrimental to tomatoes have been discovered so far [5]. Tomato spotted wilt virus (TSWV) is one of the most dangerous. TSWV cause significant reductions in tomato output and market value, as well as mortality in certain cases [6]. Usually, the entire plant is dwarfed, with necrotic streaks and dark brown flecks on the leaves, stems, and fruits. The first symptoms of seedlings inhibited the growth points and young leaves turn into copper-colored rolls, and subsequently form many small dark brown flecks, the leaves' veins will be purple. Furthermore, the virus causes growth points leaf necrosis and drooping, and the stem end will have brown necrotic streaks. The plant grows only halfway or is completely dwarfed and wilted, and causes chlorotic rings on the green fruit [7].

The *Sw-5* gene is a single dominant quality of resistance gene that protects a wide range of Tospoviruses [8], originating from *L. peruvianum*, and has been identified and introgressed in the fresh tomato cultivar (*Lycopersicon esculentum*). *Sw-5* was discovered near the telomeric area of chromosome 9, between the CT71 and CT220 RFLP markers [9], and more closely linked to the CT220 marker (about within 65 kb) [10]. *Sw-5* locus is part of a loosely clustered gene family that includes six homologous paralog genes: *Sw-5a*, *Sw-5b*, *Sw-5c*, *Sw-5d*, *Sw-5e*, and *Sw-5f* are all potential variations [11]. *Sw-5a* and *Sw-5b* genes are strongly homologous (95%), and only *Sw-5b* has universal various isolates of TSWV [12]. *Sw-5b* was also linked to Tospovirus resistance to TCSV (tomato chlorotic spot virus), TZSV (tomato zonate spot virus), and GRSV (groundnut ringspot virus) [10].

TSWV is an RNA globular virus with an envelope structure and an outer membrane with a continuous protrusion layer. Its genome consists of three separate genomic RNA strands: long (L) RNA, medium (M) RNA, and short (S) RNA, all of which code for five different proteins. RdRp (RNA-dependent RNA polymerase) is a replication-related protein that interacts with host encoding factors. L RNA is a negative-sense RNA that encodes RdRp (RNA-dependent RNA polymerase) [13]. M RNA is an antisense RNA that codes for the amino acid and carboxy-terminal location of the glycoprotein precursor (GnGc), which is required for virions to assemble, mature, and release into their host [14]. The viral nonstructural proteins (NSm), which are encoded by sense RNA, are the main promoters of TSWV infection [15]. S RNA also contains a double sense RNA, with antisense RNA encoding nucleocapsid proteins (N) and sense RNA encoding nonstructural proteins (NSs) [16]; both proteins have a crucial function in the TSWV infection cycle [17]. Tomato spotted wilt virus (TSWV) is one of the most important plant viruses in the world [18]. TSWV belongs to the species Tospovirus that infect only plants, the genus *Orthotospovirus*, the family *Tospoviridae*, and the order *Bunyaviridae* [19]. TSWV caused widespread reductions in tomato yield and commercial value, together with plant mortality. Furthermore, the fast proliferation of TSWV-carrying *Frankliniella occidentalis* (Western flower thrips) has affected tomato agriculture and output severely [6]. As a result, tomato faces a major threat from this virus. We speculated on which genes are used to defend this virus, and considered the utilization of resistant/tolerant tomato cultivars to remedy the problem. These findings may lead to a better understanding of how the PR-10 protein and the *Sw-5b* gene act in the generation of systemic necrosis by the tomato spotted wilt tospovirus, which could aid in the future development of new antiviral approaches.

## 2. Results

### 2.1. Genome-Wide Identification and Analysis of Solanum Lycopersicum PR-10 Genes

From a genome-wide investigation of tomatoes, 45 putative *PR-10* encoding gene candidates were discovered. There are no chromosomal locations for the two *PR-10* gene ID numbers. Table 1 contains basic information on the *PR-10* genes, such as protein sequence length, isoelectric points, and molecular weight. PR-10 proteins varied in length from 64 to 210 amino acids, with molecular weights ranging from 7.6 to 24.4 kDa. The bulk of PR-10 proteins was classified as acidic based on theoretical isoelectric point (pI) values. The severe acidic or basic characteristics of each *PR-10* gene may contribute to different activities. In addition, all of the PR-10 protein sequences in tomatoes were shown to have the CDS, molecular weight, pI, instability index, aliphatic index, GRAVY, and subcellular localization (Table 1).

**Table 1.** The sequence features and physicochemical properties of tomato PR-10 proteins.

| SL. No. | Chr | Start | END | Strand | Number of Amino Acids | CDS | Molecular Weight | pI | Instability Index | Aliphatic Index | GRAVY | Subcellular Localization |
|---|---|---|---|---|---|---|---|---|---|---|---|---|
| 1 | ch00 | 14,149,228 | 14,151,711 | reverse | 158 | 441 | 18,196.91 | 5.03 | 38.70 | 99.94 | −0.218 | Cytoplasmic |
| 2 | ch00 | 21,749,238 | 21,751,331 | forward | 210 | 633 | 24,401.45 | 5.15 | 43.03 | 86.62 | −0.417 | Cytoplasmic |
| 3 | ch01 | 9,342,757 | 9,343,038 | reverse | 93 | 282 | 11,233.13 | 7.65 | 50.57 | 90.11 | −0.134 | Cytoplasmic |
| 4 | ch01 | 80,454,898 | 80,456,153 | forward | 76 | 231 | 8763.32 | 6.71 | 30.32 | 86.97 | −0.051 | Cytoplasmic |
| 5 | ch03 | 66,611,609 | 66,612,183 | forward | 159 | 480 | 17,608.98 | 5.02 | 32.81 | 82.20 | −0.35 | Cytoplasmic |
| 6 | ch03 | 66,614,237 | 66,614,792 | forward | 159 | 480 | 17,636.95 | 4.90 | 34.90 | 79.12 | −0.401 | Cytoplasmic |
| 7 | ch04 | 455,936 | 457,664 | reverse | 147 | 444 | 16,774.40 | 6.29 | 35.91 | 86.19 | −0.334 | Cytoplasmic |
| 8 | ch04 | 737,430 | 738,426 | forward | 146 | 441 | 16,613.22 | 5.96 | 33.78 | 99.38 | −0.168 | Cytoplasmic |
| 9 | ch04 | 1,363,357 | 1,364,224 | forward | 95 | 288 | 10,838.61 | 9.64 | 21.65 | 109.58 | −0.226 | Cytoplasmic |
| 10 | ch04 | 1,429,473 | 1,430,430 | forward | 147 | 444 | 16,734.98 | 5.59 | 31.71 | 94.76 | −0.318 | Cytoplasmic |
| 11 | ch04 | 1,435,905 | 1,437,368 | forward | 146 | 441 | 16,583.08 | 6.03 | 25.97 | 98.01 | −0.2 | Cytoplasmic |
| 12 | ch04 | 1,456,588 | 1,458,908 | reverse | 147 | 444 | 16,595.16 | 5.96 | 42.21 | 106.05 | −0.162 | Cytoplasmic |
| 13 | ch04 | 1,466,509 | 1,468,357 | forward | 148 | 447 | 17,017.30 | 5.62 | 19.96 | 90.88 | −0.403 | Cytoplasmic |
| 14 | ch04 | 1,474,832 | 1,476,779 | forward | 147 | 444 | 16,824.05 | 5.63 | 17.31 | 88.10 | −0.407 | Cytoplasmic |
| 15 | ch04 | 1,496,628 | 1,497,874 | reverse | 147 | 444 | 16,770.01 | 5.72 | 35.06 | 87.41 | −0.365 | Cytoplasmic |
| 16 | ch04 | 1,505,284 | 1,506,417 | forward | 134 | 405 | 15,762.00 | 5.64 | 56.47 | 92.84 | −0.653 | Cytoplasmic |
| 17 | ch04 | 49,064,622 | 49,065,920 | reverse | 147 | 444 | 17,152.77 | 5.90 | 36.51 | 91.43 | −0.351 | Cytoplasmic |
| 18 | ch05 | 58,320,136 | 58,321,632 | reverse | 150 | 453 | 17,596.99 | 5.22 | 50.34 | 96.00 | −0.467 | Cytoplasmic |
| 19 | ch05 | 58,346,771 | 58,347,729 | reverse | 150 | 453 | 17,698.46 | 5.47 | 51.29 | 86.93 | −0.456 | Cytoplasmic |
| 20 | ch05 | 58,352,298 | 58,352,492 | reverse | 64 | 195 | 7600.66 | 5.18 | 43.07 | 83.75 | −0.502 | Cytoplasmic |
| 21 | ch05 | 58,352,813 | 58,353,016 | reverse | 67 | 204 | 7959.23 | 7.06 | 33.19 | 92.99 | −0.058 | Cytoplasmic |
| 22 | ch05 | 58,415,437 | 58,415,727 | reverse | 96 | 291 | 11,409.16 | 5.04 | 54.80 | 89.38 | −0.431 | Cytoplasmic |
| 23 | ch05 | 58,420,502 | 58,420,723 | reverse | 73 | 222 | 8461.61 | 7.93 | 30.44 | 77.26 | −0.458 | Cytoplasmic |
| 24 | ch05 | 64,304,632 | 64,306,501 | reverse | 162 | 489 | 18,181.90 | 4.78 | 37.17 | 89.63 | −0.188 | Cytoplasmic |
| 25 | ch06 | 1,757,744 | 1,758,229 | forward | 161 | 486 | 17,717.56 | 5.23 | 33.57 | 106.40 | 0.131 | Cytoplasmic |
| 26 | ch07 | 288,360 | 289,609 | reverse | 149 | 450 | 16,770.27 | 4.73 | 25.47 | 99.33 | −0.013 | Cytoplasmic |
| 27 | ch07 | 291,108 | 292,521 | forward | 158 | 477 | 17,981.36 | 5.23 | 37.64 | 87.47 | −0.285 | Cytoplasmic |
| 28 | ch07 | 3,683,904 | 3,685,158 | forward | 159 | 480 | 18,422.81 | 5.31 | 32.72 | 75.91 | −0.613 | Cytoplasmic |
| 29 | ch08 | 26,737,023 | 26,738,299 | reverse | 148 | 447 | 17,074.62 | 6.50 | 16.91 | 88.92 | −0.42 | Cytoplasmic |
| 30 | ch09 | 296,718 | 297,754 | reverse | 148 | 453 | 17,074.62 | 6.50 | 16.91 | 88.92 | −0.42 | Cytoplasmic |
| 31 | ch09 | 301,066 | 301,598 | reverse | 136 | 411 | 15,514.72 | 5.09 | 22.29 | 95.96 | −0.171 | Cytoplasmic |
| 32 | ch09 | 304,321 | 305,404 | reverse | 150 | 453 | 17,083.46 | 4.88 | 28.19 | 96.07 | −0.289 | Cytoplasmic |
| 33 | ch09 | 339,412 | 340,474 | forward | 151 | 456 | 17,425.73 | 5.10 | 40.68 | 94.77 | −0.438 | Cytoplasmic |
| 34 | ch09 | 6,149,317 | 6,151,058 | reverse | 150 | 453 | 17,100.71 | 5.57 | 25.66 | 92.87 | −0.223 | Cytoplasmic |
| 35 | ch09 | 6,153,756 | 6,154,061 | reverse | 101 | 306 | 11,770.43 | 8.44 | 34.15 | 73.27 | −0.716 | Cytoplasmic |
| 36 | ch09 | 6,176,738 | 6,177,747 | reverse | 152 | 459 | 17,414.09 | 5.47 | 24.29 | 88.29 | −0.266 | Cytoplasmic |
| 37 | ch09 | 6,195,377 | 6,197,648 | reverse | 74 | 225 | 8310.73 | 8.80 | 24.74 | 75.00 | −0.176 | Cytoplasmic |
| 38 | ch09 | 6,221,354 | 6,222,253 | reverse | 152 | 459 | 17,423.11 | 5.62 | 31.78 | 90.26 | −0.248 | Cytoplasmic |
| 39 | ch09 | 70,346,085 | 70,347,506 | reverse | 155 | 468 | 17,235.60 | 5.79 | 22.44 | 84.84 | −0.497 | Cytoplasmic |
| 40 | ch09 | 70,350,133 | 70,352,154 | reverse | 160 | 483 | 17,368.77 | 5.44 | 30.62 | 94.38 | −0.128 | Cytoplasmic |
| 41 | ch09 | 70,355,642 | 70,357,036 | reverse | 160 | 483 | 17,914.20 | 5.34 | 32.81 | 88.25 | −0.426 | Cytoplasmic |
| 42 | ch09 | 70,360,338 | 70,361,437 | forward | 155 | 468 | 17,352.86 | 5.67 | 27.16 | 86.06 | −0.389 | Cytoplasmic |
| 43 | ch10 | 2,466,093 | 2,467,665 | reverse | 149 | 450 | 17,272.9 | 8.74 | 25.91 | 72.01 | −0.606 | Cytoplasmic |
| 44 | ch10 | 43,660,724 | 43,661,266 | reverse | 146 | 441 | 16,685.34 | 6.37 | 13.79 | 85.48 | −0.305 | Cytoplasmic |
| 45 | ch12 | 65,664,355 | 65,665,560 | forward | 158 | 477 | 18,196.91 | 5.03 | 38.7 | 99.94 | −0.218 | Cytoplasmic |

### 2.2. Phylogenetic Analysis, Conserved Motifs, and Gene Structures

To further characterize and identify potential functional relationships between the PR-10 proteins of tomato, a phylogenetic tree was constructed and 45 gene IDs of PR-10 protein were identified (Figure 1). The neighbor-joining method, with 1000 bootstrap reconstruction and completed deletion gaps/missing data, yielded five known subfamilies.

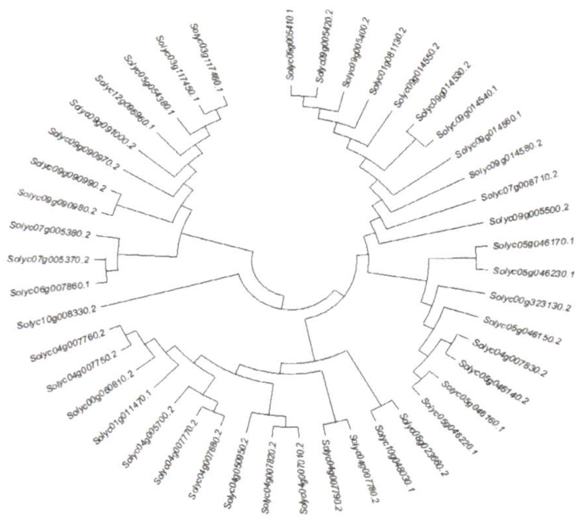

**Figure 1.** Phylogenetic tree of 45 genes of PR-10 protein in tomato. Sequence alignment was done by ClustalW and Phylogenetic tree generated by MEGA 5.1 with maximum likelihood (ML) method for 1000 replicates as the bootstrap value.

A total of nine conserved motifs were identified using the MEME server (Figure 2), and only motif 1 and motif 6 were found to be associated with the Bet_ V_ 1 (PF000407) domain. These two motifs (motif 1 and 6) were found in all SlPR-10 proteins, as expected. Remarkably, motif 9 was found only in the Solyc03g117450.1 and Solyc03g117460.1 protein sequence. The fewest motifs were identified as in Solyc01g011470.1 and Solyc01g081130.2 protein sequences. These variations of the protein motifs in number, types, and positions may be related to molecular function and roles in different tomato metabolic pathways.

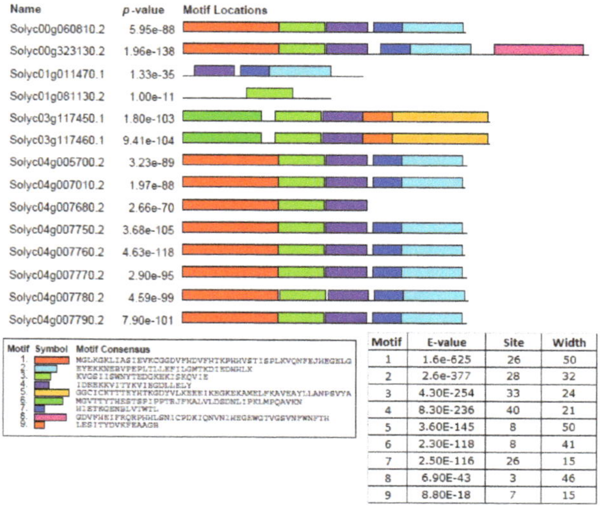


| Motif | E-value | Site | Width |
|---|---|---|---|
| 1 | 1.6e-625 | 26 | 50 |
| 2 | 2.6e-377 | 28 | 32 |
| 3 | 4.30E-254 | 33 | 24 |
| 4 | 8.30E-236 | 40 | 21 |
| 5 | 3.60E-145 | 8 | 50 |
| 6 | 2.30E-118 | 8 | 41 |
| 7 | 2.50E-116 | 26 | 15 |
| 8 | 6.90E-43 | 3 | 46 |
| 9 | 8.80E-18 | 7 | 15 |

**Figure 2.** Motif locations of PR-10 protein. Each motif is represented using a box with different colors. Motif 1, red; Motif 2, violet; Motif 3, light green; Motif 4, light blue; Motif 5, yellow; Motif 6, green; Motif 7, blue; Motif 8, purple; Motif 9, orange.

The distribution of the intronic phase and the location of exons/introns are key features for gene structure study. Introns are divided into three phases: phase 0 introns are found between two consecutive codons, phase 1 introns are found between a codon's first and second nucleotide, and phase 2 introns are found between a codon's second and third nucleotide [20]. The exon–intron and intron phase organization of tomato *PR-10* genes is shown in Figure 3. The location and length of introns–exons organization in tomato *PR-10* genes are shown in Table 1.

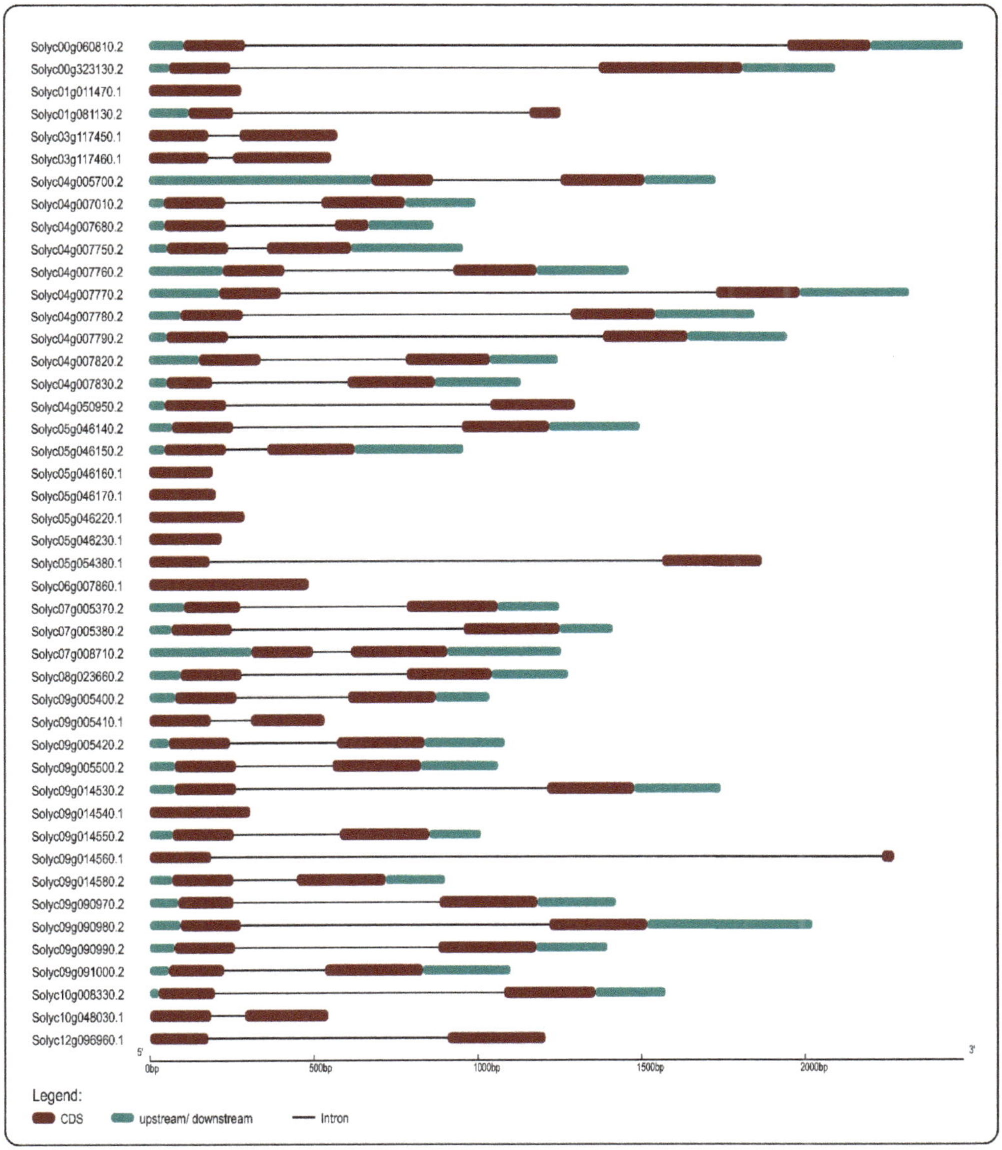


**Figure 3.** Gene structure of 45 genes of PR-10 protein in tomato. Exon, intron, upstream, and downstream regions are indicated as dark red boxes, black lines, and light blues boxes, respectively.

### *2.3. Chromosome Mapping of the PR-10 Gene*

CDD (Conserved domains database) from NCBI was used to compare the domain architecture of *PR-10* gene sequences. This study's findings include the presence of conserved domains in their sequences, indicating that they are homologs. In silico chromosomal mapping of tomato *PR-10* genes are shown in Figure 4. Tomato *PR-10* genes have been discovered on ten of the twelve chromosomes. There are 11 genes on chromosome 4 and 13 genes on chromosome 9 in tomatoes; however, PR-10 genes were not discovered on chromosomes 2 and 11.

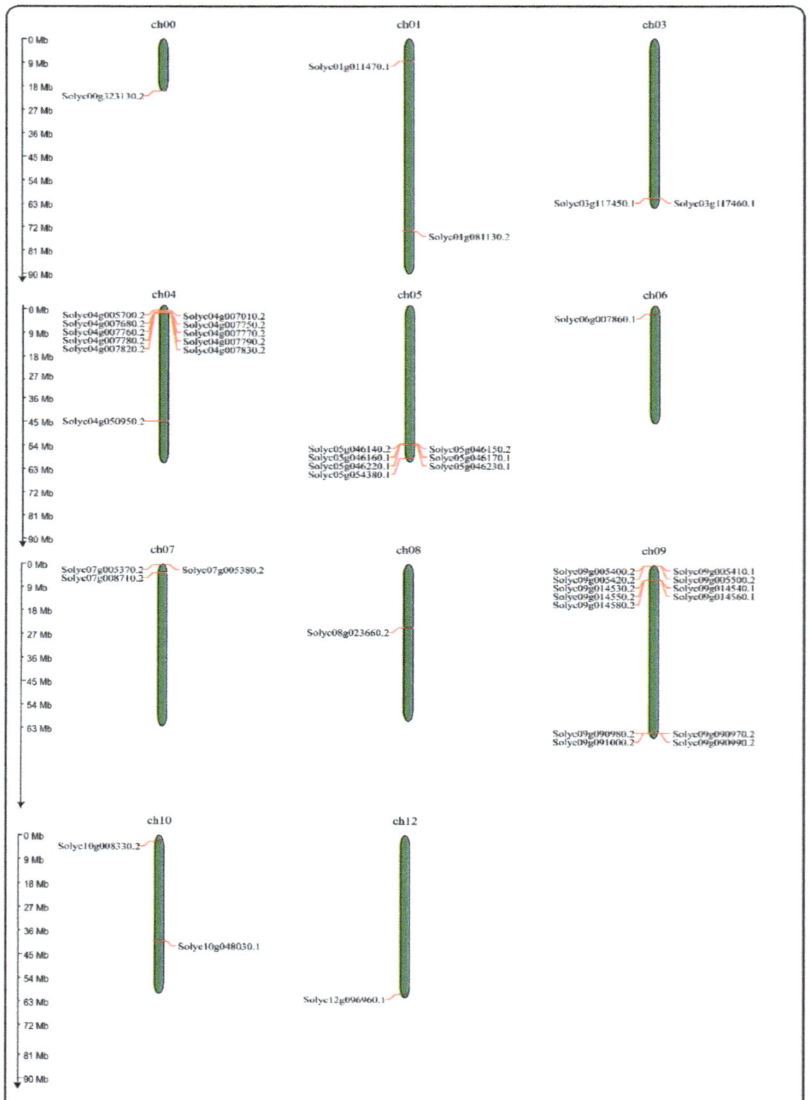


**Figure 4.** Chromosome mapping of the *PR-10* gene of tomato.

### 2.4. GO Analysis

Six biological processes and molecular activities and nine cellular components were identified via a gene ontology (GO) study, shown in Figure 5. Most PR-10 genes have a role in defensive responses and a reaction to abiotic stimulus in a biological function, according to the GO enrichment study. It exhibited protein kinase activity together with adenyl nucleotide and purine ribonucleoside binding activities at the molecular level. In the extracellular area, the cellular component had a function (Table 2).

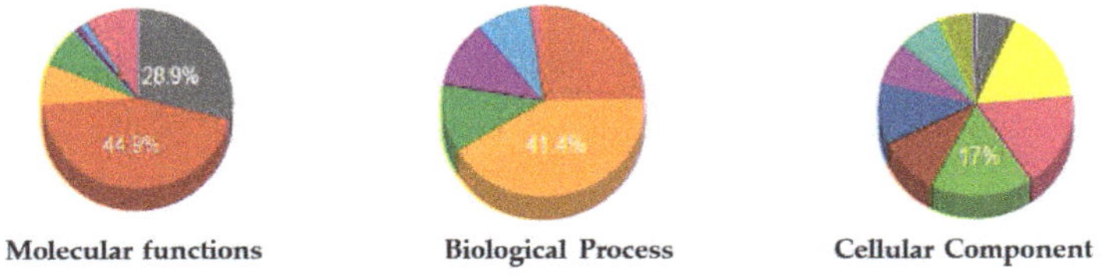


**Figure 5.** The gene ontology (GO) term distribution of PR-10 proteins of tomato.

**Table 2.** The gene ontology (GO) term distribution of PR-10 proteins of tomato.

| Molecular Functions | Biological Process | Cellular Component |
|---|---|---|
| Protein binding | Immune system process | Cytosol |
| Lipid binding 44.9% | Response to stress 41.4% | Nucleus |
| Lyase activity | Reproduction | Organelle |
| Ion binding | Anatomical structure development | Vacuole |
| Nuclease activity | Cellular nitrogen compound metabolic process | Cytoplasmic-membrane-bounded vesicle |
| Hydrolase activity | Signal transduction | Cytoplasm 17% |
| N/A 28.9% | | Cell |
| | | Intercellular |
| | | N/A |

### 2.5. Three-Dimensional Structure of the PR-10 Protein

There was a difference in the percentage of structural attributes that changed among the 45 *PR-10* gene IDs. Despite the 3D structural variances revealed among tomato PR-10 proteins, all 45 candidates have the binding proteins essential for protein interaction. For *PR-10* gene cloning and plant transformation, we chose the Solyc09g090980 gene ID of PR-10 protein, which is a phenolic, oxidative coupling protein with a length of 1424 bp and is situated on chromosome 9, as shown in Figure 6.

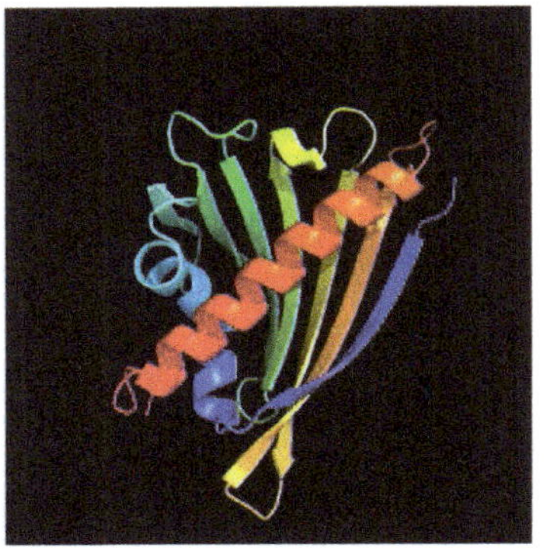

**Figure 6.** The predicted 3D structures of Solyc09g090980 of PR-10 protein, generated using the Phyre2 server, and binding pockets identified by the CASTp 3.0 server (shown in red).

### 2.6. Transcriptomic Analysis of the PR-10 Protein of Tomato

The root, stem, branch, tendon, male flower, female flower, and fully opened young leaf were all subjected to spatial expression profiling of all *PR-10* genes in tomatoes. The *PR-10* genes were expressed differently in different plant tissues, according to the heatmap analysis. However, most of the genes in tomatoes had a relatively low expression level. Furthermore, the highest level of PR-10 expression of the Solyc09g090980 gene compared to the other tissues (Figure 7).

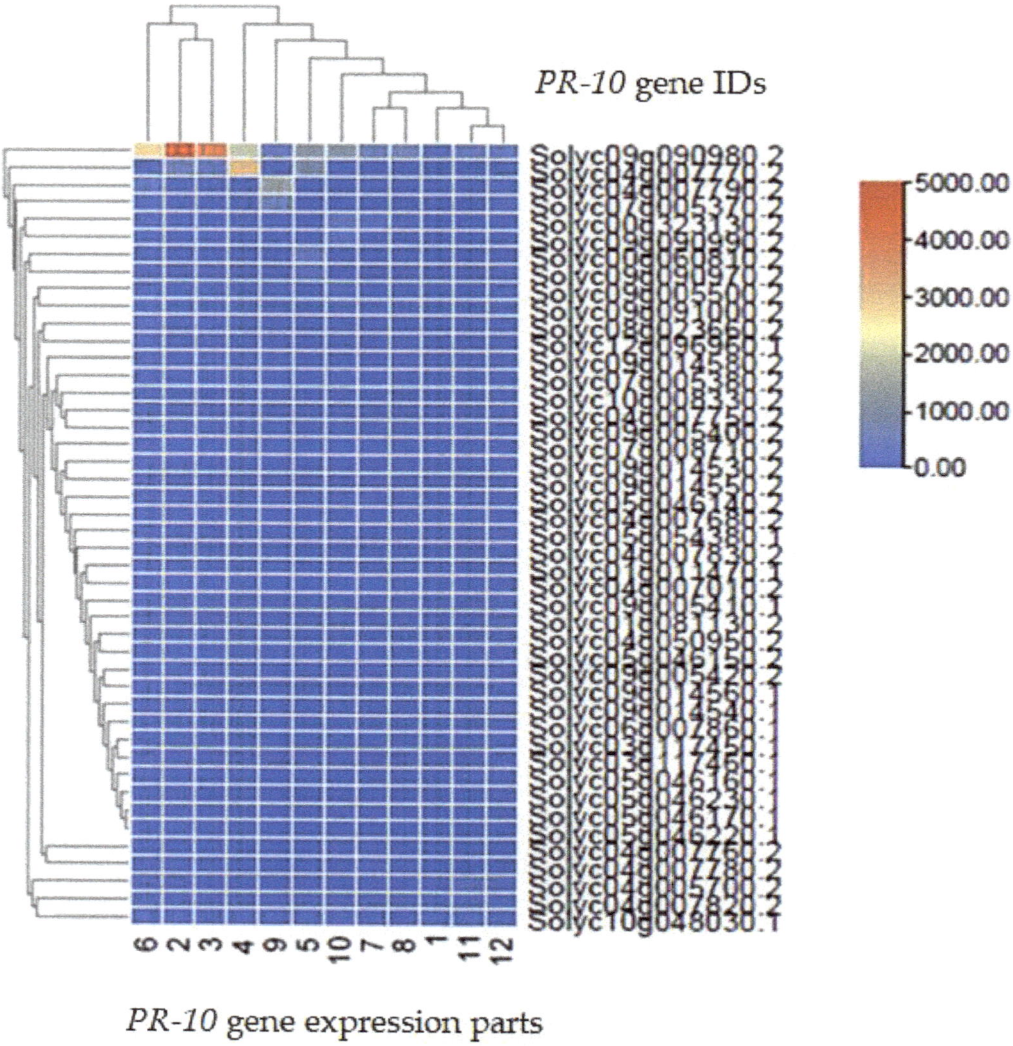


**Figure 7.** Transcriptomic analysis of the PR-10 protein in tomato. The data presented are based on at least three separate replicates. The gene expression within each family is represented using hierarchical clustering.

### 2.7. Incidence and Disease Index Level of TSWV in Each Group of Tomato Plants at Each Observation Time Point

Tomato-sensitive cultivar M82 and TSWV-resistant cultivar H8 were evaluated for their response to TSWV infections. As previously stated, plants were manually infected with TSWV and kept in an insect-proof greenhouse [21]. M82 and H8 plants treated plants with TSWV virus inoculum compared with M82 and H8 CK plants. It was found that there were no obvious symptoms of TSWV after 7 days of inoculation. After 14 days, the plants showed obvious dwarfing, and the growing point of the plant was purple; all M82 treatment groups showed purple spots in growth and wilting after 21 days. Incidence of TSWV in each CK and treatment (TR) group of M82 tomato plants showed the highest TSWV incidence rate compared to H8 plants (Table 3) after 28 days.

**Table 3.** Disease incidence and index level of TSWV in the CK and treatment (TR) groups of M82 and H8 tomato cultivars (highly resistance (0 < disease index ≤ 5); medium resistance (5 < disease index ≤ 20); resistance (20 < disease index ≤ 40); highly sensitive (60 < disease index); sensitive (40 < disease index ≤ 60)).

| Post Inoculation of TSWV | CK Group | | TR Group | |
|---|---|---|---|---|
| | cv. M82 (Sensitive) | cv. H8 (Resistant) | cv. M82 (Sensitive) | cv. H8 (Resistant) |
| 14 days | 10.0% [c]<br>5.3 [d] | 3.33% [c]<br>0.0 [e] | 16.67% [c]<br>12.5 [d] | 10.0% [c]<br>3.0 [f] |
| 21 days | 26.67% [b]<br>46.66 [b] | 10.0% [c]<br>5.4 [d] | 36.67% [b]<br>52.3 [b] | 10.0% [c]<br>8.3 [e] |
| 28 days | 40.0% [a]<br>58.88 [a] | 26.66% [b]<br>14.0 [c] | 100% [a]<br>88.88 [a] | 33.33% [c]<br>18.33 [c] |

**Note:** Same letters represent no significant difference, while different letters represent significant difference between CK and TR groups of M82 and H8 plants of TSWV disease incidence and disease index level, after TSWV inoculation at 14, 21 and 28 days. TSWV disease incidence expressed in percentages (%) and disease index express HR, MR, S, and HS respectively. Means ± SD at $p < 0.05$ using Tukey's test with three biological replicates.

The severity of TSWV symptoms was scored on a scale of 0–4 (Table 4) for each plant at 14, 21 and 28 days post inoculation (dpi) (Table 3) to calculate the TSWV disease incidence rate and disease index level.

**Table 4.** Plant disease index level parameters.

| Grade Series | TSWV Symptoms |
|---|---|
| 0 | Asymptomatic |
| 1 | Leaves are less obvious bronze (black) small spots |
| 2 | Bronze (black) necrosis spots spread in small area of leaves |
| 3 | Plant leaves on the film even into pieces, nearly 2/3 leaf area dry necrosis |
| 4 | Whole plant necrosis |

On the other hand, the plant disease index level of TSWV in each group of CK and TR; M82 showed the highest sensitivity compared to H8 plants, and H8 plants showed resistance phenotype after 28 days of infection of TSWV (Table 3).

After inoculation of TSWV, M82 and H8 plants are (shown in Figure 8) TSWV disease severity showed after 28 days of infection; the plant's leaves were purple and dark brown patches in M82 plants (shown in Figure 9).

**Figure 8.** TSWV inoculate of M82 and H8 plants of tomato; (**a**) TSWV inoculation of M82 plant, (**b**) TSWV inoculation of H8 plant, (**c**) TSWV inoculated leaves, (**d**) TSWV inoculate of M82 plants after 28 days, and (**e**) TSWV inoculate of H8 plants after 28 days.

**Figure 9.** M82 plant leaves show typical symptoms of TSWV: Grade 0, no disease; Grade 1, slightly infected; Grade 2, 50% infected; and Grade 3, 75% infected leaves.

### 2.8. PR-10 Expression in M82 and H8 Plants

The purpose of this study was to determine how PR-10 protein expression affected infection in M82 and H8 plants following TSWV virus inoculation. Both CK and TR plants were sampled at various time intervals, including 0, 3, 6, 12, 24, 48, 72, 96 h and 7, 14, 21, 28 days. The expression pattern of PR-10 in different tissues of leaves of the tomato M82 (sensitive) and H8 (resistant) plants after the TSWV virus inoculation follows: M82 inoculated plants showed a wide range of response of PR-10 protein (5.70-fold relative expression at 72 h) compared to CK and H8 inoculated plants, which showed a wide range of response of PR-10 protein relative expression compared to CK (6.74-fold at 24 h) and significantly upregulated in the leaves after the TSWV inoculation (Figure 10).

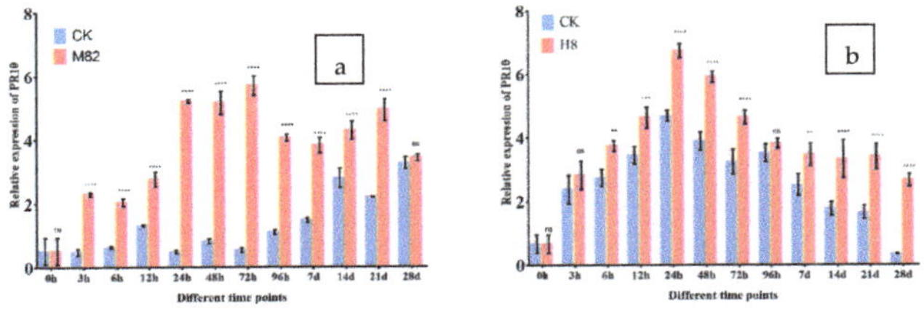


**Figure 10.** PR-10 expression in M82 and H8 plants. The relative transcription levels of TSWV-infected tomato M82 and H8 plants were normalized to the control gene *Actin*, as measured by qRT-PCR of PR-10 protein. At 0, 3, 6, 12, 24, 48, 72, and 96 h pi (hours post infection) and 7, 14, 21 and 28 dpi (days post infection), the expression levels in the leaf tissues were assessed (days post infection). The CK plants are shown by blue bars, while the M82 and H8 plants are represented by red bars. Three biological replicates' mean results (±SD) are shown as relative changes. (**a**) PR-10 relative expression in M82 plants vs. CK (not inoculated with TSWV) and (**b**) PR-10 relative expression in H8 plants compared to CK. Asterisk indicates statistically significant; ** $p = 0.01$, *** $p = 0.001$, **** $p = 0.0001$, and ns = non-significant are the $p$ values.

### *2.9. TSWV-cp Expression in M82 and H8 Plants*

To examine the effect of TSWV-cp protein expression after the TSWV virus inoculation in M82 and H8 plants. Leaf samples were taken at different time points such as 0, 3, 6, 12, 24, 48, 72, 96 h and 7, 14, 21, 28 days both CK and TR plants. The expression pattern of TSWV-cp in different tissues of leaves in the tomato M82 and H8 plants after the TSWV virus inoculation, M82 infected plants showed a wide range of response of TSWV-cp compared to CK (2.22-fold relative expression at 14D) were significantly upregulated, and H8 infected plants showed a narrow range of response of TSWV-cp compared to CK were non-significantly upregulated in leaves after the TSWV inoculation in Figure 11.

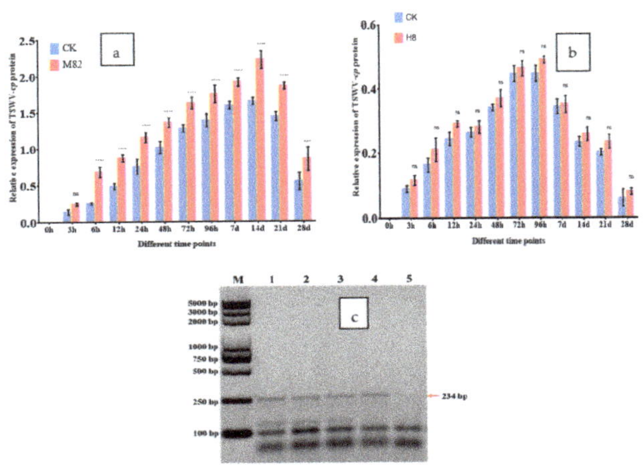


**Figure 11.** TSWV-cp expression in M82 and H8 plants. Tomato M82 and H8 plants infected with TSWV inoculum, and relative transcript levels normalized to the control gene *Actin*. At 0, 3, 6, 12, 24, 48, 72, and 96 hpi (hours post infection) and 7, 14, 21 and 28 dpi (days post infection), the expression levels in the leaf tissues were assessed (days post infection). The CK plants are shown by blue bars, while the M82 and H8 plants are represented by red bars. The mean values (±SD) from three biological replicates are provided as relative changes: (**a**) TSWV-cp relative expression in M82 plants compared to CK; (**b**) TSWV-cp relative expression in H8 plants compared to CK; and (**c**) TSWV-cp confirmation in Yangling, China, isolates 234 bp. Asterisk indicates statistically significant; *** $p = 0.001$, **** $p = 0.0001$, and ns = non-significant are the $p$ values.

## 2.10. Expression of *Sw-5b* in M82 and H8 Plants

The purpose of this study was to determine how *Sw-5b* expression changed following TSWV virus inoculation in M82 and H8 plants. Tomato leaf samples were taken from CK and TR plants at various times, including 0, 3, 6, 12, 24, 48, 72, 96 h and 7, 14, 21, 28 days. Figure 12 shows the *Sw-5b* expression pattern in different tissues of tomato M82 and H8 plants' leaves after TSWV virus inoculation; M82-infected plants showed no significant expression of *Sw-5b*, while H8-infected plants showed a wide range of relative expression of *Sw-5b* (4.14-fold at 7D) and significantly upregulated in leaves after TSWV inoculation.

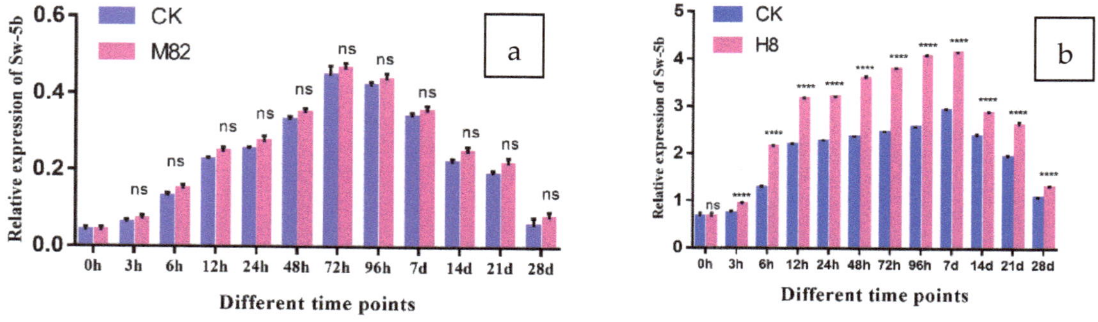


**Figure 12.** Expression of *Sw-5b* in M82 and H8 plants. *Sw-5b* relative transcription levels were assessed by qRT-PCR; tomato M82 and H8 plants were inoculated with TSWV, and relative transcript levels were normalized to the control gene *Actin*. At 0, 3, 6, 12, 24, 48, 72, and 96 hpi (hours post infection) and 7, 14, 21 and 28 dpi (days post infection), the expression levels in the leaf tissues were assessed (days post infection). The CK plants are shown by blue bars, while the M82 and H8 plants are represented by red bars. Three biological replicates' mean results (±SD) are reported as relative changes. (**a**) *Sw-5b* expression in M82 plants in comparison to CK, and (**b**) *Sw-5b* expression in H8 plants in comparison to CK. Asterisk indicates statistically significant; **** $p = 0.0001$, and ns = non-significant are the $p$ values.

## 2.11. Expression of Different Index Levels of Leaves in M82 and H8 Plants

The purpose of this study was to determine how PR-10, TSWV-cp, and *Sw-5b* expression affected different disease index levels of M82 and H8 tomato leaves following TSWV inoculation. After 28 days, tomato leaf samples were obtained from both CK and TR plant leaves. In the Section 4, different disease index levels are explained. Different disease index levels show various types of expression of PR-10, TSWV-cp, and *Sw-5b* in M82 and H8 plants. H8-infected plants showed a higher range of PR-10 protein compared to M82 plants. On the other hand, M82-infected plants showed a wide range of TSWV-cp (4.02-fold relative expression at level 3) compared to H8 plants. Additionally, M82 plants showed no significant expression of *Sw-5b*, but H8 plants showed 3.53-fold relative expression of *Sw-5b*, as shown in Figure 13.

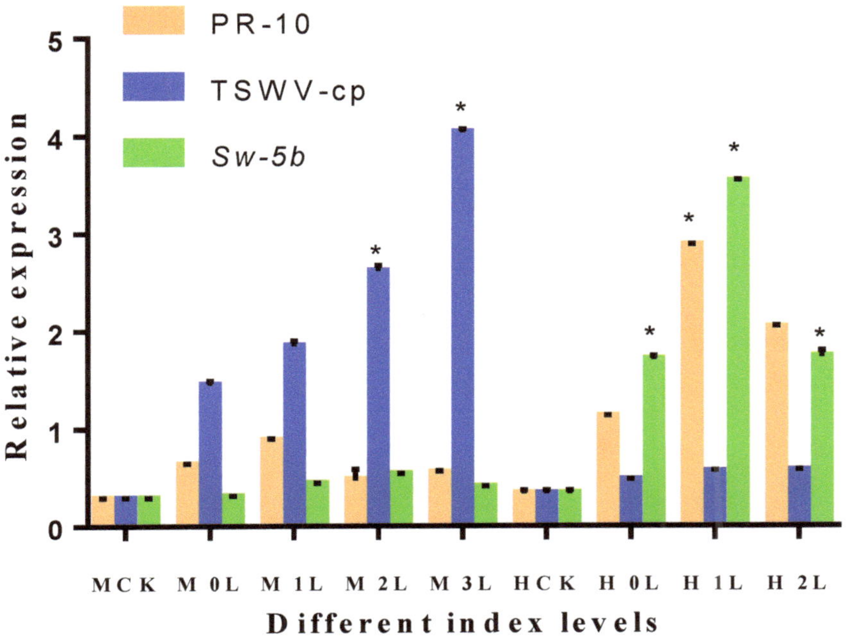


**Figure 13.** Expression of different index levels of leaves in M82 and H8 plants. After the TSWV inoculum was applied to tomato M82 and H8 plants, the relative transcript expression levels of PR-10 protein, TSWV-cp protein, and *Sw-5b* were determined using qRT-PCR, and the relative transcript levels were normalized to the control gene *Actin*. The expression levels in the leaf tissues were measured at different disease index levels. Light yellow bars represent the PR-10 proteins, blue bars represent the TSWV-cp protein, and light green bars represent the *Sw-5b*. The mean values (±SD) from three biological replicates are presented as relative changes. The least significant difference (LSD) test was used to determine if a gene showed significant induction as compared to the transcript abundance of the control group (MCK represents the M82, CK plant; MTR represents the M82-treated plant; HCK represents the H8, CK plant; and HTR represents the H8-treated plant). Asterisk indicates statistically significant; * $p = 0.05$, and ns = non-significant are the $p$ values.

### *2.12. Expression of Different Index Levels of the Tomato Plants Root*

To examine the effect of PR-10, TSWV-cp, and *Sw-5b* expression of different index levels of M82 and H8 tomato plant roots, tomato plant root samples were taken after 28 days of CK and TR plants. Different index levels show various types of expression levels of PR-10, TSWV-cp, and *Sw-5b* in M82 and H8 plants. H8-infected plants showed a higher range of PR-10 protein compared to M82 plants. Additionally, M82-infected plants showed a wide range of response of TSWV-cp (4.72-fold relative expression at level 2) and significantly upregulated, and H8 plants showed relatively lower expression of TSWV-cp. M82 plants also showed no significant expression of *Sw-5b*, but H8 plants showed 3.40-fold relative expression that was significantly upregulated (Figure 14).

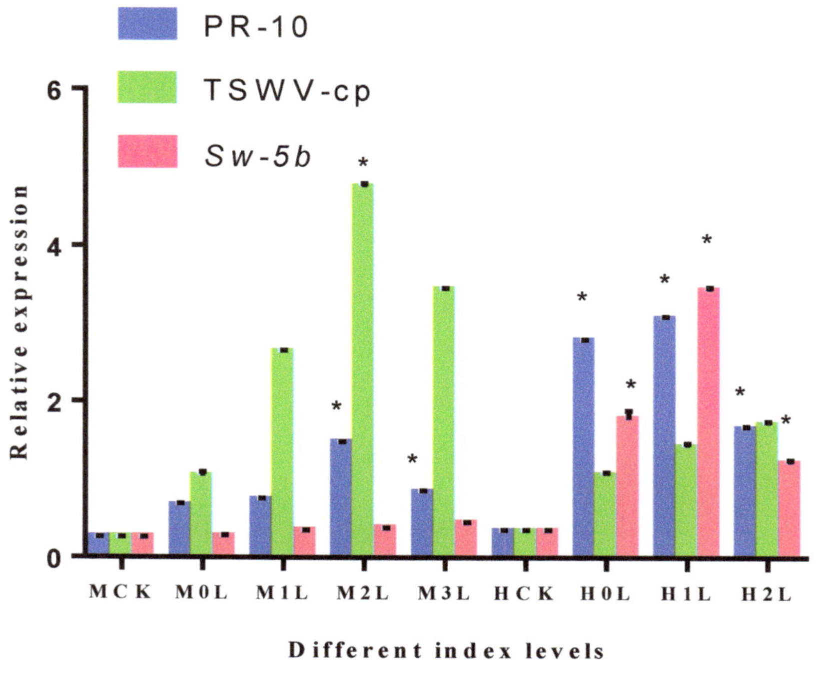


**Figure 14.** Expression of different index levels of plant roots in M82 and H8. Relative transcription levels determined by qRT-PCR of PR-10 protein, TSWV-cp protein, and *Sw-5b*; tomato M82 and H8 plants were infected with TSWV inoculum and the relative transcription levels were normalized to the control gene *Actin*. The expression levels in the root tissues were measured at different disease index levels. Blue bars represent the PR-10 proteins, light green bars represent the TSWV-cp protein, and purple bars represent *Sw-5b*. The mean values (±SD) from three biological replicates are presented as relative changes. The least significant difference (LSD) test was used to determine if a gene showed significant induction as compared to the transcript abundance of the control group (MCK represents the M82 CK plant; M represents the M82-treated plant; HCK represents the H8, CK plant; and H represents the H8-treated plant). Asterisk indicates statistically significant; * $p = 0.05$ and ns = non-significant are the $p$ values.

### *2.13. Correlation between PR-10 and Sw-5b Gene Expression after TSWV Inoculation*

Table 5 shows the results of a correlation study between the PR-10 and *Sw-5b* gene expressions following TSWV inoculation. At multiple time intervals, such as 3, 6, 12, 24, 48, 72, 96 h and 7, 14, 21, 28 days, the studies revealed extremely significant correlations between PR-10 and *Sw-5b* genes. M82 and H8 tomato plants exhibited extremely significant correlations in this study; however, H8 plants showed significant correlations when compared to M82 plants.

**Table 5.** The correlation coefficient between PR-10 protein and *Sw-5b* at different time points after TSWV inoculation.

| Index | CK | 3 h | 6 h | 12 h | 24 h | 48 h | 72 h | 96 h | 7 D | 14 D | 21 D | 28 D |
|---|---|---|---|---|---|---|---|---|---|---|---|---|
| TR | 1.000 | | | | | | | | | | | |
| 3 h | 0.495 | 1.000 | | | | | | | | | | |
| 6 h | 0.550 | 0.926 ** | 1.000 | | | | | | | | | |
| 12 h | 0.460 | 0.952 ** | 0.960 ** | 1.000 | | | | | | | | |
| 24 h | 0.123 | 0.692 ** | 0.489 | 0.606 ** | 1.000 | | | | | | | |
| 48 h | 0.516 | 0.850 ** | 0.959 ** | 0.923 ** | 0.341 | 1.000 | | | | | | |
| 72 h | 0.154 | 0.764 ** | 0.590 ** | 0.713 ** | 0.975 ** | 0.456 | 1.000 | | | | | |
| 96 h | 0.363 | 0.869 ** | 0.853 ** | 0.902 ** | 0.785 ** | 0.820 ** | 0.841 ** | 1.000 | | | | |
| 7 D | 0.314 | 0.839 ** | 0.847 ** | 0.820 ** | 0.527 | 0.822 ** | 0.602 ** | 0.777 ** | 1.000 | | | |
| 14 D | 0.322 | 0.821 ** | 0.799 ** | 0.752 ** | 0.480 | 0.717 ** | 0.565 | 0.636 ** | 0.926 ** | 1.000 | | |
| 21 D | 0.200 | 0.616 ** | 0.651 ** | 0.659 ** | 0.479 | 0.720 ** | 0.545 | 0.782 ** | 0.799 ** | 0.672 * | 1.000 | |
| 28 D | 0.323 | 0.581 ** | 0.718 ** | 0.675 ** | 0.374 | 0.772 ** | 0.447 | 0.777 ** | 0.715 ** | 0.587 * | 0.923 ** | 1.000 |

* Correlation is significant at the 0.05 level; ** correlation is significant at the 0.01 level.

### 2.14. Heat Map of PR-10 Protein and Sw-5b Gene Expression at Different Time Points

PR-10 protein and *Sw-5b* gene expression at various time periods following TSWV inoculation. TB tools software was used to create the heat map, which was then applied to two clades of M82 and H8 plants. Figure 15 shows that PR-10 protein and *Sw-5b* gene expression patterns were clustered; H8 plants demonstrated resistance to TSWV of various colors with high PR-10 protein and *Sw-5b* gene expression, whereas M82 plants showed susceptibility to TSWV. We have a comprehensive understanding of the PR-10 protein and *Sw-5b*, both of which play an equal role in TSWV resistance following TSWV inoculation. H8 plants were more resistant to *Sw-5b* than M82 plants because M82 plants lack *Sw-5b* resistant genes, but H8 plants have the resistant genes.

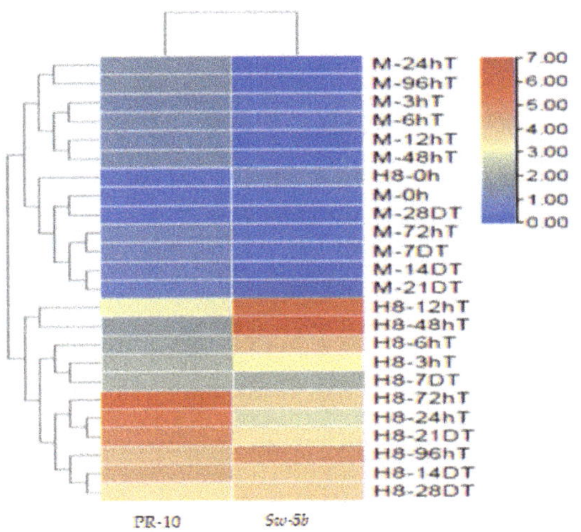


**Figure 15.** Heat map of PR-10 protein and *Sw-5b* at different time points after TSWV inoculation. *PR-10* and *Sw-5b* genes are highly expressed in H8 plants compared to M82 plants.

## 3. Discussion

To confirm that the PR-10 proteins found in tomatoes are related to pathogenesis-related proteins from the PR-10 family, a general NCBI-BLAST was performed using their deduced amino acid (a.a) sequence as the query of FASTA search. PR-10 protein sequences are near in molecular weight (17–18 kDa) and amino acid length (157–166 a.a) with a pI value of 4.69–6.17 and a Bet_ V_ 1 (PF000407) domain structure. The subcellular localization of the PR-10 protein was predicted to be cytoplasmic (Table 1).

To characterize and identify potential functional relationships between the PR-10 proteins of tomato, a phylogenetic tree was constructed and 45 gene IDs of PR-10 protein identified. The neighbor-joining method, with 1000 bootstrap reconstruction and completed deletion gaps/missing data, yielded five known subfamilies. The subsequent phylogenetic analysis of all forty-five PR-10 nucleotide and protein sequences were carried out using the maximum-likelihood method with 1000 bootstraps (Figure 1).

Although PR-10 proteins are mostly recognized for their roles in plant defense in response to biotic and abiotic stress, the overall biological function of many PR-10 members are yet unknown. We investigated the *PR-10* genes to better understand the evolutionary relationship between tomato genes and proteins. The findings of this work serve as a theoretical foundation for future research into the structural relationships that define pathogenesis-related proteins [22]. To fight broad pathogenic activity, plants have evolved several defensive mechanisms. Among other strategies, they create antibiotic chemicals such as phytoalexins [23]; and multiple pathogens such as viruses, bacteria, and fungi induce the expression of several genes, as well as the formation of ethylene and salicylic acid in the plant after pathogen infection, resulting in a stress response [24]. The induced genes inform the plants to generate pathogenesis-related (PR) proteins, which are involved in a defensive mechanism [25]. The level at which new cases of a disease appear in a plant population within a given period is referred to as incidence and the disease intensity is calculated using an interval scale measure, which has been used to quantify a disease severity index (DSI) on a percentage basis [26]. We measured different types of diseased leaf samples at different time points and estimated disease incidence and index level (Table 3) with the information contained in Table 4.

The PR-10 family is a critical element of pathogenesis-related proteins, which aid plant defense against biotic and abiotic stress. For example, several tobacco PRs have been identified as chitinases and β-1-3 glucanases with antifungal activity [27]. Multigene families code for PR-10 proteins. This may be the reason for the multifunctional existence of these ancient proteins, which spent time in a phase known as protein promiscuity for mutations and functions [28]. The cAMP-dependent protein kinase, casein kinase II, and protein kinase C also have phosphorylation sites on several of the PR-10 proteins [29].

Tomato spotted wilt virus (TSWV) is transmitted by thrips and has become one of the most important viral vector complexes for agriculture and food defense [30]. One of the most effective approaches to decrease viral diseases is to grow virus resistant cultivars. This technique's effectiveness is dependent on the presence of resistant genes in either cultivated or wild relatives. The most prevalent mechanism of natural plant resistance to viral infection is the hypersensitive response [31]. HR causes viral invasion-associated cells to die quickly, reducing viral cell-to-cell multiplication and the virus's subsequent spread throughout the plant. The hypersensitive reaction is caused by the virus's exact identification, which is based on comparable plant and viral gene products (HR). In Tospoviruses, particularly TSWV, the dominant genes of *Sw-5* are the principal sources of HR-based resistance [30]. *Sw-5* clustered proteins were discovered to be part of the resistance (R) gene family, encoded by the amino-terminal *Solanaceae* domain (SD) and coiled-coil domain (CC), a core nucleotide binding-adapter shared by APAF-1, R proteins, and a CED-4 (NB-ARC), and a leucine-rich repeat (LRR) domain [32]. Multiple homologs have been found in the tomato genome; nevertheless, *Sw-5b* provides broad and long-lasting resistance and has been extensively researched because of this functionality [11]. The NSm encoded in the TSWV M segment is the TSWV product that triggers the resistance response (*Avr* Determinant) of *Sw-5b*-mediated resistance [11,32,33]. *Sw-5* containing tomato cultivar have L and S segments, as well as the M segment form a TSWV resistance-inducing (RI) isolate and a TSWV resistance-breaking (RB) isolate [34]. Because H8 plants had *Sw-5b* resistant genes, PR-10 protein expression was higher in H8 plants compared to M82 plants (Figure 10); nevertheless, TSWV-cp protein expression was lower in H8 plants, as seen in Figure 11. However, due to these resistance genes (*Sw-5b*) (Figure 12), *Sw-5b* was substantially expressed in H8 plants.

The viral small interfering RNAs (vsiRNAs) originating from the TSWV genome have been discovered to potentially target host genes involved in a variety of processes, including plant–pathogen interactions [35]. As a result, they experience much plant protection and autoimmunity. Plant pattern-recognition receptors (PRRs) sense the virus that has infected the plants by recognizing pathogen-associated molecular patterns (PAMPs). The plants' first line of defense against an immunological response is PAMP-triggered immunity (PTI). On the other hand, rapid pathogen effectors hinder the PTI response. PAMP-triggered immunity (PTI) and effector-triggered immunity (ETI) are two types of immune systems found in plants. PTI is elicited by pathogen/microbe-associated molecular patterns (PAMPs/MAMPs), while ETI is elicited by disease resistance plant proteins (R), which are highly effective for disease resistance reactions upon precise identification of pathogen effectors, and both PTI and ETI cause local immune responses. During virus invasion, nucleotide-binding leucine-rich repeat receptors (NLRs) recognize specific pathogen effectors and induce effector-triggered immunity (ETI), which triggers a long-distance protective mechanism known as SAR (systemic acquired resistance) in plants [36].

NLRs (nucleotide-binding leucine-rich repeat receptors) identify unique pathogen effectors during virus invasion and cause effector-triggered immunity (ETI) in the plant [37]. Because of their distinct N-terminal configurations, plant NLRs are known as coiled–coil (CC)-NLRs (CNLs) or Toll/interleukin-1 (TIR)-NLRs (TNLs) [38]. The NLRs specifically or implicitly recognize pathogen effectors and initiate a hypersensitive cell death response (HR) aimed at limiting TSWV infection to the site of infection [39]. TSWV-cp protein was highly expressed in M82 sensitive plants, but PR-10 protein and *Sw-5b* were highly expressed in H8 plants leave and roots in different types of disease index levels (Figures 13 and 14), and PR-10 protein and *Sw-5b* are highly correlated for TSWV disease resistance in the plants (Figure 15). The tomato *Sw-5b* belongs to the coiled–coil leucine-rich repeats (CNLs) and also auto-activated. *Sw-5b* is a typical class of proteins, CC-NB-ARC protein, and *Sw-5b* has broad-spectrum resistance that its SD domain can specifically recognize a 21- amino acid (21-aa) in NSm of TSWV. The region of NSm is highly conserved in the American-type Tospoviruses only, and not in Euro-Asian-type Tospoviruses [39]. At the moment, the *Sw-5b* genes have been discovered in various tomato germplasm materials by genome sequencing, and the sequence variations between the genes of different tomato plants are very large [40]. TSWV-specific NSm attaches to the expanded SD domain of the *Sw-5b* protein, activating the switch that triggers the receptor and HR, resulting in a vigorous defensive reaction against Tospoviruses [40]. The CNLs (including *Sw-5b* protein) from tomato, on the other hand, detect NSm and NSs with robust effector-triggered immunity (ETI) and activate the hypersensitive cell death reaction (HR). Salicylic acid (SA), jasmonic acid (JA), and ethylene (ET) play critical roles in PTI and ETI immunity, along with helping plants to establish systemic acquired resistance (SAR) [41]. The SA signaling pathway is essential in tomato plants for basal protection against TSWV [42]. The SA accumulates in the contaminated regions, where it then triggers the rapid activation of transcriptional resistance (R) genes [43]. These findings suggest that PR-10 protein and its involvement with *Sw-5b* have a resistance mechanism against TSWV to protect against these devastating viruses in tomatoes, which also establish systemic acquired resistance (SAR) in the plants.

## 4. Materials and Methods

### *4.1. Materials*

Forty-five amino acid sequences of PR-10 encoding genes were used to analyze the genome-wide identification of the PR-10 superfamily. M82 and H8 tomato seeds were used in this experiment as a sensitive and resistant material because M82 was included as a TSWV susceptible cultivar, as reported by [44], and H8 was included as a TSWV resistant cultivar, as reported by the Laboratory of Tomato Quality and Stress Tolerance Regulation Mechanism and Genetic Improvement, College of Horticulture, Northwest A and F University, China.

## 4.2. Methods

### 4.2.1. Identification and Sequence Analysis

The PR-10 protein amino acid sequences were used as a reference point for a number of database searches against the Phytozome database proteome files. (https://phytozome-next.jgi.doe.gov/) accessed on 23 May 2021. The National Center for Biotechnology Information (NCBI) provided isolated versions of BLASTP, which were utilized with an e-value threshold of $1e^{-10}$ [45]. The candidate sequences were further screened by searching for the PR-10 domain in tomato by using PFAM (http://pfam.xfam.org/) accessed on 23 May 2021 and SMART (http://smart.embl-heidelberg.de/) accessed on 23 May 2021. The characteristic of PR-10 protein, including protein length, isoelectric point (pI), molecular weight ($M_W$), and grand average of hydropathicity (GRAVY) were predicted by ExPASy ProtParam (http://web.expasy.org/protparam/) accessed on 23 May 2021. The signal peptides and transmembrane (TM) domains were predicted with SignalP 4.1 (http://www.cbs.dtu.dk/services/SignalP/) accessed on 23 May 2021.

### 4.2.2. Phylogenetic Analysis

To compare the evolutionary relationships and identify the subfamilies of PR-10, proteins were used to construct the phylogenetic tree using MEGA-X with the neighbor-joining (NJ) method [46]. The phylogenetic tree was then visualized by iTol (https://itol.embl.de/) accessed on 24 May 2021.

### 4.2.3. Gene Structure, Conserved Motifs, and Chromosome Mapping Analysis

Gene Structure Display Server 2.0 software (http://gsds.cbi.pku.edu.cn/) accessed on 23 May 2021 was used to investigate the exon–intron organizations of PR-10 genes based on their information given in the Phytozome database. The novel motifs of PR proteins were searched using MEME 5.0.3 (http://meme-suite.org/tools/meme) accessed on 23 May 2021. The parameters were set as follows: the site distribution was any number of repetitions (anr); the number of motifs was 20; the width of the motif was limited to between 10 and 30, and other optional parameters remained as the default. The combination of gene structures, motifs, and the phylogenetic tree was then generated using the iTol tool. The distributions of *PR-10* genes on tomato chromosome mapping were illustrated with MapInspect 1.0 (http://mapinspect.software.informer.com/) accessed on 24 May 2021.

### 4.2.4. GO Analysis

Biological processes, cellular components, and molecular function are the three categories of gene ontology (GO). Using the PANNZER2 web server, the *PR-10* genes were investigated for their role in GO (http://ekhidna2.biocenter.helsinki.fi/sanspanz/) accessed on 23 May 2021.

### 4.2.5. Three-Dimensional Structure of the PR-10 Protein

A Protein Homology/Analogy Recognition Engine v2 (Phyre2) server was used to generate the anticipated 3D structures. (http://www.sbg.bio.ic.ac.uk/~phyre2/html/page.cgi?id=index) accessed on 23 May 2021.

### 4.2.6. Transcriptomic Analysis

Genome-wide expression data from tomatoes were utilized to reveal the expression patterns of the PR-10 protein family in different tissues and developmental periods. Transcriptomic data from several different plant tissues, including leaves, stems, roots, and various developmental stages of tomato fruit, were downloaded from the Sol genomics network (https://solgenomics.net/) accessed on 30 May 2021, and then TB tools software was used to analyze the expression levels of *PR-10* genes in various tissues of tomato plants by using 45 gene IDs of PR-10 protein.

#### 4.2.7. TSWV Virus Solution Stock Preparation

TSWV virus solution stock was prepared according to the method of [30]. A 0.5 g diseased leaf sample was ground in a mortar and pestle and mixed with a buffer solution (pH 7.0) of 10 mL (0.1 mol/LH24 $PO_4$)/$Na_2HPO_4$; 2% poly-vinyl pyrolidon; and 0.2% $Na_2SO_3$. The TSWV virus solution was kept at −80 °C prior to further analysis.

#### 4.2.8. Plant Materials and Growth Conditions

To minimize pests and disease incidence during seedling growth and development, the experiment was conducted in the greenhouse at the College of Horticulture Northwest Agriculture and Forestry University in Yangling, China. The seeds were germinated for 3 days on moist filter paper in Petri dishes at 28 °C in the dark. The seedlings were transported to a growth chamber with a 16 h light/8 h dark photoperiod and a temperature cycle of 25/16 °C. The plants were moved to a solar greenhouse at Northwest A and F University in Yangling, Shaanxi Province, China, at the four true-leaf stage. The plants were cultivated under a 22/18 °C day/night temperature during the TSWV inoculation.

#### 4.2.9. Virus Inoculation

The TSWV virus was inoculated into six-week-old seedlings. A syringe was used to inject the TSWV virus. The injections were made with a 1.0 mL syringe at three sites on the leaf: border, middle, and base, covering around 0.5 $cm^2$ of leaf area. The presence of the virus was determined visually and confirmed by PCR.

#### 4.2.10. Treatment and Sampling

Inoculated treatment (TR) and non-inoculated (CK) seedlings were used to split tomato cultivars into two groups. At 0, 3, 6, 12, 24, 48, 72, 96 h, and 7, 14, 21, 28 days, samples were taken from the control and treatment groups. TSWV infection of tomato leaves was identified by PCR using TSWV-$_{CP}$ specific primers. At 14, 21, and 28 days after inoculation, the incidence and disease index of TSWV were evaluated, along with the disease symptoms of susceptible and resistant plants, to determine the relative expression of TSWV-cp, PR-10, and *Sw-5b*. There were 20 plants in each treatment. For this experiment, three biological replicates were used. Viral disease classification standard records disease index level in Table 4 [47].

#### 4.2.11. Investigation of Incidence Rate and Disease Index

The incidence of TSWV was observed at 14, 21, and 28 days after inoculation, and the disease symptoms of susceptible and resistant plants were observed.

Incidence rate (%) = (number of infected plants / total number of treated plants) × 100.

Measurement of Disease index = (number of disease plants at all levels × number of representatives at all levels)/(total number of investigated plants × maximum series)] × 100 (at 14, 21, and 28 days after inoculation).

#### 4.2.12. RNA Extraction and qRT-PCR

Leaf samples were collected from three TSWV-inoculated and three uninoculated plants. The leaves were directly frozen in liquid nitrogen and kept at −80 °C for further analysis. Total RNA was derived from tomato leaves using an Omega plant RNA kit, and cDNA was generated using an M-MuLV reverse transcriptase kit (Thermo Scientific, USA). For TSWV-cp protein confirmation, the cDNA samples were amplified using PCR: 95 °C for 5 min, 35 cycles of 95 °C for 30 s, 53 °C for 30 s, 72 °C for 3.5 min, and 72 °C for 10 min. Another cDNA was also synthesized for qPCR using the Evo M-MLV RT Kit (Accurate Biotechnology, Hunan). PCR was used to amplify the cDNA samples: 42 °C for 2 min, 37 °C for 15 min, and 85 °C for 5 s. Thermocycler iQ5 Real-Time PCR Detection Device (BIO-RAD Corp., Hercules, CA, USA) and SYBR Green Pro Taq HS qPCR (Accurate Biotechnology, Hunan) were then used to operate qRT-PCR according to manufacturer guidance. The PCR conditions were as follows: 95 °C for 30 s, accompanied by 40 cycles of 95 °C for 5 s, 60 °C for 30 s, and 72 °C for 20 s. As an internal control, the tomato *actin* gene was used. The qRT-PCR data were analyzed using the $2-\Delta\Delta_{CT}$ method [48]. The primers are listed in Table S1.

#### 4.2.13. Statistical Analysis

GraphPad Prism version 7.00 for Windows (one-way ANOVA followed by Dunnett test) GraphPad Software, San Diego, California, USA, www.graphpad.com, accessed on 23 May 2021, was used to plot and statistically analyze the graphs. A value of $p > 0.05$ was regarded as statistically significant; * $p = 0.05$, ** $p = 0.01$, *** $p = 0.001$, **** $p = 0.0001$, and ns = non-significant. Correlation coefficients were determined by SPSS 25.0 (IBM, Armonk, NY, USA).

## 5. Conclusions

We found 45 PR-10 protein superfamily candidate genes with lengths ranging from 64 to 210 amino acid residues and molecular weights ranging from 7.6 to 24.4 kDa. The 5 subfamilies of PR-10 protein was discovered, as well as 9 conserved motifs. On ten of the twelve chromosomes, *PR-10* genes have been found. Gene ontology research revealed six biological processes and molecular activities, as well as nine cellular components, as well as the greatest level of expression in the Solyc09g090980 *PR-10* gene compared to other tomato tissues. TSWV infection strongly induced *PR-10* and *Sw-5b* gene transcription and activity in tomato leaves, H8 plants having the highest significant expression of PR-10 at 24 h (6.74-fold) and *Sw-5b* gene at 7 days (4.14-fold) compared to control after the TSWV inoculation, TSWV inoculated M82 plants showed significant expression of *PR-10* gene at 72 h (5.70-fold) and no significant expression of *Sw-5b* gene all the time, compared to control. The expressions of *PR-10* and *Sw-5b* genes showed highly significant correlations in H8 plants after the inoculation of TSWV at different time points and also heat map showed that these two genes may participate in regulating the defense response after the inoculation of TSWV in tomatoes. Thus, we conclude that *PR-10* and *Sw-5b* genes have an important role in the defense response after the infection of TSWV.

**Supplementary Materials:** The following supporting information can be downloaded at: https://www.mdpi.com/article/10.3390/ijms23031502/s1.

**Author Contributions:** Conceptualization, M.M.I. and Y.L.; writing and original draft preparation, M.M.I.; editing, S.Q., S.Z., B.A., V.Y., A.H.E.-S. and F.Z.; supervision, Y.L.; project administration, Y.L. All authors have read and agreed to the published version of the manuscript.

**Funding:** This work was supported by the National Key Research and Development Program of Stem Cell and Translational Research (2016YFD0101703) and the Shaanxi Province Key Research and Development Plan Project (2019ZDLNY03-05).

**Institutional Review Board Statement:** Not applicable.

**Informed Consent Statement:** Not applicable.

**Data Availability Statement:** The data presented in this study are available on request from the corresponding author.

**Acknowledgments:** The authors are extremely thankful to the Liang Yan, College of Horticulture, Northwest A and F University, Yangling, Shaanxi, 712100, China, for providing necessary facilities during research work.

**Conflicts of Interest:** The authors declare no conflict of interest.

## References


1. Ali, S.; Mir, Z.A.; Tyagi, A.; Bhat, J.A.; Chandrashekar, N.; Papolu, P.K.; Rawat, S.; Grover, A. Identification and comparative analysis of Brassica juncea pathogenesis-related genes in response to hormonal, biotic and abiotic stresses. *Acta Physiol. Plant.* **2017**, *39*, 268. [CrossRef]
2. Boccardo, N.A.; Segretin, M.E.; Hernandez, I.; Mirkin, F.G.; Chacón, O.; Lopez, Y.; Borrás-Hidalgo, O.; Bravo-Almonacid, F.F. Expression of pathogenesis-related proteins in transplastomic tobacco plants confers resistance to filamentous pathogens under field trials. *Sci. Rep.* **2019**, *9*, 2791. [CrossRef]
3. Sinha, R.K.; Verma, S.S.; Rastogi, A. Role of Pathogen-Related Protein 10 (PR 10) under abiotic and biotic stresses in plants. *Phyton* **2020**, *89*, 167. [CrossRef]
4. Jain, S.; Kumar, A. The pathogenesis related class 10 proteins in plant defense against biotic and abiotic stresses. *Adv. Plants Agric. Res.* **2015**, *3*, 00077. [CrossRef]
5. Hanssen, I.M.; Lapidot, M.; Thomma, B.P. Emerging viral diseases of tomato crops. *Mol. Plant Microbe Interact.* **2010**, *23*, 539–548. [CrossRef] [PubMed]
6. Sevik, M.A.; Arli-Sokmen, M. Estimation of the effect of Tomato spotted wilt virus (TSWV) infection on some yield components of tomato. *Phytoparasitica* **2012**, *40*, 87–93. [CrossRef]
7. Newman, S.; Pottorff, L. *Recognizing Tomato Problems*; Colorado State University Libraries: Fort Collins, CO, USA, 2013.
8. Padmanabhan, C.; Ma, Q.; Shekasteband, R.; Stewart, K.S.; Hutton, S.F.; Scott, J.W.; Fei, Z.; Ling, K.-S. Comprehensive transcriptome analysis and functional characterization of PR-5 for its involvement in tomato Sw-7 resistance to tomato spotted wilt tospovirus. *Sci. Rep.* **2019**, *9*, 7673. [CrossRef]
9. Stevens, M.; Lamb, E.; Rhoads, D. Mapping the Sw-5 locus for tomato spotted wilt virus resistance in tomatoes using RAPD and RFLP analyses. *Theor. Appl. Genet.* **1995**, *90*, 451–456. [CrossRef] [PubMed]
10. Spassova, M.I.; Prins, T.W.; Folkertsma, R.T.; Klein-Lankhorst, R.M.; Hille, J.; Goldbach, R.W.; Prins, M. The tomato gene Sw5 is a member of the coiled coil, nucleotide binding, leucine-rich repeat class of plant resistance genes and confers resistance to TSWV in tobacco. *Mol. Breed.* **2001**, *7*, 151–161. [CrossRef]
11. De Oliveira, A.S.; Boiteux, L.S.; Kormelink, R.; Resende, R.O. The Sw-5 gene cluster: Tomato breeding and research toward orthotospovirus disease control. *Front. Plant Sci.* **2018**, *9*, 1055. [CrossRef]
12. Roselló, S.; Díez, M.J.; Nuez, F. Genetics of tomato spotted wilt virus resistance coming from Lycopersicon peruvianum. *Eur. J. Plant Pathol.* **1998**, *104*, 499–509. [CrossRef]
13. Kim, J.-H.; Kim, Y.-S.; Jang, S.-W.; Jeon, Y.-H. Complete genome sequence of Tomato spotted wilt virus from paprika in Korea. *Int. J. Phytopathol.* **2013**, *2*, 121–136. [CrossRef]
14. Gupta, R.; Kwon, S.-Y.; Kim, S.T. An insight into the tomato spotted wilt virus (TSWV), tomato and thrips interaction. *Plant Biotechnol. Rep.* **2018**, *12*, 157–163. [CrossRef]
15. Leventhal, S.S.; Wilson, D.; Feldmann, H.; Hawman, D.W. A Look into Bunyavirales Genomes: Functions of Non-Structural (NS) Proteins. *Viruses* **2021**, *13*, 314. [CrossRef]
16. Hedil, M.; Kormelink, R. Viral RNA silencing suppression: The enigma of bunyavirus NSs proteins. *Viruses* **2016**, *8*, 208. [CrossRef]
17. Guo, Y.; Liu, B.; Ding, Z.; Li, G.; Liu, M.; Zhu, D.; Sun, Y.; Dong, S.; Lou, Z. Distinct mechanism for the formation of the ribonucleoprotein complex of tomato spotted wilt virus. *J. Virol.* **2017**, *91*, 23. [CrossRef]
18. Rybicki, E.P. A Top Ten list for economically important plant viruses. *Arch. Virol.* **2015**, *160*, 17–20. [CrossRef]
19. Adams, M.J.; Lefkowitz, E.J.; King, A.M.; Harrach, B.; Harrison, R.L.; Knowles, N.J.; Kropinski, A.M.; Krupovic, M.; Kuhn, J.H.; Mushegian, A.R. Changes to taxonomy and the International Code of Virus Classification and Nomenclature ratified by the International Committee on Taxonomy of Viruses (2017). *Arch. Virol.* **2017**, *162*, 2505–2538. [CrossRef]
20. Kerényi, Z.; Mérai, Z.; Hiripi, L.; Benkovics, A.; Gyula, P.; Lacomme, C.; Barta, E.; Nagy, F.; Silhavy, D. Inter-kingdom conservation of mechanism of nonsense-mediated mRNA decay. *EMBO J.* **2008**, *27*, 1585–1595. [CrossRef]
21. Anfoka, G.H.; Abhary, M.; Stevens, M.R. Occurrence of Tomato spotted wilt virus (TSWV) in Jordan. *EPPO Bull.* **2006**, *36*, 517–522. [CrossRef]
22. Swoboda, I.; Hoffmann-Sommergruber, K.; O'Ríordáin, G.; Scheiner, O.; Heberle-Bors, E.; Vicente, O. Bet v 1 proteins, the major birch pollen allergens and members of a family of conserved pathogenesis-related proteins, show ribonuclease activity in vitro. *Physiol. Plant.* **1996**, *96*, 433–438. [CrossRef]
23. Hofius, D.; Tsitsigiannis, D.I.; Jones, J.D.; Mundy, J. *Inducible Cell Death in Plant Immunity, Seminars in Cancer Biology, 2007*; Elsevier: Amsterdam, The Netherlands, 2007; pp. 166–187.
24. Hammond-Kosack, K.E.; Jones, J. Resistance gene-dependent plant defense responses. *Plant Cell* **1996**, *8*, 1773. [PubMed]
25. Marković-Housley, Z.; Degano, M.; Lamba, D.; von Roepenack-Lahaye, E.; Clemens, S.; Susani, M.; Ferreira, F.; Scheiner, O.; Breiteneder, H. Crystal structure of a hypoallergenic isoform of the major birch pollen allergen Bet v 1 and its likely biological function as a plant steroid carrier. *J. Mol. Biol.* **2003**, *325*, 123–133. [CrossRef]
26. Yu, H.; Zhang, Q.; Sun, P.; Song, C. Impact of droughts on winter wheat yield in different growth stages during 2001–2016 in Eastern China. *Int. J. Disaster Risk Sci.* **2018**, *9*, 376–391. [CrossRef]
27. Takahashi, M.; Shigeto, J.; Izumi, S.; Yoshizato, K.; Morikawa, H. Nitration is exclusive to defense-related PR-1, PR-3 and PR-5 proteins in tobacco leaves. *Plant Signal. Behav.* **2016**, *11*, e1197464. [CrossRef]

28. Tokuriki, N.; Tawfik, D.S. Stability effects of mutations and protein evolvability. *Curr. Opin. Struct. Biol.* **2009**, *19*, 596–604. [CrossRef]
29. Ziadi, S.L.; Poupard, P.; Brisset, M.-N.; Paulin, J.-P.; Simoneau, P. Characterization in apple leaves of two subclasses of PR-10 transcripts inducible by acibenzolar-S-methyl, a functional analogue of salicylic acid. *Physiol. Mol. Plant Pathol.* **2001**, *59*, 33–43. [CrossRef]
30. Olaya, C.; Fletcher, S.J.; Zhai, Y.; Peters, J.; Margaria, P.; Winter, S.; Mitter, N.; Pappu, H.R. The Tomato spotted wilt virus (TSWV) Genome is Differentially Targeted in TSWV-Infected Tomato (Solanum lycopersicum) with or without Sw-5 Gene. *Viruses* **2020**, *12*, 363. [CrossRef]
31. Goldbach, R.; Bucher, E.; Prins, M. Resistance mechanisms to plant viruses: An overview. *Virus Res.* **2003**, *92*, 207–212. [CrossRef]
32. De Oliveira, A.S.; Koolhaas, I.; Boiteux, L.S.; Caldararu, O.F.; Petrescu, A.J.; Oliveira Resende, R.; Kormelink, R. Cell death triggering and effector recognition by Sw-5 SD-CNL proteins from resistant and susceptible tomato isolines to Tomato spotted wilt virus. *Mol. Plant Pathol.* **2016**, *17*, 1442–1454. [CrossRef]
33. Hallwass, M.; De Oliveira, A.S.; de Campos Dianese, E.; Lohuis, D.; Boiteux, L.S.; Inoue-Nagata, A.K.; Resende, R.O.; Kormelink, R. The T omato spotted wilt virus cell-to-cell movement protein (NSM) triggers a hypersensitive response in S w-5-containing resistant tomato lines and in N icotiana benthamiana transformed with the functional S w-5b resistance gene copy. *Mol. Plant Pathol.* **2014**, *15*, 871–880. [CrossRef] [PubMed]
34. Hoffmann, K.; Qiu, W.; Moyer, J. Overcoming host-and pathogen-mediated resistance in tomato and tobacco maps to the M RNA of Tomato spotted wilt virus. *Mol. Plant Microbe Interact.* **2001**, *14*, 242–249. [CrossRef] [PubMed]
35. Ramesh, S.V.; Williams, S.; Kappagantu, M.; Mitter, N.; Pappu, H.R. Transcriptome-wide identification of host genes targeted by tomato spotted wilt virus-derived small interfering RNAs. *Virus Res.* **2017**, *238*, 13–23. [CrossRef] [PubMed]
36. Poltronieri, P.; Brutus, A.; Reca, I.B.; Francocci, F.; Cheng, X.; Stigliano, E. Engineering plant leucine rich repeat-receptors for enhanced pattern-triggered immunity (PTI) and effector-triggered immunity (ETI). In *Applied Plant Biotechnology for Improving Resistance to Biotic Stress*; Elsevier: Amsterdam, The Netherlands, 2020; pp. 1–31.
37. Panigrahi, G.K.; Sahoo, A.; Satapathy, K.B. Insights to Plant Immunity: Defense Signaling to Epigenetics. *Physiol. Mol. Plant Pathol.* **2020**, 101568. [CrossRef]
38. Saile, S.C.; Jacob, P.; Castel, B.; Jubic, L.M.; Salas-Gonzalez, I.; Bäcker, M.; Jones, J.D.; Dangl, J.L.; El Kasmi, F. Two unequally redundant "helper" immune receptor families mediate Arabidopsis thaliana intracellular "sensor" immune receptor functions. *PLoS Biol.* **2020**, *18*, e3000783. [CrossRef]
39. Zhu, M.; Jiang, L.; Bai, B.; Zhao, W.; Chen, X.; Li, J.; Liu, Y.; Chen, Z.; Wang, B.; Wang, C. The intracellular immune receptor *Sw-5b* confers broad-spectrum resistance to tospoviruses through recognition of a conserved 21-amino acid viral effector epitope. *Plant Cell* **2017**, *29*, 2214–2232. [CrossRef]
40. Li, J.; Huang, H.; Zhu, M.; Huang, S.; Zhang, W.; Dinesh-Kumar, S.P.; Tao, X. A plant immune receptor adopts a two-step recognition mechanism to enhance viral effector perception. *Mol. Plant* **2019**, *12*, 248–262. [CrossRef]
41. Klessig, D.F.; Choi, H.W.; Dempsey, D.M.A. Systemic acquired resistance and salicylic acid: Past, present, and future. *Mol. Plant Microbe Interact.* **2018**, *31*, 871–888. [CrossRef]
42. Zheng, X.; Chen, Y.; Zhao, L.; Chen, Y.; Zheng, L.; Zheng, K.; Mu, Y.; Zhao, X.; Gao, Y.; Zhang, J. Tripartite interactions between jasmonic/salicylic acid pathways, western flower thrips, and thrips-transmitted tomato zonate spot virus infection in Capsicuum annuum. *Arthropod Plant Interact.* **2019**, *13*, 289–297. [CrossRef]
43. Yang, L.; Li, B.; Zheng, X.-y.; Li, J.; Yang, M.; Dong, X.; He, G.; An, C.; Deng, X.W. Salicylic acid biosynthesis is enhanced and contributes to increased biotrophic pathogen resistance in Arabidopsis hybrids. *Nat. Commun.* **2015**, *6*, 7309. [CrossRef]
44. Lee, H.J.; Kim, B.; Bae, C.; Kang, W.-H.; Kang, B.-C.; Yeam, I.; Oh, C.-S. Development of a single-nucleotide polymorphism marker for the *Sw-5b* gene conferring disease resistance to Tomato spotted wilt virus in tomato. *Hortic. Sci. Technol.* **2015**, *33*, 730–736. [CrossRef]
45. Altschul, S.F.; Madden, T.L.; Schäffer, A.A.; Zhang, J.; Zhang, Z.; Miller, W.; Lipman, D.J. Gapped BLAST and PSI-BLAST: A new generation of protein database search programs. *Nucleic Acids Res.* **1997**, *25*, 3389–3402. [CrossRef] [PubMed]
46. Kumar, S.; Stecher, G.; Li, M.; Knyaz, C.; Tamura, K. MEGA X: Molecular evolutionary genetics analysis across computing platforms. *Mol. Biol. Evol.* **2018**, *35*, 1547. [CrossRef] [PubMed]
47. Mo, Y.; Bao, J.; Zhu, H.; Zhao, K. Advance in research on resistance breeding of tomato against Tomato spotted wilt virus. *J. Yunnan Agric. Univ.* **2016**, *31*, 733–737.
48. Livak, K.J.; Schmittgen, T.D. Analysis of relative gene expression data using real-time quantitative PCR and the $2-\Delta\Delta CT$ method. *Methods* **2001**, *25*, 402–408. [CrossRef] [PubMed]

International Journal of
*Molecular Sciences*

MDPI

*Article*

# Combined Transcriptomics and Metabolomics Analysis Reveals the Molecular Mechanism of Salt Tolerance of Huayouza 62, an Elite Cultivar in Rapeseed (*Brassica napus* L.)

**Heping Wan [1,2,†], Jiali Qian [1,†], Hao Zhang [1], Hongchen Lu [1], Ouqi Li [1], Rihui Li [1], Yi Yu [2], Jing Wen [1], Lun Zhao [1], Bin Yi [1], Tingdong Fu [1] and Jinxiong Shen [1,*]**

1 National Key Laboratory of Crop Genetic Improvement, National Engineering Research Center of Rapeseed, College of Plant Science and Technology, Huazhong Agricultural University, Wuhan 430070, China; wanheping@jhun.edu.cn (H.W.); qianjiali@vazyme.com (J.Q.); haozhang1108@163.com (H.Z.); zjluhongchen@163.com (H.L.); oql1119@163.com (O.L.); lirh20212021@163.com (R.L.); wenjing@mail.hzau.edu.cn (J.W.); zhaolun@mail.hzau.edu.cn (L.Z.); yibin@mail.hzau.edu.cn (B.Y.); futing@mail.hzau.edu.cn (T.F.)

2 Hubei Engineering Research Center for Protection and Utilization of Special Biological Resources in the Hanjiang River Basin, School of Life Science, Jianghan University, Wuhan 430056, China; yuyixxx@yeah.net

* Correspondence: jxshen@mail.hzau.edu.cn; Tel.: +86-27-8728-1507; Fax: +86-27-8728-0009

† These authors have contributed equally to this work.

**Abstract:** Soil salinity is one of the most significant abiotic stresses affecting crop yield around the world. To explore the molecular mechanism of salt tolerance in rapeseed (*Brassica napus* L.), the transcriptome analysis and metabolomics analysis were used to dissect the differentially expressed genes and metabolites in two rapeseed varieties with significant differences in salt tolerance; one is an elite rapeseed cultivar, Huayouza 62. A total of 103 key differentially expressed metabolites (DEMs) and 53 key differentials expressed genes (DEGs) that might be related to salt stress were identified through metabolomics and transcriptomics analysis. GO and KEGG analysis revealed that the DEGs were mainly involved in ion transport, reactive oxygen scavenging, osmotic regulation substance synthesis, and macromolecular protein synthesis. The DEMs were involved in TCA cycle, proline metabolism, inositol metabolism, carbohydrate metabolic processes, and oxidation-reduction processes. In addition, overexpression of *BnLTP3*, which was one of the key DEGs, could increase tolerance to salt stress in *Arabidopsis* plants. This study reveals that the regulation mechanism of salt tolerance in rapeseed at the transcriptome and metabolism level and provides abundant data for further in-depth identification of essential salt tolerance genes.

**Keywords:** *Brassica napus*; transcriptomics; metabolomics; salt stress; *BnLTP3*

**Citation:** Wan, H.; Qian, J.; Zhang, H.; Lu, H.; Li, O.; Li, R.; Yu, Y.; Wen, J.; Zhao, L.; Yi, B.; et al. Combined Transcriptomics and Metabolomics Analysis Reveals the Molecular Mechanism of Salt Tolerance of Huayouza 62, an Elite Cultivar in Rapeseed (*Brassica napus* L.). *Int. J. Mol. Sci.* **2022**, *23*, 1279. https://doi.org/10.3390/ijms23031279

Academic Editors: Isabel Marques, Ana I Ribeiro-Barros and José C. Ramalho

Received: 31 December 2021
Accepted: 20 January 2022
Published: 24 January 2022

**Publisher's Note:** MDPI stays neutral with regard to jurisdictional claims in published maps and institutional affiliations.



## 1. Introduction

Soil salinity is one of the most significant abiotic stresses affecting seed germination, crop growth, and productivity [1]. According to a rough estimate, approximately 280 million hectares of agricultural land are affected by salt stress, and this problem continues to worsen [2]. It is predicted that over 50% of all agricultural land will be threatened by 2050 [3]. Thus, understanding the crop salt tolerance mechanisms and breeding high salt-tolerant crops has been of great significance to the sustainable development of world agriculture.

Soil salinity can result in early-occurring osmotic stress and the accumulation of toxic ions in plants [4]. In order to reduce the salt damage, plants have four main mechanisms to cope with salt stress. The first category is osmotic regulation, and there are two osmotic adjustment methods in plants. One is to absorb and accumulate inorganic ions such as $Na^+$, $K^+$, $Cl^-$, $Ca^{2+}$, and $Mg^{2+}$ in the cell, and the other is to accumulate small molecules such as proline, betaine, polyols, sugars, and other osmotic substances in cells [5]. The second category is the ion balance regulation, which regulates the ion balance through ion transporters, and channels, and maintains the stable state of ions in cells and tissues [6]. The third category is that plants produce and accumulate specific macromolecular proteins that are induced and expressed by high salt, such as common osmolarity (OSM) and aquaporins (AQP) and late embryogenesis-abundant protein (LEA) [5]. The fourth category is active oxygen scavenging methods. The scavenging of excessive ROS under high salinity may be attributed to non-enzymatic antioxidant metabolites, including ascorbate, glutathione, and tocopherols, and enzymatic agents, such as catalases (CAT), superoxide dismutase (SOD), ascorbate peroxidase (APX), and glutathione reductase (GR) [7]. A large number of salt-responsive genes have been identified, such as *SNC1*, which encoded a high-affinity $K^+$ transporter (HKT)-type sodium transporter [6], and salt-response genes include transcription factors (TFs), such as ERF and WRKY, and signal-related protein kinase [8]. By combining the interaction mode among the mentioned genes above, several pathways that improved the salt-stress signal have been revealed, such as the salt overly sensitive (SOS) pathway [7–10], the calcium-dependent protein kinase (CDPK) pathway [11], and the mitogen-activated protein kinase (MAPK) pathway [12].

Most crop species, such as rice, maize, and barley, are not salt-tolerant and are adversely affected by high salt stress [1]. Rapeseed (*Brassica napus* L.), one of the most important oil crops all over the world, is regarded as a middle salt-tolerant crop [13]. A series of research advances have been made in researching the genetic mechanism of salt tolerance in rapeseed, and many QTLs and genes related to salt tolerance in rapeseed have been identified. For example, the expression levels of *BnBDC1, BnLEA4, BnMPK3*, and *BnNAC2* were up-regulated under salt stress [14]. Over-expression of *BnaAOX1b* significantly improved seed germination under salt stress [15]. Although a number of genes related to salt tolerance in rapeseed have been identified, the function and regulatory network of the genes are still unclear. Therefore, a more comprehensive and in-depth study on the molecular mechanism of rapeseed salt tolerance is needed.

With the rapid development of high-throughput sequencing technology, transcriptomics analysis (RNA-seq) has become been successfully employed to explore the molecular mechanism of salt tolerance in different crops, such as rice [16,17] and wheat [18]. To date, limited transcriptome information in response to salt stress in rapeseed has been reported. For example, Long [19] performed RNA-seq to identify 163 DEGs at 0, 3, 12, and 24 h after NaCl treatments on rapeseed roots at the germination stage, such as the glycine-rich protein, ERD family proteins, glycosyltransferase family, and ubiquitin-protein ligase. Yong [20] used RNA-seq to perform comparative transcriptome analysis of leaves and roots in response to salt stress in rapeseed, and a total of 582 transcription factors and 438 transporter genes were differentially regulated in both organs in response to salt stress. The successful application of transcriptomics analysis in the gene mining of salt tolerance traits in rapeseed has shown excellent efficiency in gene detection.

Metabolomics is a quantitative analysis of all metabolites in living organisms and finds the relative relationship between metabolites and physiological and pathological changes [21,22]. Metabolomic analysis can detect small molecular substances and exogenous substances, where small molecular substances include endogenous substances in tissues or organs, and the relative molecular mass is <1000 [23]. Metabolites are the final products of cell-regulation processes, and their levels can be regarded as the final response of biological systems to genetic or environmental changes [24]. In recent years, metabolomics research methods have been successfully applied to analyze the salt tolerance mechanism of different crops, such as maize [25,26], barley [27], and peanut [28], and a large number of metabolites have been identified, such as alanine, glutamate, asparagine, glycine-betaine, sucrose, malic acid, trans-aconitic acid, and glucose. For example, when plants respond to salt stress, the osmotic regulators like proline content and soluble sugars such as glucose, fructose, and sucrose in plants increases significantly, and the plant accumulates more energy metabolites such as phenylalanine, aspartic acid, citric acid, and citramalic acid under salt stress [25–28]. However, the metabolites in response to salt stress in rapeseed are still uncle ar.

The Huayouza62 (H62) cultivar, which is considered a salt-tolerant rapeseed variety, has been widely planted in the main production areas of rapeseed in the Yangtze River basin of China, and in saline-alkali land in the Gansu, Xinjiang, and Inner Mongolia provinces of China. In addition, H62 has been extended to countries such as Outer Mongolia and Russia. However, the molecular mechanism of salt tolerance in H62 rapeseed is still unclear. Zhongshuang 11 (ZS11) is a conventional rapeseed variety, which is sensitive to salt stress [29]. In this study, the salt-tolerant variety (H62) and salt-sensitive variety (ZS11) were selected, and RNA-seq and metabolomics analysis methods were employed to analyze key salt stress response genes and metabolites in rapeseed under salt stress at the germination stage. This study provides plenty of valuable information about how rapeseed responds to salt stress at the transcriptome and metabolism level.

## 2. Results

### 2.1. Effects of Different NaCl Concentrations on the Germination of Rapeseed Seeds

The germination of rapeseed was affected under different NaCl concentrations from 0 to 200 mM (0, 100, 150, and 200 mM) (Figure 1A). The majority of seeds, both ZS11 and H62, normally germinated under low NaCl (100 and 150 mM)-concentration treatments. When the concentration reached 200 mM, the seeds of ZS11 did not germinate, while the germination rate of H62 reached 90%. The root length (RL) and hypocotyl length (HL) under a low NaCl concentration (100 mM) were higher than the RL and HL of the control sample (Figure 1B). When NaCl concentration reached 150 mM and 200 mM, the RL and HL of seedlings were significantly lower than the RL and HL of the control (Figure 1C). All the above results suggested that there are significant differences in salt tolerance between the H62 and ZS11. Additionally, H62 could be regarded as a salt-tolerant variety, and ZS11 could be regarded as a salt-sensitive variety.

### 2.2. DEMs Description of Two Rapeseed Varieties under Salt Stress

PCA analysis and PLS-DA analysis were performed to analyze the mass spectrometry data of the treatment group. The results showed that the seven biological repeats data points in the treatment group could basically be collected together and the data points of the samples of the two varieties of the treatment groups could be clearly distinguished in space, indicating that the metabolites of each group of samples were different in terms of species, quantity, and concentration (Figure S1).

Conditions for screening differentially expressed metabolites (DEMs): (1) VIP $\geq 1$; (2) Fold-change $\geq 1.2$ or Fold-change $\leq 0.833$; (3) q-value $< 0.05$. 912, 1251, and 757 differentially expressed metabolites (DEMs) were detected in three groups (H62_T1 vs. H62_CK, H62_T2 vs. H62_CK and ZS11_T1 vs. ZS11_CK), respectively, and most DEMs were uniquely associated with a specific salt treatment (Figure 2A). For example, compared with H62_CK, 196 and 295 DEMs were only specifically up-regulated in H62_T1 and H62_T2, respectively, and 222 DEMs were up-regulated in both H62_T1 and H62_T2. Considering the common DEMs in H62_T1, H62_T2, and ZS11_T1, 124 and 178 DEMs were up-regulated and down-regulated in all the salt treatment including H62_T1, H62_T2 and ZS11_T1, respectively, and these DEMs were considered as the salt-related metabolites in rapeseed (Figure 2B,C). The enrichment analysis of the KEGG pathway showed that most of the DEMs were mainly enriched in metabolic pathways, and the biosynthesis of secondary metabolites (Table S2).

Combining the pathway function enrichment analysis results, the result of DEMs annotation, and the fold-change levels of the two varieties under different salt concentrations, 103 DEMs were screened out (Table S3), and some DEMs are listed in Figure 3. These DEMs can be divided into several categories. The first group is amino acid, including Proline, Arginine, Ornithine, Leucine, Threonine, Aspartate, and Hygroline. Except for Proline, Ornithine, and Hygroline, which were highly induced by salt stress, the content of other amino acids was significantly down-regulated, suggesting that salt stress could inhibit the normal metabolic process of amino acids in rapeseed. The second group was carbohydrate, including Glucose, Fructan, Inositol, Sucrose, Mannose, and Oligoglucan; most of these were up-regulated by salt stress in two rapeseed. The third group was plant hormone, including ABA, NAA, GA, and JA. NAA was up-regulated and ABA, and GA and JA were down-regulated in H62_T1, H62_T2, and ZS11_T1. Several important metabolic intermediates in the TCA cycle were all up-regulated, such as Citrate, Aconitare, and Isocitrate, indicating that rapeseed cloud enhances its energy metabolism pathway under salt stress.

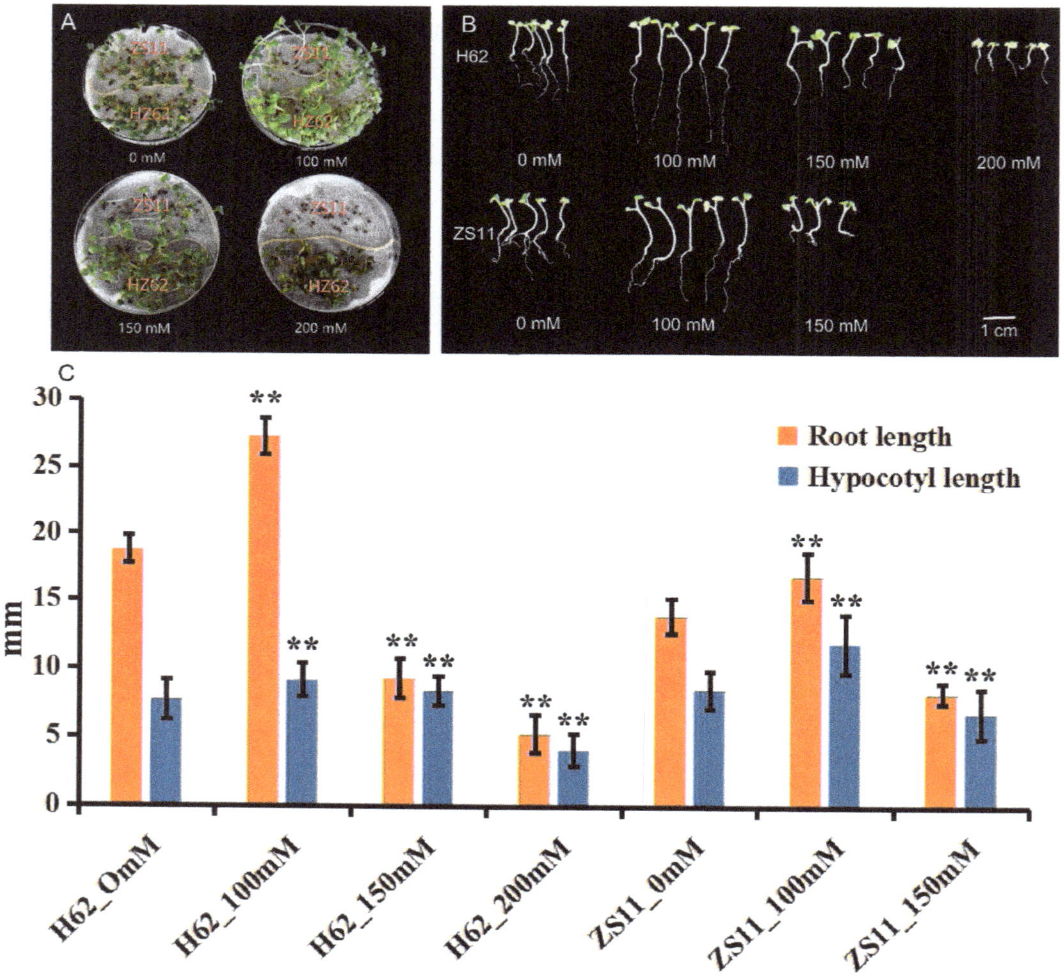


**Figure 1.** The germination and seedlings grow of two rapeseed varieties (H62 and ZS11) under different NaCl concentrations (0, 100, 150 and 200 mM) on the eighth day. (**A**) Comparison of germination of H62 and ZS11 under different salt concentrations; (**B**) Comparison of early seedlings of H62 and ZS11 under different salt concentrations; (**C**) Hypocotyl length and root length of early seedlings of H62 and ZS11 at different salt concentrations. ** indicates that the root length and hypocotyl length of each variety (H62 and ZS11) under salt stress were significantly different from those of the control condition at $p < 0.01$ (Student's $t$ test).

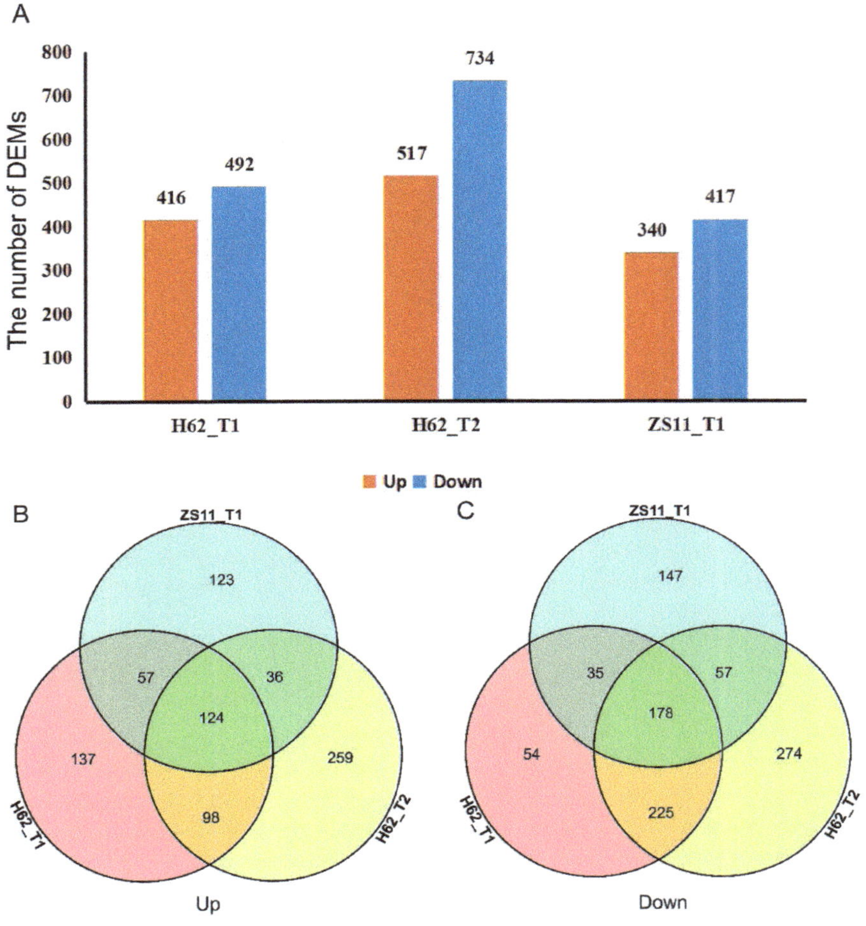


**Figure 2.** The differentially expressed metabolites (DEMs) among the groups for H62 and ZS11 under salt stress (T) and normal (C) conditions at different NaCl concentrations (T1, 150 mM, T2, 200 mM). (**A**) The number of DEMs in three groups. (**B**) Venn diagram of up-regulated DEMs in three groups. (**C**) Venn diagram of down-regulated DEMs in three groups.

### *2.3. Sequencing Output and Assembly*

Under treatment and control conditions, 15 sequencing libraries were constructed from *Brassica napus* H62 and ZS11 samples at three concentrations (0, 150, 200 mM NaCl) (Table 1). 54.03 Mb to 57.14 Mb of total raw reads were obtained from each of the 15 libraries. The software SOAPnuke (v1.5.2) (https://github.com/BGI-flexlab/SOAPnuke, accessed on 25 March 2019) was used to filter reads, followed as: (1) Remove reads with adaptors; (2) Remove reads in which unknown bases (N) are more than 5%; (3) Remove low-quality reads (the percentage of base whose quality is lesser than 15 is greater than 20% in a read). After filtering, it was confirmed that >80% of the sequences were clean reads. The proportion of filtered reads (CRR) exceeded 75%, and most of them exceeded 80%. The proportion of clean reads (TMR) of the reference genome on the comparison exceeded 70%, and the total number of genes (TGN) exceeded 70,000, so the sequence data were sufficient for gene expression analysis.

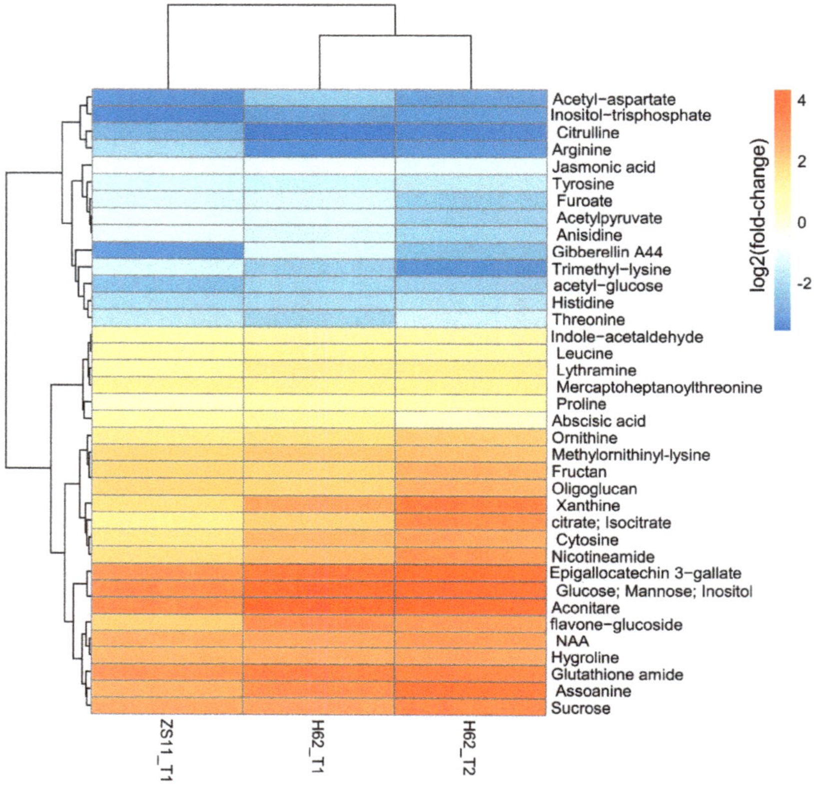


**Figure 3.** The heat map of the part of DEMs related to salt stress base on the $\log_2$Foldchange among the groups for H62 and ZS11 under salt stress (T) and normal (C) conditions at different NaCl concentrations (T1, 150 mM, T2, 200 mM).

**Table 1.** Summary of the sequencing results.

| Sample | TRR | TCR | CRR | TMR | UMR | TGN | KGN |
|---|---|---|---|---|---|---|---|
| H62_CK_1 | 54.04 | 45.39 | 84.01 | 73.44 | 21.63 | 77,996 | 72,412 |
| H62_CK_2 | 54.04 | 44.55 | 82.44 | 71.43 | 21.90 | 79,639 | 73,968 |
| H62_CK_3 | 54.03 | 44.55 | 82.44 | 71.71 | 21.35 | 77,173 | 71,656 |
| H62_T1_1 | 54.03 | 44.18 | 81.76 | 71.6 | 21.01 | 77,378 | 71,886 |
| H62_T1_2 | 54.04 | 44.35 | 82.08 | 73.06 | 20.64 | 77,245 | 71,684 |
| H62_T1_3 | 55.67 | 44.71 | 80.30 | 74.1 | 20.98 | 75,820 | 70,371 |
| H62_T2_1 | 54.03 | 44.84 | 82.98 | 76.88 | 19.52 | 72,893 | 67,757 |
| H62_T2_2 | 54.03 | 44.76 | 82.84 | 75.4 | 21.7 | 75,294 | 69,933 |
| H62_T2_3 | 54.03 | 44.22 | 81.83 | 75.19 | 20.19 | 75,088 | 69,839 |
| ZS11_CK_1 | 54.03 | 44.56 | 82.46 | 74.48 | 21.96 | 76,629 | 71,160 |
| ZS11_CK_2 | 54.04 | 45.22 | 83.69 | 74.13 | 22.2 | 77,841 | 72,342 |
| ZS11_CK_3 | 54.03 | 44.28 | 81.94 | 73.32 | 21.53 | 76,138 | 70,811 |
| ZS11_T1_1 | 54.03 | 44.13 | 81.68 | 75.23 | 22.35 | 75,167 | 69,754 |
| ZS11_T1_2 | 57.14 | 45.15 | 79.00 | 76.26 | 22.34 | 72,940 | 67,636 |
| ZS11_T1_3 | 57.14 | 45.30 | 79.28 | 75.21 | 22.64 | 73,875 | 68,597 |

TRR: Total Raw Reads (Mb); TCR: Total Clean Reads (Mb); CRR: Clean Reads Ratio (%); TMR: Total Mapping Ratio; UMR: Uniquely Mapping Ratio; TGN: Total Gene Number; KGN: Known Gene Number; Seedlings of two genotypes with differential salt tolerance, H62 (tolerant) and ZS11 (sensitive), were sampled for RNA sequencing after exposure to 0 (CK), 150 mM (T1), 200 mM NaCl (T2) for 8 days.

*2.4. General Gene-Expression Description under Salt Stress*

Adjust *p*-value $\leq 0.05$ and a $|\log_2 \text{FoldChange}| \geq 1$ were set as the thresholds to determine the significance of the gene-expression difference between samples. Compared with H62_CK, there were 4277 (Up 1480 and Down 2797) and 10,592 (Up 4166 and Down 6426) DEGs at the treatments of H62_T1 and H62_T2, displaying a rising tendency in DEGs numbers. Compared with ZS11_CK, there were 7745 (Up 3380 and Down 4365) DEGs at the treatments of ZS11_T1 (Figure 4A). Down-regulated genes had greater numbers than those of the up-regulated ones, H62 and ZS11, under different NaCl concentrations. Comparing the number of DEGs of H62 and ZS11 under two salt-concentration conditions, the number of DEGs of H62_T2 and ZS11_T1 were significantly higher than that of H62_T2, providing evidence of gene transcription level for the difference in salt tolerance between H62 and ZS11.

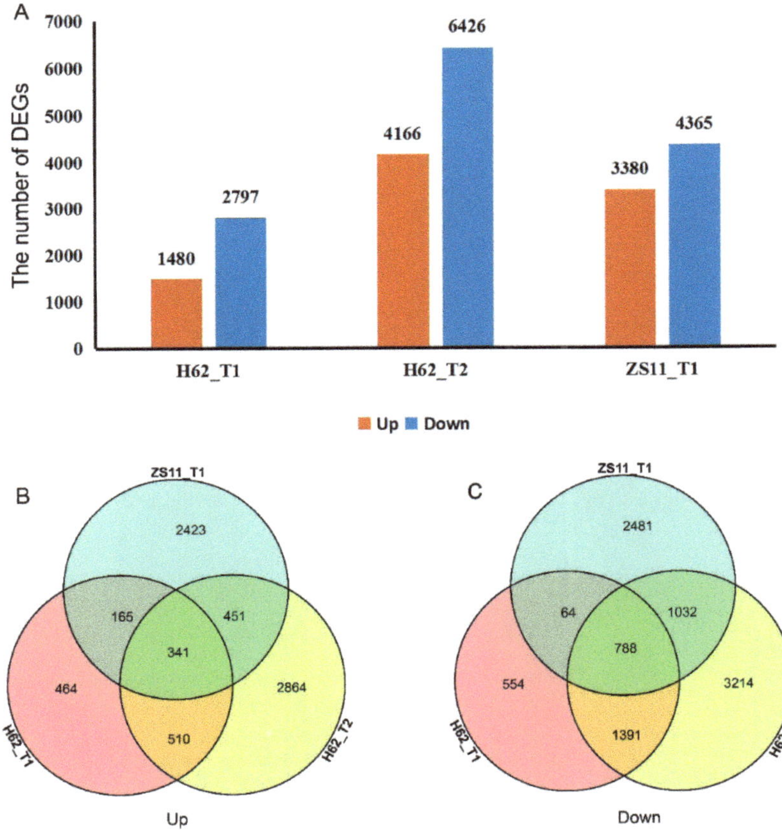


**Figure 4.** The differentially expressed genes (DEGs) among the groups for H62 and ZS11 under salt stress (T) and normal (C) conditions at different NaCl concentrations (T1, 150 mM, T2, 200 mM). (**A**) The number of DEGs in three groups. (**B**) Venn diagram of up-regulated DEGs in three groups. (**C**) Venn diagram of down-regulated DEGs in three groups.

Most DEGs were uniquely associated with a specific salt treatment. For example, compared with H62_CK, 464 and 2864 DEGs were only specifically up-regulated in H62_T1 and H62_T2, respectively, and 851 DEGs were up-regulated in both H62_T1 and H62_T2 (Figure 4B). Considering the common DEGs in H62_T1, H62_T2, and ZS11_T1,

341 and 788 DEGs were up-regulated and down-regulated in all the salt treatments including H62_T1, H62_T2, and ZS11_T1 (Figure 4A–C), and these DEGs were considered as the possible salt-related genes.

### 2.5. GO and KEGG Analysis of DEGs Two Rapeseed Varieties under Different NaCl Condition

Gene Ontology (GO) assignments were used to classify the functions of the DEGs responding to salt stress. Three non-mutually exclusive GO categories, biological process (BP), cellular component (CC), and molecular function (MF), were well represented. The most represented GO terms are presented in Figure 5. In the BP category, the most abundant GO terms were "metabolic process", "cellular process", and "response to stimulus". In the CC category, "cell" was the most abundant, followed by "cell part", "organelle", and "membrane". In the MF category, "catalytic activity" was the most abundant category, followed by "binding", "nucleic acid binding transcription factor activity", "transporter activity", and "antioxidant activity". Notably, some genes were assigned to more than one category.

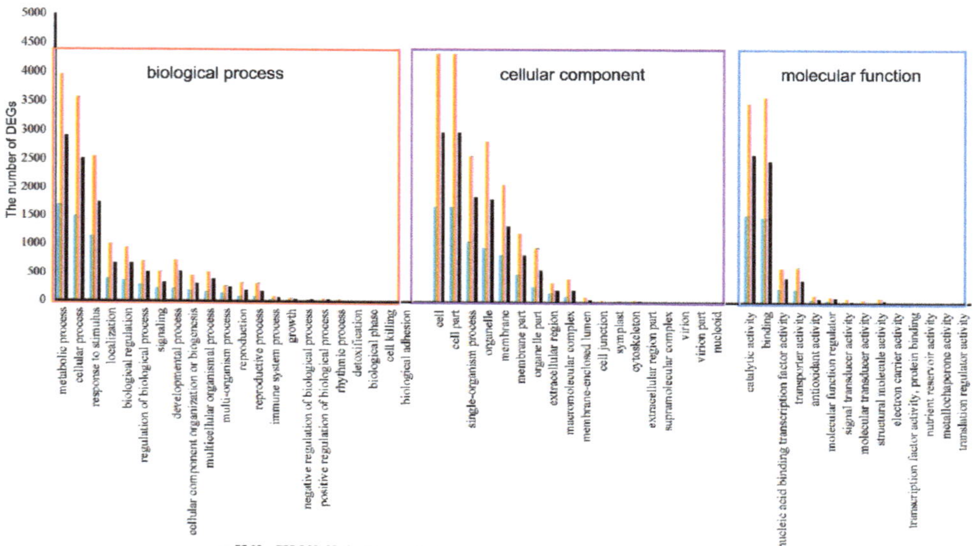


**Figure 5.** Gene Ontology (GO) analysis of common DEGs of two rapeseed varieties under different NaCl conditions. The DEGs were assigned into three groups, including biological process, cellular components, and molecular function. The *x*-axis represents the most abundant categories of each group, and the *y*-axis represents the percentages of the total genes in each category. Only partial results are shown.

According to the results of KEGG, a total of 3261, 7897, and 5856 DEGs could be aligned to the KEGG pathways in H62_T1 vs. H62_CK, H62_T2 vs. H62_CK, and ZS11_T1 vs. ZS11_CK, respectively (Table S4). The pathways with more mapped genes were the metabolic pathways, biosynthesis of secondary metabolites, and starch and sucrose biosynthesis (Figures 6 and S2). In addition, the pathways with more mapped genes were the alanine, aspartate and glutamate metabolism, inositol phosphate metabolism, and fructose, MAPK signaling pathway-plant, and mannose metabolism (Figures 6 and S2).

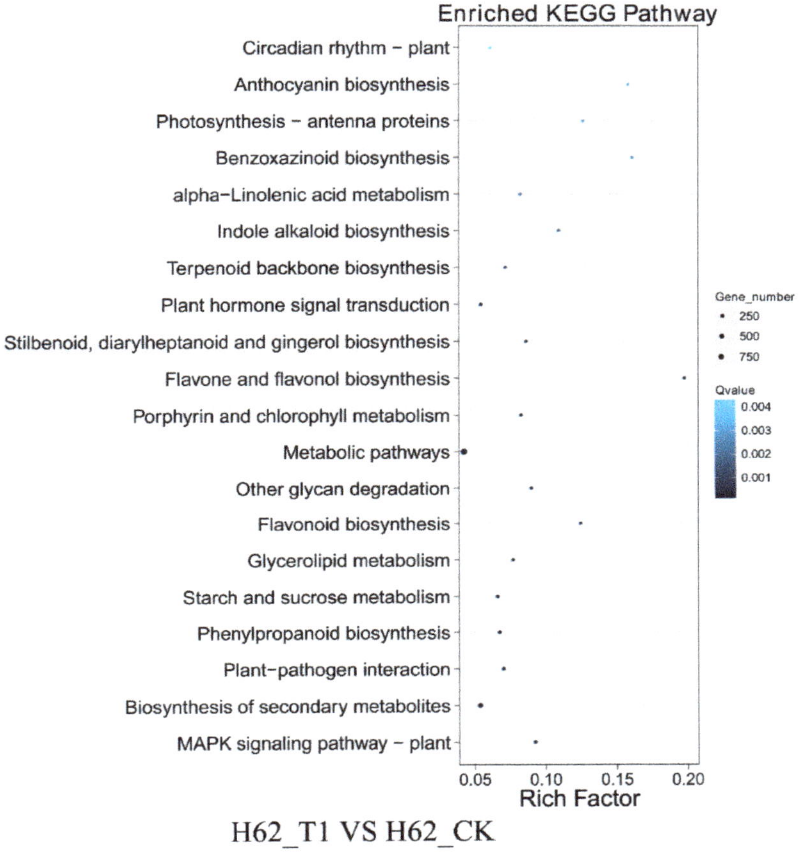


**Figure 6.** Kyoto Encyclopedia of Genes and Genomes (KEGG) pathway analysis of the DEGs in H62_T1 vs. H62_CK. Only partial results are shown in Figure 6.

### 2.6. Identification of Genes Responding to Salt Stress

A total of 341 genes were up-regulated in H62_T1, H62_T2, and ZS11_T1, and a total of 788 genes were down-regulated in H62_T1, H62_T2, and ZS11_T1 (Figure 4). These genes can be considered as responses to salt stress in rapeseed. Combining the pathway function enrichment analysis results, the result of DEGs function annotation, and the expression levels of the two varieties under different salt concentrations, 53 candidate genes were screened out (Table S5), and some candidate genes are listed in Figure 7. These genes can be divided into several categories, such as ion-channel protein gene (*HKT1, CLIC5, NHX2, NHX4, KAT1*), osmotic regulation-related genes (*P5CS, P5C reductase, ProDH1, Sus3, MIPS*), transcription factor (*ERF023, ERF018, ABI3, ETC3, WRKY29, MYB39, bHLH122*), macromolecular protein genes (*RD22, DIR1, PIP1-4, TIP3-2, LEA1, LEA76, Rab18*), reactive oxygen-scavenging enzyme gene (*SOD1, POD34, PER1*), and functional enzyme (*GS1, SAG12, ADH*) (Figure 7, Table S5).

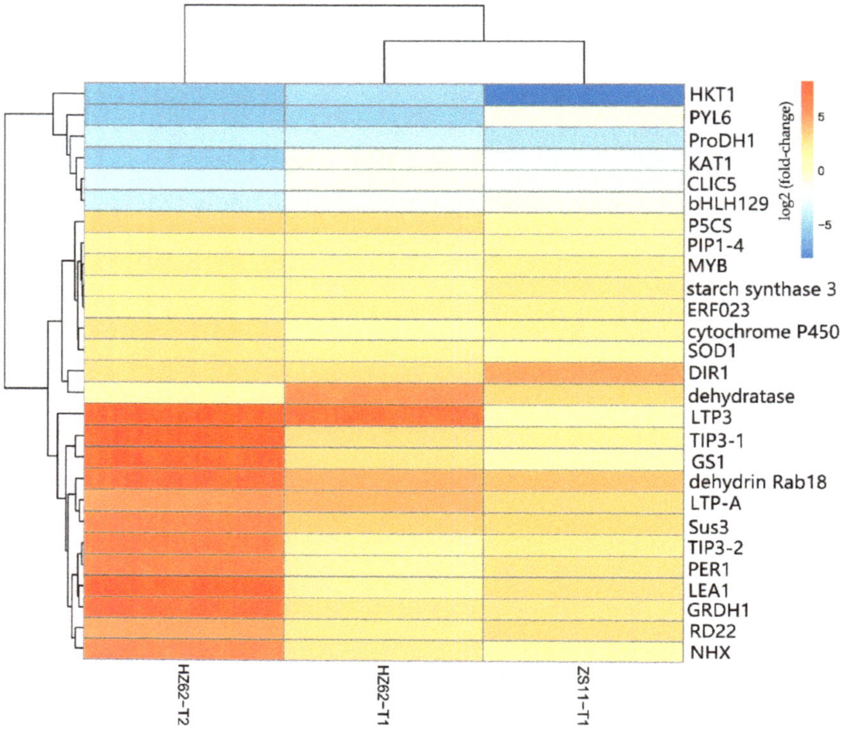


**Figure 7.** The heat map of part of the candidate genes related to salt stress base on the $log_2$Foldchange among the groups for H62 and ZS11 under salt stress (T) and normal (C) conditions at different NaCl concentrations (T1, 150 mM, T2, 200 mM).

### *2.7. Confirmation of DEG Profiles by qPCR Analysis*

In order to validate the reliability of our RNA-seq data, we measured mRNA abundance using qRT-PCR for 6 DEGs. All the six tested genes by qRT-PCR were significantly changed between H62 and ZS11 rapeseed under the different NaCl treatments, which was similar to the result observed through RNA-seq analysis (Figure 8). Although the change folds were not exactly the same as those revealed by the transcriptome-profiling data, all the validated genes showed similar expression patterns considering the DEGs data.

### *2.8. Over-Expression of BnLTP3 Increased Salt Tolerance in Arabidopsis thaliana*

A number of candidate genes homologous to *LTPs* were also identified. The expression levels of lots of *LTPs* were up-regulated significantly as the salt concentration increased in salt-tolerant H62, but was not significantly up-regulated in ZS11 (Figure S3). In order to determine that the *BnLTP3* is related to salt tolerance, it was transferred to *Arabidopsis thaliana*. The result showed no significant difference between the root length of the wild-type and over-expression lines under 0 mM NaCl (Figures 9 and S4). While, the root lengths of the over-expression lines were significantly higher than that of the wild-type under salt stress conditions (50 mM and 100 mM), indicating that *BnLTP3* has the function of improving salt tolerance (Figure 9).

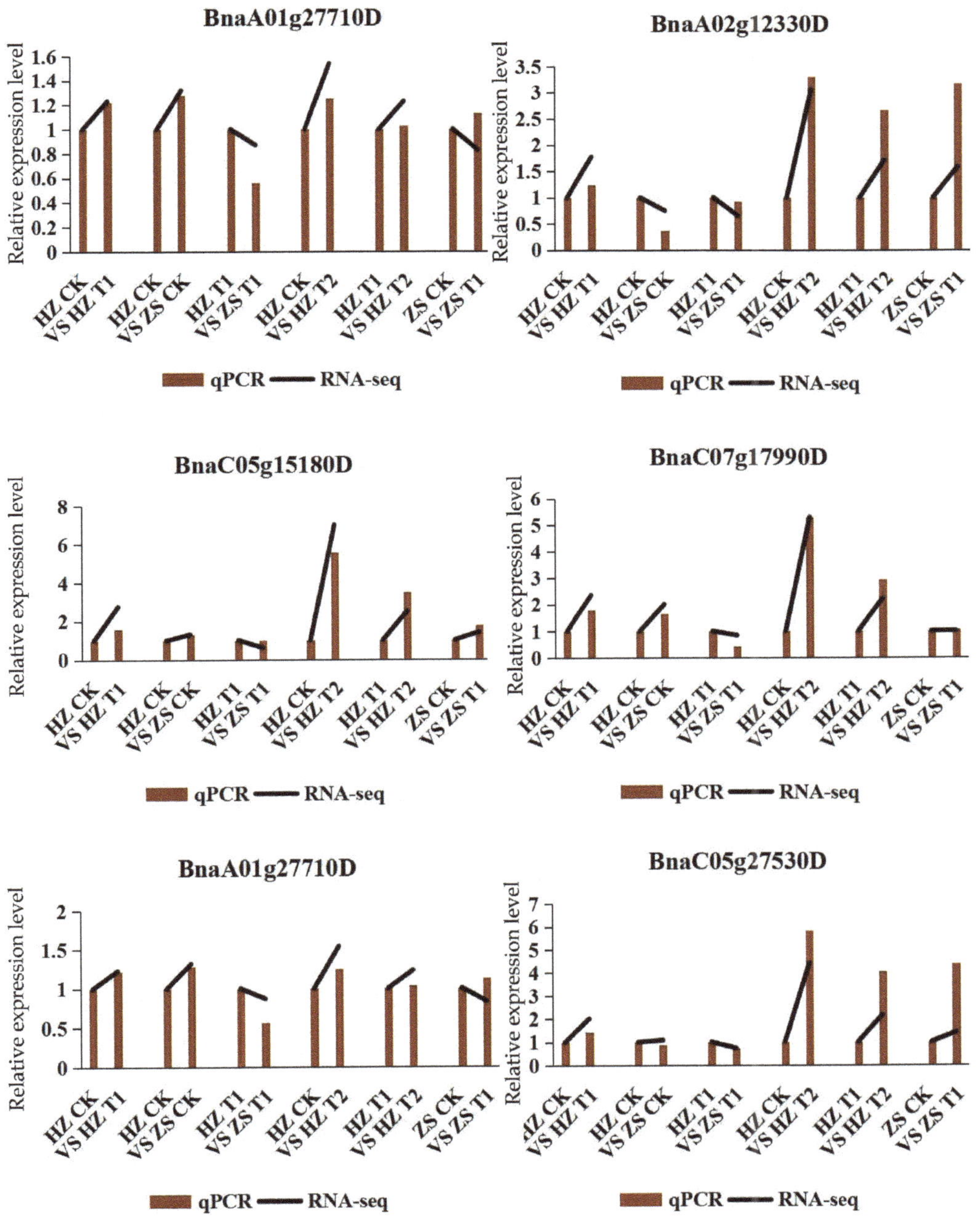


**Figure 8.** The relative expression of qRT-PCR and RNA-seq of six DEGs.

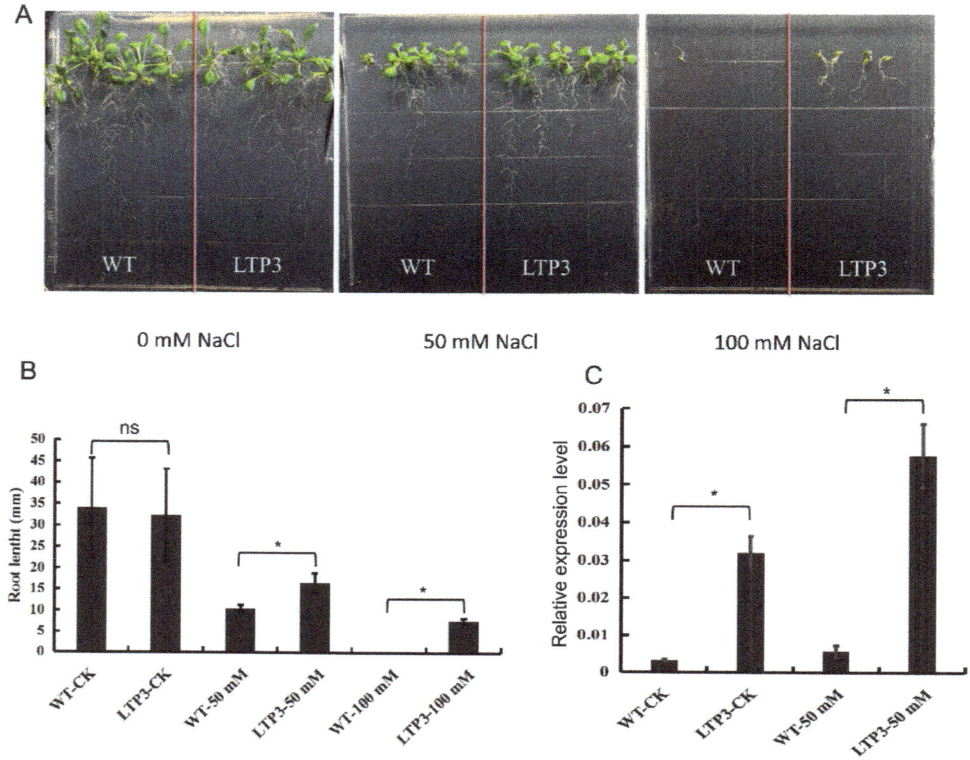


**Figure 9.** Over-expression of *BnLTP3* in *Arabidopsis thaliana.* (**A**) Phenotypes of WT (wild-type) and *LTP3*-overexpressing plants grown on MS and MS + NaCl (50, 100 mM) media. (**B**) Root lengths of WT and *LTP3* overexpressing plants on MS + NaCl medium (50, 100 mM). Values are means ± SD ($n$ = 3); * $p < 0.05$ (Student's $t$ test); ns not significant. (**C**) *LTP3* gene expression level of WT and *LTP3*-overexpressing plants grown on MS and MS + NaCl (50, 100 mM) media.

## 3. Discussion

With the rapid development of modern molecular biology, the research on the mechanism of plant salt tolerance has reached the level of transcriptome, proteome, metabolome, and ionome [30]. The "omics" research provides a powerful method for identifying salt-tolerant genes and the mining of marker metabolites [31]. Unique genetic resources are the genetic basis for omics research. In our study, two distinctive rapeseed varieties (Huayouza 62 and Zhongshuang 11) were selected for transcriptome and metabolome analysis. Huayouza 62 is an elite hybrid rapeseed variety, successfully promoted and planted in China's saline-alkali land due to its relatively strong salt tolerance. Zhongshuang 11 is an elite conventional rapeseed variety; however, it is very sensitive to salt stress [29]. Our results showed a significant difference between H62 and ZS11 in salt tolerance at the early-seedling stage (Figure 1). Therefore, these two rapeseed varieties are ideal resources for identifying salt-tolerant genes and studying salt-tolerant mechanisms in rapeseed.

Plant accumulate inorganic ions such as $Na^+$, $K^+$, and $Ca^{2+}$ in the cell, and accumulate small molecules such as proline, betaine, polyols, and sugars in cells under salt stress [5]. Among them, proline has been studied most in salt stress. The overexpression of *P5CS1*, which is closely related to the proline synthesis pathway, can significantly improved the salt tolerance of plants [32]. In our study, *BnP5CS* (*BnaA03g18760D*) was induced by salt stress in the two rapeseed varieties and the expression of *BnProDH1* (*BnaAnng07910D*), which encodes proline dehydrogenase, was significantly suppressed under salt stress. The down-regulation of *ProDH* and up-regulation of *P5CS* would make glutamate flow more towards proline, which is in agreement with other studies [33,34]. It is worth noting that more proline was detected in the salt-tolerant variety H62 under salt stress, while no significant difference in proline content in the sensitive variety ZS11 was detected, suggesting that the difference in proline synthesis may lead to the different salt tolerance between H62 and ZS11. In addition, increasing the concentration of sucrose in cells could also regulate osmotic stress [35]. The results of transcriptomics and metabolomics analysis showed that the concentration of sucrose increased significantly under salt stress, and the expression of *Sus3* (sucrose synthase 3) was significantly induced by salt stress in the two rapeseed varieties, suggesting that rapeseed can reduce salt damage by synthesizing sucrose in cells. Fructans were also detected to be synthesized in large amounts under salt-stress conditions, and the increase in salt-tolerant material H62 was more than that of ZS11. Inositol is also a small molecule that regulates osmosis in cells. The metabolome results showed that the content of inositol was significantly increased, and the increase in inositol content in H62 was significantly higher than that of ZS11. Myo-inositol-3-phosphate synthase (*MIPS*) catalyzes the first step of myo-inositol biosynthesis, and its over-expression enhances salt stress tolerance in rice [36]. The transcriptome results showed that *BnaC01g00680D* (*MIPS*) was increased significantly by salt stress.

The balance of intracellular $Na^+/K^+$ concentration is the key to ensuring plants' normal physiological metabolism under salt stress [37]. At present, several types of ion transporters and ion channels in plants have been confirmed to be involved in the regulation of plant salt stress, such as the NHX family of vacuolar $Na^+/H^+$ antiporters [38], potassium/sodium transporter (HKT-type) [5], and the potassium channel (KAT) [39]. In this study, some DEGs related to ion balance were identified, such as *NHX4* (*BnaC09g19230D*), *HKT1* (*BnaC02g29240D*), and *KAT1* (*BnaC09g19230D*). Among them, the expression of *HKT1* in two rapeseed varieties under salt stress were both down-regulated greatly, which can limit the absorption of $Na^+$ by the cell membrane. Under NaCl stress, the expression of *NHX4* significantly increased, which can force cells to transport more $Na^+$ into the vacuole to reduce ion toxicity. The expression of *KAT1* in the two rapeseed varieties was reduced, which may be due to the presence of excessive $Na^+$ outside the cell [40]. The expression of *NHX4* of H62 was up-regulated under 150 mM and 200 mM NaCl, which were significantly higher than that of ZS11, indicating that the difference of *NHX4* expression may be one of the reasons for the difference salt tolerance of the two varieties.

Osmomodulin (OSM) is widely present in various tissues of different plants. It is a type of protein synthesized in order to adapt to osmotic pressure. When plant cells are subjected to osmotic stress, the osmomodulin in their cells can absorb water and reduce excessive water loss by changing the membrane's water permeability. Aquaporins (AQPs) are a class of integral membrane proteins that efficiently and specifically transport water molecules and play an important role in plant water transfer [41], including Plasma membrane intrinsic proteins (PIPs), Tonoplast intrinsic proteins (TIPs), NLM proteins (Nodulin 26-like MIPs, NLMs), and Small and basic intrinsic proteins (SIPs) [42–44]. Many studies have shown that aquaporin can be upregulated by salt stress, and the over-expression of related genes can significantly improve the salt tolerance of plants [45–47]. In our study, one candidate gene *BnaC03g32130D* (*PIP1-1*) and three candidate genes (*BnaA06g12030D, BnaA07g30640D, BnaA09g44820D*) homologous to aquaporin *TIP3-1* were identified, all of which can be highly induced by salt stress, and their expression was up-regulated more as the salt concentration increased. The LEA protein (Late-embryogenesis-abundant protein) is widely present in higher plants and accumulates largely during the later stage of seed embryogenesis [48]. Under osmotic stress, the accumulation of a large amount of LEA protein in plants can alleviate cell damage caused by reduced water potential [49,50]. For example, studies have transferred exogenous LEA genes into tobacco, and the salt tolerance and drought resistance of transgenic plants have been identified and the salt tolerance and drought resistance of transgenic tobacco plants have been improved [51]. In our study,

*BnaA01g10880D* (*LEA1*), *BnaA09g43150D* (*LEA*), and *BnaC01g35900D* (*LEA76*) were highly induced by salt stress, and their expression was up-regulated more as the salt concentration increased. Dehydrins, part of LEA family, are a group of proteins believed to play a fundamental role in plant response and adaptation to abiotic stresses that lead to cellular dehydration [52]. Therefore, they can alleviate the impact caused by osmotic water loss when exposed to salt stress during the seed-germination stage. In this study, the expression of dehydrohydrin *Rab18* (*BnaC09g08130D*) was highly induced by salt stress, and it was more up-regulated in H62. Non-specific lipid transporter (nsLTP) is an important type of small alkaline-secreted protein in plants, which plays an important role in plant resistance to adversity and stress [53]. In this study, a number of candidate genes homologous to *LTPs* were also identified, and expression levels of *LTPs* were up-regulated significantly as the salt concentration increased in the salt-tolerant H62 (Figure S3). In order to verify the function of the family genes, *BnLTP3* was selected and transferred to *Arabidopsis thaliana*. The results showed that overexpression of *BnLTP3* could increase tolerance to salt stress in *Arabidopsis* plants, but whether the *BnLTP3* has the same function in rapeseed needs further verification. It is worth noting that the root length of over-expressed *Arabidopsis thaliana* under salt stress was significantly higher than that of the wild type under salt stress (Figures 9 and S4), which is consistent with the phenotypes of H62 and ZS11 (Figure 1), indicating that *LTP3* may be involved in promoting root development under salt stress.

Enhancement of antioxidant systems such as catalases (CAT), superoxide dismutase (SOD), ascorbate peroxidase (APX), and glutathione reductase (GR) can increase salt-stress tolerance in plants [7]. In this study, *SOD1* (LOC106346305) was increased in H62_T1, H62_T2, and ZS_T1, and the amount of up-regulation in H62 was significantly higher than that in ZS11. *POD34* (LOC106447805) was increased in H62 and ZS11 under salt stress, and the amount of up-regulation in H62 was significantly higher than that in ZS11. 1-cys peroxiredoxin (*PER1*) is a seed-specific antioxidant that eliminates ROS with cysteine residues [54,55]. *BnaAnng40060D*(*PER1*) was increased in both H62 and ZS11 under salt stress. The results showed that increasing the activity of antioxidant enzymes to resist salt stress is also one of the important mechanisms for rapeseed to resist salt stress, and the expression levels of related genes in salt-tolerant rapeseed were significantly higher than salt-sensitive rapeseed.

Transcription factors (TFs) also play a vital role in response to salt stress in plants, such as *bZIP* [56], *NAC* [57], *WARY* [58], *MYB* [59], and *ERFs* [60]. In this study, several transcription factors induced by salt stress were also identified, such as *BnaC03g31750D* (*MYB-like ETC3*), *BnaA07g21980D* (*ERF018*), and *BnaC03g44820D* (B3 domain-containing transcription factor *ABI3*). These transcription factors can be induced by salt stress and may play a role in the salt tolerance mechanism of rapeseed. Ethylene-responsive factors (ERFs), within a subgroup of the AP2/ERF transcription factor family, are involved in diverse plant reactions to biotic or abiotic stresses [61]. Over-expression of *SlERF1* in tomato plants enhanced salt tolerance during tomato seedling root development, and *SlERF1* activated the expression of stress-related genes, including *LEA*, *P5CS* in tomato plants under salt stress [62]. Overexpression of *BrERF4* from *Brassica rapa* increased tolerance to salt and drought in *Arabidopsis* plants [60]. In this study, *BnaC05g00680D* (*ERF023*) and *BnaA07g21980D* (*ERF018*) were highly induced by salt stress, and were more up-regulated in H62 under salt stress, implying that these gene have a certain relationship with the difference in salt tolerance between the two rapeseed varieties.

In addition, there are some other genes and metabolites also involved in the regulation of salt tolerance. For example, NADP-malic enzyme functions in many different pathways in plant and may be involved in plant defense, such as salt stress [63]. Transgenic *Arabidopsis* plants over-expressing rice *NADP-ME1* had a greater salt tolerance at the seedling stage than wild-type plants [64]. In our study, *NADP-ME1* (*BnaA07g00860D*) was significantly up-regulated in two rapeseed varieties under salt stress, and the amount of up-regulation in H62 was significantly higher than that in ZS11. Flavonoids are a large group of secondary plant metabolites, playing diverse roles in plant growth, development, and responses to stress. Epigallocatechin-3-gallate (EGCG, a bioactive flavonoid) was found to enhance tolerance to salt stress [65], and EGCG could alleviate salt stress-induced inhibition in seed germination and root growth in tomato [66]. In our study, the content of EGCG was significantly increased in both H62 and ZS11, and the increase in EGCG content in H62 was significantly higher than that of ZS11.

Based on the data results of the correlation analysis of key DEMs and DEGs (Figure S5), some putative mechanisms of rapeseed salt tolerance were (Figure 10): (1) High concentrations of NaCl in the environment decreased the water potential, which made it difficult for the plant cells to absorb external water. In order to reduce the harm of osmotic stress, the genes related to osmotic regulation in rapeseed is regulated by related transcription factors (MYB, ERF, bHLH) in the nucleus. Salt stress can promote the TCA cycle of rapeseed and produce more ATP for rapeseed metabolic activity. The up-regulation of *P5CS* and down-regulation of *ProDH* increased proline synthesis, the up-regulation of *Sus3* and *MIPs* increased Sucrose and Inositol synthesis, and macromolecular proteins, such as LEA, Dehydrins, began to accumulate in large quantities, which could balance the water potential inside and outside the cells to some extent. In addition, the PIP protein located on the cell membrane and the TIP protein located on the vacuole membrane can also regulate the water balance in the cell and the vacuole. (2) High concentrations of $Na^+$ inhibit the activity of *KAT1*, thereby inhibiting the absorption of $K^+$. A large amount of $Na^+$ enters the cells, and then causes ion toxicity in the cells. The down-regulation of *HKT1* can limit the absorption of $Na^+$ by the cell membrane, and the up-regulation of NHXs can force cells to transport more $Na^+$ into the vacuole to reduce ion toxicity. (3) Large amounts of ROS were generated in cells under salt stress, which may cause oxidative damage to cells. The ROS scavenging system may neutralize the excess ROS. The up-regulation of the ROS scavenging system-related genes (such as *POD*, *SOD*, and *PER1*) decreased ROS content in the cell to reduce its damage to the cell. In addition, the enhanced TCA cycle provides energy for gene transcription, translation, and substance synthesis under salt stress. The functional genes mentioned above need further functional verification.

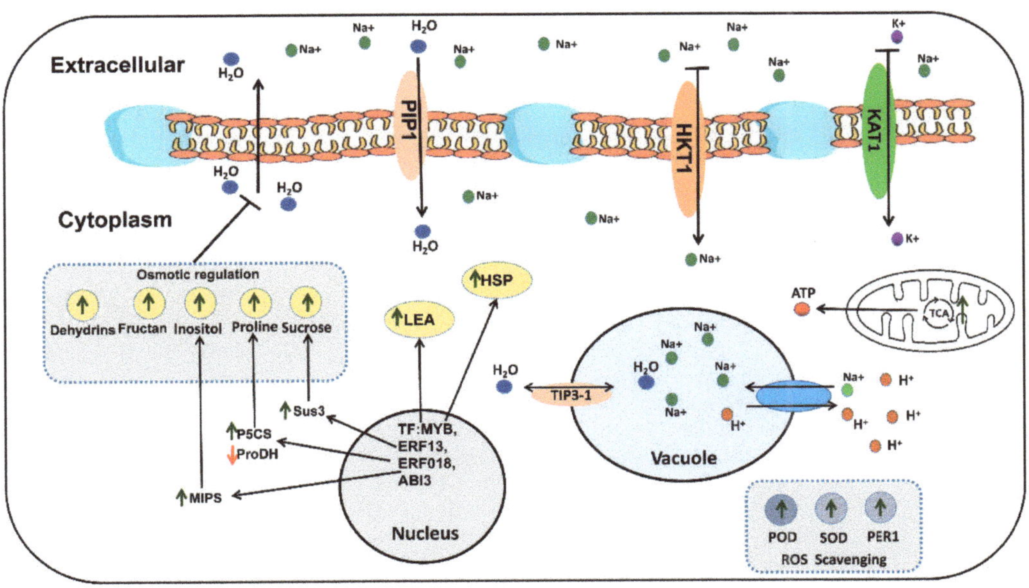


**Figure 10.** Possible molecular mechanism of salt tolerance in rapeseed.

## 4. Materials and Methods

### 4.1. Plant Material, Salt Stress Treatment, and Sample Collection

The experimental materials included Huayouza 62 (H62) and Zhongshuang 11 (ZS11). The seed germination experiment method was performed as previously described [67]. Thirty healthy seeds of H62 and ZS11, full and of the same size, were selected, surface-sterilized with 0.1% HgCl solution for 3 min, rinsed with distilled water three times, and equally distributed into Petri dishes (9 cm in diameter) containing two sheets of medical gauze. 15 mL of the NaCl solutions (0, 100, 150, and 200 mM) were added to the respective Petri dishes. The seed germination experiment was performed in a greenhouse (20 °C), with an 8 h day, 16 h night photoperiod.

On the 8th day after sowing, 8–10 seedlings with relatively consistent growth were selected for each treatment, and the hypocotyl length and main root length were measured; 3 seedlings were selected for each treatment and mixed into a biological sample, which were stored in the refrigerator at −80 °C by liquid nitrogen quick freezing. Finally, Huayouza 62 took three groups of samples, named H62_CK, H62_T1, and H62_T2, and Zhongshuang 11 took two groups of samples, named ZS_CK and ZS_T1, in which CK was 0 mM, T1 was 150 mM, and T2 was 200 mM. The samples were determined and analyzed by BGI. Finally, there were 3 and 7 biological repeats for the transcriptomics and metabolomics analysis, respectively.

### 4.2. LC-MS for Metabolite Determination

The tissue samples were stored at −80 °C for metabolomics analysis. We weighed 25 mg tissue samples into Eppendorf Micro Test Tubes (EP tubes), added 800 μL of pre-cooled (4 °C) methanol/water (1:1) buffer solution to each EP tube, added two tiny steel balls to each EP tube, and placed the sample in TissueLyser Medium grinding (QIAGEN, Düsseldorf, Germany), setting the parameter to 50 HZ, 4 min; then, we removed the steel balls after grinding, placed the centrifuge tube in a refrigerator at −20 °C for overnight precipitation, and centrifuged at 30,000 *g* for 20 min (4 °C). We then carefully removed the EP tube from the centrifuge, and drew 550 μL of each sample; GC-MS analysis was performed in a new EP tube.

### 4.3. LC-MS (Gas Chromatography-Mass Spectrometry) Analysis

The LC-MS acquired all samples, and the system followed machine orders. Firstly, all chromatographic separations were performed using an ultra-performance liquid chromatography (UPLC) system (Waters, UK). An ACQUITY UPLC HSS T3 column (100 mm * 2.1 mm, 1.8 μm, Waters, UK) was used to reverse phase separation. The column oven was maintained at 50 °C. The flow rate was 0.4 mL/min and the mobile phase consisted of solvent A (water + 0.1% formic acid) and solvent B (acetonitrile + 0.1% formic acid). Gradient elution conditions were set as follows: 0~2 min, 100% phase A; 2~11 min, 0% to 100% B; 11~13 min, 100% B; 13~15 min, 0% to 100% A. The injection volume for each sample was 10 μL. A high-resolution tandem mass spectrometer Xevo G2 XS QTOF (Waters, UK) was used to detect metabolites eluted from the column. The Q-TOF was operated in both positive and negative ion modes. The mass spectrometry data were acquired in Centroid MSE mode. The TOF mass range was from 50 to 1200 Da and the scan time was 0.2 s. For the MS/MS detection, all precursors were fragmented using 20–40 eV, and the scan time was 0.2 s. During the acquisition, the LE signal was acquired every 3 s to calibrate the mass accuracy. Furthermore, to evaluate the LC-MS stability during the whole acquisition, a quality control sample (pool of all samples) was acquired after every 10 samples. Original data were preprocessed, peak list information was extracted, and data correction was performed. Peak extraction was mainly achieved through the commercial software Progenesis QI (version 2.2), which includes peak alignment, peak extraction, normalization, deconvolution, and compound identification. Local polynomial regression fitting signal correction (QC-RSC) based on QC sample information for real sample signals is a more effective data correction method for omics analysis in the field of metabolomics [68].

*4.4. Metabolite Data Analysis to Obtain Differential Metabolites*

The main purpose of metabolomics analysis was to screen statistically and biologically significant metabolites from the mass metabolites detected and clarify the metabolic process and the changing mechanism of the organism based on this. Single and multi-dimensional methods were used to analyze from different angles according to the characteristics of the data. Univariate analysis was performed using *t*-test and fold change analysis (FC analysis). Further FDR correction was performed in the statistical analysis process on the *p*-value produced by the statistical test to obtain the q-value. Multivariate analysis was performed using principal component analysis (PCA) and partial optimal multiplication-discrimination analysis (PLS-DA). PCA was mainly used to observe the trend of separation between groups in the experimental model, specifically, whether any abnormal points appeared and reflected the variability between and within groups from the original data. PLS-DA uses partial maximum regression to establish a relationship model between metabolite expression and sample categories to achieve model prediction of sample categories. At the same time, the variable projection important degree (Variable Important for the Projection, VIP) was used to measure the influence intensity and explanatory power of each metabolite expression pattern on the classification and discrimination of each group of samples, thereby assisting the selection of metabolic markers. In this study, the ion identification data in positive mode were selected. Conditions for screening differentially expressed metabolites (DEMs): (1) VIP $\geq 1$; (2) Fold-change $\geq 1.2$ or Fold-change $\leq 0.833$; (3) q-value $< 0.05$. The three were taken to intersect, and the obtained substance was regarded as a substance with a significant change in content. Determination of metabolites and analysis of metabolic pathways were based on the KEGG database.

*4.5. Transcriptome Sequencing Analysis*

Fifteen libraries representing the plant from the two lines collected at three NaCl concentrations and in three replicates were constructed for transcriptome sequencing. Total RNA was extracted using a TIANGEN RNA Prep Pure Plant kit (Tiangen Biotech Co., Ltd., Beijing, China) and purified with the DNase I, RNase-free, and Thermo Scientific (Waltham, MA, USA) RebertAid First Strand cDNA Synthesis Kit. RNA concentration was measured using Agilent 2100 Bioanalyzer (Agilent Technologies, Santa Clara, CA, USA) and NanoDropTM (Thermo Scientific, Waltham, MA, USA) evaluated the purity of RNA. Sequencing was performed using an Illumina HiSeqTM paired-end sequencing system (San Diego, California, USA). It contains crucial steps such as mRNA purification, mRNA fragmentation, adaptor addition, reverse transcription, and library validation.

*4.6. Sequencing Analysis*

The sequencing reads, which contain low-quality, adaptor-polluted and high content of unknown base (N) reads, should be processed to be removed before downstream analysis. After reads filtering, the clean reads were mapped to the reference genome (ZS11: https://www.ncbi.nlm.nih.gov/assembly/GCF_000686985.2/, accessed on 10 April 2019) using HISAT2 (v2.0.4) software [69]. Fragments per kilobase per million mapped fragments (FPKM) were calculated to estimate the expression level of genes in each sample [70]. DESeq2 was used to estimate the expression level of each gene in each sample [71]. In this experiment, *p*-value $\leq 0.05$ and $|\log_2\text{FoldChange}| \geq 1$ were set as the thresholds to determine the significance of the gene-expression difference between samples. Notably, gene-expression comparisons of the samples at the 0, 150 mM, and 200 mM concentrations (H62_T1/H62_CK, H62_T2/H62_CK, and ZS11_T1/ZS11_CK) were performed. Gene Ontology (GO) annotation (http://geneontology.org/, accessed on 28 May 2019) and Kyoto Encyclopedia of Genes and Genomes (KEGG) reference pathway (https://www.genome.jp/kegg/kegg.html, accessed on 4 June 2019) analyses were performed for DEGs.

### 4.7. RT-PCR and Real-Time Quantitative PCR

In order to validate the reliability of our RNA-seq data, we measured mRNA abundance using qRT-PCR for 6 DEGs (Table S1). Three biological replications of RNA from the samples (H62 and ZS11) at the 0, 150 mM, and 200 mM concentrations were used for qPCR analysis. The gene-specific primers for real-time PCR analysis were designed using Primer 3 by applying the parameters described by Thornton and Basu [72]. The first-strand cDNAs were synthesized from 1 μg of total RNAs using SuperScript III Reverse Transcriptase (Invitrogen, Carlsbad, CA, USA). Ten microliters of PCR samples containing 1 μL of first-strand cDNAs and 5 pmol of primers were then subjected to 30 cycles of 30 s denaturing at 94 °C, 30 s annealing at 60 °C, and 30 s extending at 72 °C. The PCR products were electrophoresed on 1.5% agarose gel. Real-time PCR was performed on CFX Connect Real-Time PCR Detection System (Bio-Rad, Hercules, CA, USA) using 1 μL of cDNAs and SsoAdvanced SYBR Green Supermix (Bio-Rad). The thermal conditions were set at 95 °C for 3 min denaturation, followed by 40 cycles of 95 °C for 1 s and 60 °C for 30 s. Following denaturation at 95 °C for 30 s and cooling to 65 °C for 30 s, a melting curve was generated by heating from 65 °C to 95 °C in 0.5 °C increments with a dwell time at each temperature of 2 s while continuously monitoring the fluorescence. All of the reactions were performed in triplicate and the average expression value was calculated. The relative expression level for each gene was calculated using the $2^{-\Delta\Delta C_T}$ method with normalization to the internal control gene [73].

### 4.8. BnLTP3 Gene Function Verification

The Col-0 ecotype of *Arabidopsis thaliana* was used as the WT (wild-type). Plants were grown at 23 °C with a long-day light cycle (16 h light/8 h dark). *BnLTP3* (*BnaC03g12050D*) cDNA sequences were cloned, and the overexpression vector pC2300 were driven using a CaMV 35S promoter. The vectors were introduced into the *Agrobacterium tumefaciensstrain* GV3101 for genetic transformation into Col-0 ecotype of *Arabidopsis thaliana*. Homozygous mutants were confirmed by PCR-based genotyping. The T2 generations *Arabidopsis* seeds were certified, and uniform and healthy seeds were surface-sterilized with sodium hypochlorite solution and then rinsed in sterile distilled water. The seeds were germinated on 1/2 MS and 1/2 MS + NaCl (50, 100 mM) basal medium containing 2% (*w*/*v*) sucrose. After 10 days, 3 relatively uniform seedlings were selected to measure root length and RNA was extracted to detect the expression of *LTP3* gene (F:GTTCCTCCTCCGTGTTGTG, R:TTGTCGCAGTTAGTGCTCAC).

## 5. Conclusions

A total of 103 key differentially expressed metabolites (DEMs) involved in TCA cycle, proline metabolism, inositol metabolism, carbohydrate metabolic processes, and oxidation-reduction processes and 53 key differentials expressed genes (DEGs) involved in ion transport, reactive oxygen scavenging, osmotic regulation substance synthesis, and macromolecular protein synthesis were identified. The overexpression of *BnLTP3* could increase tolerance to salt stress in *Arabidopsis* plants.

**Supplementary Materials:** The following supporting information can be downloaded at: https://www.mdpi.com/article/10.3390/ijms23031279/s1.

**Author Contributions:** J.S. and T.F. were responsible for providing ideas and experimental guidance; H.W. and J.Q. were responsible for data collection; H.Z., H.L., O.L., R.L. and Y.Y. were responsible for data analysis; H.W. and J.Q. were responsible for writing the original draft; J.S., L.Z., B.Y. and J.W. were responsible for review and editing. All authors have read and agreed to the published version of the manuscript.

**Funding:** This research was financially supported by the Hubei Hongshan Laboratory research funding (grant No. 2021HSZD004) and the Program for Modern Agricultural Industrial Technology System of China (grant No. CARS-12).

**Institutional Review Board Statement:** Not applicable.

**Informed Consent Statement:** Not applicable.

**Data Availability Statement:** Not applicable.

**Conflicts of Interest:** The authors declare no conflict of interest. This project was submitted to NCBI BioProject with BioProject, ID: PRJNA757136. The raw reads were deposited in NCBI SRA (Short Read Archive) with submission number SUB10231030.

## References


1. Zhu, J.K. Abiotic stress signaling and responses in plants. *Cell* **2016**, *167*, 313–324. [CrossRef]
2. Ruan, C.J.; da Silva, J.A.T.; Mopper, S.; Qin, P.; Lutts, S. Halophyte improvement for a salinized world. *Crit. Rev. Plant Sci.* **2010**, *6*, 329–359. [CrossRef]
3. Flowers, T.J. Improving crop salt tolerance. *J. Exp. Bot.* **2004**, *396*, 307–319. [CrossRef]
4. Munns, R.; Tester, M. Mechanisms of salinity tolerance. *Annu. Rev. Plant Biol.* **2008**, *59*, 651–681. [CrossRef]
5. Zhao, C.; Zhang, H.; Song, C.; Zhu, J.K.; Shabala, S. Mechanisms of plant responses and adaptation to soil salinity. *Innovation* **2020**, *1*, 100017. [CrossRef]
6. Ren, Z.H.; Gao, J.P.; Li, G.; Cai, X.L.; Huang, W.; Chao, D.Y.; Zhu, M.; Wang, Z.; Luan, S.; Lin, H. A rice quantitative trait locus for salt tolerance encodes a sodium transporter. *Nat. Genet.* **2005**, *37*, 1141–1146. [CrossRef]
7. Moez, H.; Chantal, E.; Mariama, N.; Laurent, L.; Khaled, M. New insights on plant salt tolerance mechanisms and their potential use for breeding. *Front. Plant Sci.* **2016**, *7*, 1787. [CrossRef]
8. Jiang, Y.Q.; Deyholos, M.K. Comprehensive transcriptional profiling of NaCl-stressed *Arabidopsis* roots reveals novel classes of responsive genes. *BMC Plant Biol.* **2006**, *6*, 25. [CrossRef] [PubMed]
9. Mahajan, S.; Pandey, G.K.; Tuteja, N. Calcium- and salt-stress signaling in plants: Shedding light on SOS pathway. *Arch. Biochem. Biophys.* **2008**, *471*, 146–158. [CrossRef]
10. Sánchez-Barrena, M.J.; Martínez-Ripoll, M.; Zhu, J.K.; Albert, A. The structure of the *Arabidopsis thaliana* SOS3: Molecular mechanism of sensing calcium for salt stress response. *J. Mol. Biol.* **2015**, *345*, 1253–1264. [CrossRef] [PubMed]
11. Vivek, P.J.; Tuteja, N.; Soniya, E.V. *CDPK1* from ginger promotes salinity and drought stress tolerance without yield penalty by improving growth and photosynthesis in *Nicotiana tabacum*. *PLoS ONE* **2013**, *8*, e76392. [CrossRef] [PubMed]
12. Liu, J.; Wang, X.; Yang, L.; Nan, W.; Bi, Y. Involvement of active MKK9-MAPK3/MAPK6 in increasing respiration in salt-treated *Arabidopsis* callus. *Protoplasma* **2020**, *257*, 965–977. [CrossRef] [PubMed]
13. Purty, R.S.; Kumar, G.; Singla-Pareek, S.L.; Pareek, A. Towards salinity tolerance in *Brassica*: An overview. *Physiol. Mol. Biol. Plants* **2008**, *14*, 39–49. [CrossRef]
14. Hu, Y.F.; Guo, S.X.; Li, X.X.; Ren, X.Q. Comparative analysis of salt-responsive phosphoproteins in maize leaves using Ti4+-IMAC enrichment and ESI-Q-TOF MS. *Electrophoresis* **2013**, *34*, 485–492. [CrossRef]
15. Yang, H.; Deng, L.; Liu, H.; Fan, S.; Hua, W.; Liu, J. Overexpression of *BnaAOX1b* confers tolerance to osmotic and salt stress in rapeseed. *G3-Genes Genom. Genet.* **2019**, *9*, 3501–3511. [CrossRef]
16. Zhou, Y.; Yang, P.; Cui, F.; Zhang, F.; Luo, X.; Xie, J. Transcriptome analysis of salt stress responsiveness in the seedlings of Dongxiang wild rice (*Oryza rufipogon* Griff.). *PLoS ONE* **2016**, *11*, e0146242. [CrossRef]
17. Chandran, A.N.; Kim, J.W.; Yoo, Y.H.; Park, H.L.; Kim, Y.J.; Cho, M.H.; Jung, K.H. Transcriptome analysis of rice-seedling roots under soil-salt stress using RNA-Seq method. *Plant Biotechnol. Rep.* **2019**, *13*, 567–578. [CrossRef]
18. Amirbakhtiar, N.; Ismaili, A.; Ghaffari, M.R.; Firouzabadi, F.N.; Shobbar, Z.S. Transcriptome response of roots to salt stress in a salinity-tolerant bread wheat cultivar. *PLoS ONE* **2019**, *14*, e0213305. [CrossRef]
19. Long, W.; Zou, X.; Zhang, X. Transcriptome analysis of canola (*Brassica napus*) under salt stress at the germination stage. *PLoS ONE* **2015**, *10*, e0116217. [CrossRef]
20. Yong, H.Y.; Zou, Z.; Kok, E.P.; Kwan, B.H.; Chow, K.; Nasu, S.; Nanzyo, M.; Kitashiba, H.; Nishio, T. Comparative transcriptome analysis of leaves and roots in response to sudden increase in salinity in *Brassica napus* by RNA-seq. *J. Biomed. Biotechnol.* **2014**, *2014*, 467395. [CrossRef]
21. Nicholson, J.K.; Lindon, J.C.; Holmes, E. Metabonomics: Understanding the metabolic responses of living systems to pathophysiological stimuli via multivariate statistical analysis of biological nmr spectroscopic data. *Xenobiotica* **1999**, *29*, 1181–1189. [CrossRef]
22. Fiehn, O. Metabolomics—The link between genotypes and phenotypes. *Plant Mol. Biol.* **2002**, *48*, 155–171. [CrossRef]
23. Wishart, D.S. Current progress in computational metabolomics. *Brief. Bioinform.* **2007**, *5*, 279–293. [CrossRef]
24. Fernie, A.R.; Schauer, N. Metabolomics-assisted breeding: A viable option for crop improvement? *Trends Genet.* **2009**, *25*, 39–48. [CrossRef]
25. Gavaghan, C.L.; Jia, V.L.; Hadfield, S.T.; Hole, S.; Nicholson, J.K.; Wilson, I.D.; Howe, P.W.; Stanley, P.D.; Holmes, E. Application of NMR-based metabolomics to the investigation of salt stress in maize (*Zea mays*). *Phytochem. Anal.* **2011**, *22*, 214–224. [CrossRef]
26. Richter, J.A.; Erban, A.; Kopka, J.; Zörb, C. Metabolic contribution to salt stress in two maize hybrids with contrasting resistance. *Plant Sci.* **2015**, *233*, 107–115. [CrossRef]

27. Widodo; Patterson, J.H.; Newbigin, E.; Tester, M.; Bacic, A.; Roessner, U. Metabolic responses to salt stress of barley (*Hordeum vulgare* L.) cultivars, sahara and clipper, which differ in salinity tolerance. *J. Exp. Bot.* **2009**, *60*, 4089–4103. [CrossRef]
28. Cui, F.; Sui, N.; Duan, G.; Liu, Y.; Han, Y.; Liu, S.; Wan, S.; Li, G. Identification of Metabolites and Transcripts Involved in Salt Stress and Recovery in Peanut. *Front. Plant Sci.* **2018**, *9*, 217. [CrossRef]
29. Mohamed, I.A.A.; Shalby, N.; El-Badri, A.M.; Batool, M.; Wang, C.; Wang, Z.; Salah, A.; Rady, M.; Jie, K.; Wang, B.; et al. NA-seq analysis revealed key genes associated with salt tolerance in rapeseed germination through carbohydrate metabolism, hormone, and MAPK signaling pathways. *Ind. Crop. Prod.* **2022**, *176*, 114262. [CrossRef]
30. Zhang, J.; Zhang, Y.; Du, S.; Chen, S.; Tang, H. Dynamic metabonomic responses of tobacco (*Nicotiana tabacum*) plants to salt stress. *J. Proteome Res.* **2011**, *10*, 1904–1914. [CrossRef]
31. Urano, K.; Kurihara, Y.; Seki, M.; Shinozaki, K. 'Omics' analyses of regulatory networks in plant abiotic stress responses. *Curr. Opin. Plant Biol.* **2010**, *13*, 132–138. [CrossRef] [PubMed]
32. Wang, H.; Ding, Q.; Shao, H.; Wang, H. Overexpression of *KvP5CS1* increases salt tolerance in transgenic tobacco. *Pak. J. Bot.* **2019**, *51*, 831–836. [CrossRef]
33. Xue, X.; Liu, A.; Hua, X. Proline accumulation and transcriptional regulation of proline biosynthesis and degradation in *Brassica napus*. *BMB Rep.* **2009**, *42*, 28–34. [CrossRef]
34. Saadia, M.; Jamil, A.; Akram, N.A.; Ashraf, M.A. Study of proline metabolism in canola (*Brassica napus* L.) seedlings under salt stress. *Molecules* **2012**, *17*, 5803–5815. [CrossRef] [PubMed]
35. Fernandes, F.M.; Arrabaça, M.C.; Carvalho, L.M.M. Sucrose Metabolism in *Lupinus albus* L. under salt stress. *Biol. Plant.* **2004**, *48*, 317–319. [CrossRef]
36. Kusuda, H.; Koga, W.; Kusano, M.; Oikawa, A.; Saito, K.; Hirai, M.Y.; Yoshida, K. Ectopic expression of myo-inositol 3-phosphate synthase induces a wide range of metabolic changes and confers salt tolerance in rice. *Plant Sci.* **2015**, *232*, 49–56. [CrossRef]
37. Zhang, Y.; Fang, J.; Wu, X.; Dong, L. $Na^+/K^+$ balance and transport regulatory mechanisms in weedy and cultivated rice (*Oryza sativa* L.) under salt stress. *BMC Plant Biol.* **2018**, *18*, 375. [CrossRef]
38. Jarvis, D.E.; Ryu, C.H.; Beilstein, M.A.; Schumaker, K.S. Distinct roles for SOS1 in the convergent evolution of salt tolerance in *Eutrema salsugineum* and *Schrenkiella parvula*. *Mol. Biol. Evol.* **2014**, *31*, 2094–2107. [CrossRef]
39. Abdelaziz, M.E.; Kim, D.; Ali, S.; Fedoroff, N.V.; Al-Babili, S. The endophytic fungus *Piriformospora* indica enhances *Arabidopsis thaliana* growth and modulates $Na^+/K^+$ homeostasis under salt stress conditions. *Plant Sci.* **2017**, *263*, 107–115. [CrossRef]
40. Obata, T.; Kitamoto, H.K.; Nakamura, A.; Fukuda, A.; Tanaka, Y. Rice Shaker potassium channel *OsKAT1* confers tolerance to salinity stress on yeast and rice cells. *Plant Physiol.* **2007**, *144*, 1978–1985. [CrossRef]
41. Tyerman, S.D.; Niemietz, C.M.; Bramley, H.H. Plant aquaporins: Multifunctional water and solute channels with expanding roles. *Plant Cell Environ.* **2010**, *25*, 173–194. [CrossRef] [PubMed]
42. Fotiadis, D.; Jeno, P.; Mini, T.; Wirtz, S.; Muller, S.A.; Fraysse, L.; Kjellbom, P.; Engel, A. Structural characterization of two aquaporins isolated from native spinach leaf plasma membranes. *J. Biol. Chem.* **2001**, *276*, 1707–1714. [CrossRef] [PubMed]
43. Santoni, V.; Gerbeau, P.; Javot, H.; Maurel, C. The high diversity of aquaporins reveals novel facets of plant membrane functions. *Curr. Opin. Plant Biol.* **2000**, *3*, 476–481. [CrossRef]
44. Chaumont, F.; Barrieu, F.; Wojcik, E.; Chrispeels, M.; Jung, R. Aquaporins constitute a large and highly fivergent protein family in maize. *Plant Physiol.* **2001**, *125*, 1206–1215. [CrossRef] [PubMed]
45. Hu, W.; Yuan, Q.; Wang, Y.; Cai, R.; Deng, X.; Wang, J.S.; Zhou, S.; Chen, M.; Chen, L.; Huang, C.; et al. Overexpression of a wheat aquaporin gene, *TaAQP8*, enhances salt stress tolerance in transgenic tobacco. *Plant Cell Physiol.* **2012**, *53*, 2127–2141. [CrossRef]
46. Li, R.; Wang, J.; Li, S.; Zhang, L.; Qi, C.; Weeda, S.L.; Guo, Y.D. Conferring enhanced drought stress tolerance in tomato. *Sci. Rep.* **2016**, *6*, 31814. [CrossRef]
47. Ayadi, M.; Brini, F.; Masmoudi, K. Overexpression of a wheat aquaporin gene, *Td*PIP2;1, enhances salt and drought tolerance in transgenic durum wheat cv. Maali. *Int. J. Mol. Sci.* **2019**, *20*, 2389. [CrossRef]
48. Dure, L.; Crouch, M.; Harada, J.; Ho, T.H.D.; Mundy, J.; Quatrano, R.; Thomas, T.; Sung, Z.R. Common amino acid sequence domains among the LEA proteins of higher plants. *Plant Mol. Biol.* **1989**, *12*, 475–486. [CrossRef]
49. Hu, T.Z. Oslea3, a late embryogenesis abundant protein gene from rice, confers tolerance to water deficit and salt stress to transgenic rice. Russ. *J. Plant Physiol.* **2008**, *55*, 530–537. [CrossRef]
50. Bies-Ethève, N.; Gaubier-Comella, P.; Debures, A.; Lasserre, E.; Jobet, E.; Raynal, M.; Cooke, R.; Delseny, M. Inventory, evolution and expression profiling diversity of the LEA (late embryogenesis abundant) protein gene family in *Arabidopsis thaliana*. *Plant Mol. Biol.* **2008**, *67*, 107–124. [CrossRef]
51. Yan, L.; Pei, L.; Zhao, S.X.; Zhang, R.Y.; Liu, L.; Li, L.C.; Chen, M.; Ye, X.G.; Chen, Y.F.; Ma, Y.Z. Overexpression of *W6* gene tncreases salt tolerance in transgenic tobacco plants. *Acta Agron. Sinica* **2008**, *34*, 984–990. [CrossRef]
52. Liu, Y.; Song, Q.P.; Li, D.X.; Yang, X.H.; Li, D.Q. Multifunctional roles of plant dehydrins in response to environmental stresses. *Front. Plant Sci.* **2017**, *8*, 1018. [CrossRef]
53. Lindorff-Larsen, K.; Winther, J.R. Surprisingly high stability of barley lipid transfer protein, *LTP1*, towards denaturant, heat and proteases. *FEBS Lett.* **2001**, *488*, 145–148. [CrossRef]
54. Haslekås, C.; Stacy, R.A.P.; Nygaard, V.; Culiáñez-Macià, F.A.; Aalen, R.B. The expression of a peroxiredoxin antioxidant gene, *atper1*, in *Arabidopsis thaliana* is seed-specific and related to dormancy. *Plant Mol. Biol.* **1998**, *36*, 833–845. [CrossRef] [PubMed]

55. Chen, H.H.; Chu, P.; Zhou, Y.L.; Ding, Y.; Li, Y.; Liu, J.; Jiang, L.; Huang, S.Z. Ectopic expression of *NnPER1*, a Nelumbo nucifera 1-cysteine peroxiredoxin antioxidant, enhances seed longevity and stress tolerance in *Arabidopsis*. *Plant J.* **2016**, *88*, 608–619. [CrossRef]
56. Sun, X.; Li, Y.; Cai, H.; Bai, X.; Ji, W.; Ding, X.; Zhu, Y. The *Arabidopsis AtbZIP1* transcription factor is a positive regulator of plant tolerance to salt, osmotic and drought stresses. *J. Plant Res.* **2012**, *125*, 429–438. [CrossRef]
57. Zheng, X.; Bo, C.; Lu, G.; Han, B. Overexpression of a NAC transcription factor enhances rice drought and salt tolerance. *Biochem. Bioph. Res. Commun.* **2009**, *379*, 985–989. [CrossRef]
58. Zhou, L.; Wang, N.N.; Gong, S.Y.; Lu, R.; Li, Y.; Li, X.B. Overexpression of a cotton (*Gossypium hirsutum*) WRKY gene, *GhWRKY34*, in *Arabidopsis* enhances salt-tolerance of the transgenic plants. *Plant Physiol. Biochem.* **2015**, *96*, 311–320. [CrossRef]
59. Tominaga-Wada, R.; Nukumizu, Y.; Wada, T. Tomato (*Solanum lycopersicum*) Homologs of TRIPTYCHON(SlTRY) and GLABRA3 (SlGL3) are involved in anthocyanin accumulation. *Plant Signal. Behav.* **2013**, *8*, e24575. [CrossRef]
60. Long, L.; Yang, W.W.; Liao, P.; Guo, Y.W.; Kumar, A.; Gao, W. Transcriptome analysis reveals differentially expressed erf transcription factors associated with salt response in cotton. *Plant Sci.* **2019**, *281*, 72–81. [CrossRef]
61. Seo, Y.J.; Park, J.B.; Cho, Y.J.; Jung, C.; Seo, H.S.; Park, S.K.; Nahm, B.; Song, J. Overexpression of the ethylene-responsive factor gene *BrERF4* from *Brassica rapa* increases tolerance to salt and drought in *Arabidopsis* plants. *Mol. Cells* **2010**, *30*, 271–277. [CrossRef]
62. Lu, C.W.; Shao, Y.; Li, L.; Chen, A.J.; Xu, W.Q.; Wu, K.J.; Luo, Y.; Zhu, B. Overexpression of *SlERF1* tomato gene encoding an ERF-type transcription activator enhances salt tolerance. *Russ. J. Plant Physiol.* **2011**, *58*, 118–125. [CrossRef]
63. Drincovich, M.A.F.; Casati, P.; Andreo, C.S. NADP-malic enzyme from plants: A ubiquitous enzyme involved in different metabolic pathways. *FEBS Lett.* **2001**, *490*, 1–6. [CrossRef]
64. Cheng, Y.; Long, M.A. cytosolic nadp-malic enzyme gene from rice (*Oryza sativa* L.) confers salt tolerance in transgenic arabidopsis. *Biotechnol. Lett.* **2007**, *29*, 1129–1134. [CrossRef]
65. Li, X.; Li, Y.; Ahammed, G.J.; Zhang, X.L.; Ying, L.; Zhang, L.; Li, Q.; Han, W. RBOH1-dependent apoplastic $H_2O_2$ mediates epigallocatechin-3-gallate-induced abiotic stress tolerance in *Solanum lycopersicum* L. *Environ. Exp. Bot.* **2019**, *161*, 357–366. [CrossRef]
66. Jalal, A.G.; Yang, L.; Li, X.; Han, W.Y.; Chen, S. Epigallocatechin-3-Gallate alleviates salinity-retarded seed germination and oxidative stress in tomato. *J. Plant Growth Regul.* **2018**, *37*, 1349–1356. [CrossRef]
67. Wan, H.; Wei, Y.; Qian, J.; Gao, Y.; Wen, J.; Yi, B.; Ma, C.; Tu, J.; Fu, T.; Shen, J. Association mapping of salt tolerance traits at germination stage of rapeseed (*Brassica napus* L.). *Euphytica* **2018**, *214*, 190. [CrossRef]
68. Dunn, W.B.; Broadhurst, D.; Begley, P.; Zelena, E.; Francis-Mcintyre, S.; Anderson, N.; Brown, M.; Knowles, J.D.; Halsall, A.; Haselden, J.N.; et al. Procedures for large-scale metabolic profiling of serum and plasma using gas chromatography and liquid chromatography coupled to mass spectrometry. *Nat. Protoc.* **2011**, *6*, 1060–1083. [CrossRef]
69. Kim, D.; Paggi, J.M.; Park, C.; Bennett, C.; Salzberg, S.L. Graph-based genome alignment and genotyping with HISAT2 and HISAT-genotype. *Nat. Biotechnol.* **2019**, *37*, 907–915. [CrossRef]
70. Trapnell, C.; Williams, B.A.; Pertea, G.; Mortazavi, A.; Kwan, G.; van Baren, M.J.; Salzberg, S.L.; Wold, B.J.; Pachter, L. Transcript assembly and quantification by RNAseq reveals unannotated transcripts and isoform switching during cell differentiation. *Nat. Biotechnol.* **2010**, *28*, 511–515. [CrossRef]
71. Love, M.I.; Huber, W.; Anders, S. Moderated estimation of fold change and dispersion for RNA-seq data with DESeq2. *Genome Biol.* **2014**, *15*, 550. [CrossRef] [PubMed]
72. Thornton, B.; Basu, C. Real-Time PCR (qPCR) primer design using free online software. *Biochem. Mol. Biol. Educ.* **2011**, *39*, 145–154. [CrossRef]
73. Livak, K.J.; Schmittgen, T.D. Analysis of relative gene expression data using real-time quantitative PCR and the $2^{-\Delta\Delta C_T}$ method. *Methods* **2011**, *25*, 402–408. [CrossRef]

International Journal of
*Molecular Sciences*

MDPI

*Article*

# Transcriptome Analysis Reveals Roles of Anthocyanin- and Jasmonic Acid-Biosynthetic Pathways in Rapeseed in Response to High Light Stress

**Yuxiu Luo [1], Shoulian Teng [1], Hengxia Yin [2,*], Shengping Zhang [3], Xiaoyun Tuo [1] and Lam-Son Phan Tran [4,5,*]**

1 College of Eco-Environmental Engineering, Qinghai University, Xining 810016, China; 1992990033@qhu.edu.cn (Y.L.); y200713000525@qhu.edu.cn (S.T.); 1800601019@qhu.edu.cn (X.T.)
2 State Key Laboratory of Plateau Ecology and Agriculture, Qinghai University, Xining 810016, China
3 Qinghai Academy of Agriculture and Forestry, Qinghai University, Xining 810016, China; 2019990075@qhu.edu.cn
4 Institute of Research and Development, Duy Tan University, 03 Quang Trung, Da Nang 550000, Vietnam
5 Institute of Genomics for Crop Abiotic Stress Tolerance, Department of Plant and Soil Science, Texas Tech University, Lubbock, TX 79409, USA
* Correspondence: hengxiayin@qhu.edu.cn (H.Y.); tranplamson@duytan.edu.vn or son.tran@ttu.edu (L.-S.P.T.); Tel.: +86-971-531-0086 (H.Y.)

**Abstract:** Rapeseed (*Brassica napus*) is one of the major important oil crops worldwide and is largely cultivated in the Qinghai-Tibetan plateau (QTP), where long and strong solar-radiation is well-known. However, the molecular mechanisms underlying rapeseed's response to light stress are largely unknown. In the present study, the color of rapeseed seedlings changed from green to purple under high light (HL) stress conditions. Therefore, changes in anthocyanin metabolism and the transcriptome of rapeseed seedlings cultured under normal light (NL) and HL conditions were analyzed to dissect how rapeseed responds to HL at the molecular level. Results indicated that the contents of anthocyanins, especially glucosides of cyanidin, delphinidin, and petunidin, which were determined by liquid chromatography-mass spectrometry (LC-MS), increased by 9.6-, 4.2-, and 59.7-fold in rapeseed seedlings exposed to HL conditions, respectively. Next, RNA-sequencing analysis identified 7390 differentially expressed genes (DEGs), which included 4393 up-regulated and 2997 down-regulated genes. Among the up-regulated genes, many genes related to the anthocyanin-biosynthetic pathway were enriched. For example, genes encoding dihydroflavonol reductase (*BnDFR*) and anthocyanin synthase (*BnANS*) were especially induced by HL conditions, which was also confirmed by RT-qPCR analysis. In addition, two *PRODUCTION OF ANTHOCYANIN PIGMENTATION 2* (*BnPAP2*) and *GLABRA 3* (*BnGL3*) genes encoding *MYB*-type and *bHLH*-type transcription factors, respectively, whose expression was also up-regulated by HL stress, were found to be associated with the changes in anthocyanin biosynthesis. Many genes involved in the jasmonic acid (JA)-biosynthetic pathway were also up-regulated under HL conditions. This finding, which is in agreement with the well-known positive regulatory role of JA in anthocyanin biosynthesis, suggests that the JA may also play a key role in the responses of rapeseed seedlings to HL. Collectively, these data indicate that anthocyanin biosynthesis-related and JA biosynthesis-related pathways mediate HL responses in rapeseed. These findings collectively provide mechanistic insights into the mechanisms involved in the response of rapeseed to HL stress, and the identified key genes may potentially be used to improve HL tolerance of rapeseed cultivars through genetic engineering or breeding strategies.

**Keywords:** anthocyanin biosynthesis; *Brassica napus*; high light; jasmonic acid pathway; transcriptome analysis

**Citation:** Luo, Y.; Teng, S.; Yin, H.; Zhang, S.; Tuo, X.; Tran, L.-S.P. Transcriptome Analysis Reveals Roles of Anthocyanin- and Jasmonic Acid-Biosynthetic Pathways in Rapeseed in Response to High Light Stress. *Int. J. Mol. Sci.* **2021**, *22*, 13027. https://doi.org/10.3390/ijms222313027

Academic Editors: Ana I. Ribeiro-Barros, José C. Ramalho and Isabel Marques

Received: 15 October 2021
Accepted: 27 November 2021
Published: 1 December 2021

## 1. Introduction

Rapeseed (*Brassica napus* L., AACC, $n = 19$) is a hybrid allotetraploid species derived from *B. rapa* (AA, $n = 10$) and *B. oleracea* (CC, $n = 9$) [1]. As the third most important industrial oil crop after soybean (*Glycine max*) and palm (*Trachycarpus fortunei*) [2], rapeseed is an important source of edible oil, protein-rich animal feed, as well as biodiesel [3]. Many studies in rapeseed plants have characterized their physiological and molecular responses to abiotic stresses, such as salt [2], drought [4,5], heat [6], freezing temperatures [7], as well as cadmium stress [7,8], using transcriptome sequencing. This approach has successfully identified numerous stress response- and/or tolerance-related genes, providing information that has been dissected to better understand the underlying mechanisms associated with stress tolerance in rapeseed. To date, however, the molecular response to high light (HL) stress in rapeseed has not been fully elucidated.

As an essential environmental factor, light plays a crucial role in the growth and development of plants throughout their life cycle. However, plants will be stressful when being exposed to high intensity light, and a wide range of HL stress responses will be activated for their adaptation [9]. Previous studies mainly focused on underling photosynthetic and physiological variations, e.g., activation of non-photochemical quenching (NPQ) [10,11], changes in chloroplast and leaf avoidance movements [12,13], and leaf phenotypic changes (e.g., leaves tend to be thicken) [14]. In addition, numerous studies reported HL stress also triggered transcriptional and metabolic responses in plants [15], e.g., activation of various genes encoding enzymes responsible for reactive oxygen species (ROS) scavenging [16–18], and reconfiguration of primary and energy metabolism (e.g., changes in the content of various carbohydrates) [19–21]. However, it should be noted that the responses of plants to HL stress are diverse partly due to species-specific responses [22,23].

Anthocyanins, present in most plant species, are flavonoid secondary metabolites that serve in many cases as the source of pigmentation of a variety of plant organs, including seeds, leaves, flowers, and fruits [24–26]. Anthocyanins are also considered a nutraceutical and have human health-protective effects against a variety of chronic diseases, such as diabetes, allergies, viral infections, and cancer [27–29]. Anthocyanin accumulation in plants is frequently induced by diverse environmental stresses [30–33]. For example, anthocyanin biosynthesis is activated under HL conditions and anthocyanins function in photoprotection [34,35]. Although there have been a few studies on the roles of anthocyanin metabolism in plants under HL stress, the production of anthocyanin in rapeseed in response to HL stress has not been yet investigated.

Anthocyanins are derived from phenylalanine in a branch of the flavonoid-biosynthesis pathway. The key enzymes involved in anthocyanin biosynthesis are primarily phenylalanine ammonium lyase (PAL), cinnamic acid 4-hydroxylase (C4H), 4-coumarate-CoA ligase (4CL), chalcone synthase (CHS), chalcone isomerase (CHI), flavonoid-3′,5′-hydroxylase (F3′5′H), flavanone-3-hydroxylase (F3H), flavonoid-3′-hydroxylase (F3′H), flavonol synthase (FLS), anthocyanin reductase (ANR), dihydroflavonol reductase (DFR), anthocyanin synthase (ANS), and uridine diphosphate glucose-flavonoid- 3-*O*-glycosyltranferase (UFGT) [25,36]. Among these enzymes, DFR, ANS, ANR, and UFGT catalyze the production of anthocyanins and their glucosides in plants [36]. The predominant anthocyanins in plants are mainly derived from cyanidin, delphinidin, and petunidin [37], which may contribute to the pigmentation of plant organs [38].

It has been well documented that anthocyanin biosynthesis is primarily regulated by three types of transcription factors (TFs), namely *MYB*, *bHLH* (basic helix-loop-helix), and *WD40* that form a MBW complex, in which *MYB* TFs play a crucial role [25,39,40]. R2R3-type *MYB*s are predominantly responsible for activation of anthocyanin biosynthesis, and many *R2R3-MYB*s have been well characterized in various plant species, including *Arabidopsis thaliana* [41], grape (*Vitis vinifera*) [42], *Medicago truncatula* [43], strawberry (*Fragaria* × *ananassa*) [44], and apple (*Malus* × *domestica*) [45]. For example, *PRODUCTION OF ANTHOCYANIN PIGMENTATION 1* (*PAP1*)/*MYB75*, *PAP2*/*MYB90*, *MYB113*, and *MYB114* (*R2R3-MYB* family members) have been identified and well-studied in

*Arabidopsis* [41]. *bHLH* TFs also act as important regulators in anthocyanin biosynthesis by mediating the transcriptional expression of the anthocyanin biosynthesis-associated genes *DFR*, *ANS* and *UFGT* [41]. More specifically, in *Arabidopsis*, three *bHLH* TF-encoding genes, *GLABRA 3* (*GL3*), *ENHANCER OF GLABRA 3* (*EGL3*), and *TRANSPARENT TESTA 8* (*TT8*), were demonstrated to participate in regulating anthocyanin biosynthesis [46]. Recently, the functions of *bHLHs* in regulating the anthocyanin-biosynthetic pathway have been investigated in various plant species, including apple, grape (*V. davidii*) [47], sheepgrass (*Leymus chinensis*) [48], kiwifruit (*Actinidia* spp.) [49], and lotus (*Nelumbo nucifera*) [50].

Phytohormones also play an important role in regulating plant responses to a wide variety of abiotic stresses [51]. In particular, jasmonic acid (JA) and its conjugate form, jasmonoyl-L-isoleucine (JA-Ile), have been shown to be implicated in plant responses to various abiotic stresses, such as UV, osmotic stress, drought, salinity, cold, heat, and heavy metals [52,53]. A report on the ultra-fast transcriptomic response of *Arabidopsis* to HL stress showed that ~12% of its transcripts that were induced within seconds of HL stress exposure were JA-responsive ones [54], demonstrating a potential role of this hormone in the rapid response to HL stress in plants. Additionally, JA has been reported to induce anthocyanin production in *Arabidopsis* [55,56].

In the present study, the potential roles of anthocyanins and JA in the responses of rapeseed seedlings to HL were characterized by phenotypic analysis, anthocyanin profiling, and transcriptome analysis with the main focus on anthocyanin-biosynthetic and JA-biosynthetic pathways. In addition, the expression patterns of key associated genes were validated by RT-qPCR analysis. The results of our study provide the first in-depth insight into the molecular mechanisms underlying the response of rapeseed to HL, and may potentially contribute to breeding and genetic engineering for the development of HL-tolerant rapeseed varieties.

## 2. Results

### 2.1. Changes in the Appearance and Level of Anthocyanins in Leaves of Rapeseed Seedlings in Response to HL Treatment

In comparison with plants exposed to a normal light (NL) intensity (300 μmol $m^{-2}$ $s^{-1}$), rapeseed seedlings (four-leave stage) exposed to a HL intensity (600 μmol $m^{-2}$ $s^{-1}$) resulted in plants with a purple appearance in the aboveground organs, including the hypocotyls, cotyledons, petioles and leaves, after 16 h of exposure (Figure 1A,B). Due to the close relationship between the pigmentation and anthocyanin accumulation in plants [57], the levels of total anthocyanins and six major anthocyanins, namely cyanidin, delphinidin, malvidin, pelargonidin, peonidin, and petunidin, in NL and HL rapeseed leaves were determined using both colorimetric assay and liquid chromatography-mass spectrometry (LC-MS). Our data revealed that the contents of total anthocyanins were 9.32 mg $g^{-1}$ dry weight (DW) in the HL group and 6.51 mg $g^{-1}$ DW in the NL group, indicating that a significant induction of total anthocyanins had occurred in the HL leaves (Figure 1C; Supplementary Table S1). The level of flavonoids, as precursors of anthocyanins, will reflect the potential for anthocyanin accumulation [58]. Therefore, the contents of total flavonoids in the two groups were also measured. Rapeseed leaves exposed to HL were found to have significantly higher total flavonoid contents than those exposed to NL (Figure 1C; Supplementary Table S1). Among the six anthocyanins that were measured, the levels of cyanidin, delphinidin, and petunidin were significantly higher in the HL group than in the NL group. The content of petunidin increased the most in response to HL exposure among the anthocyanins measured, exhibiting a level of 272.67 μg $g^{-1}$ DW in the HL group, compared with 4.57 μg $g^{-1}$ DW in leaves of the NL group (Figure 1D; Supplementary Table S1). These results indicated that anthocyanin biosynthesis was indeed involved in the response of rapeseed to HL, and that anthocyanin accumulation induced by HL might contribute to the color change in HL-exposed rapeseed seedlings.

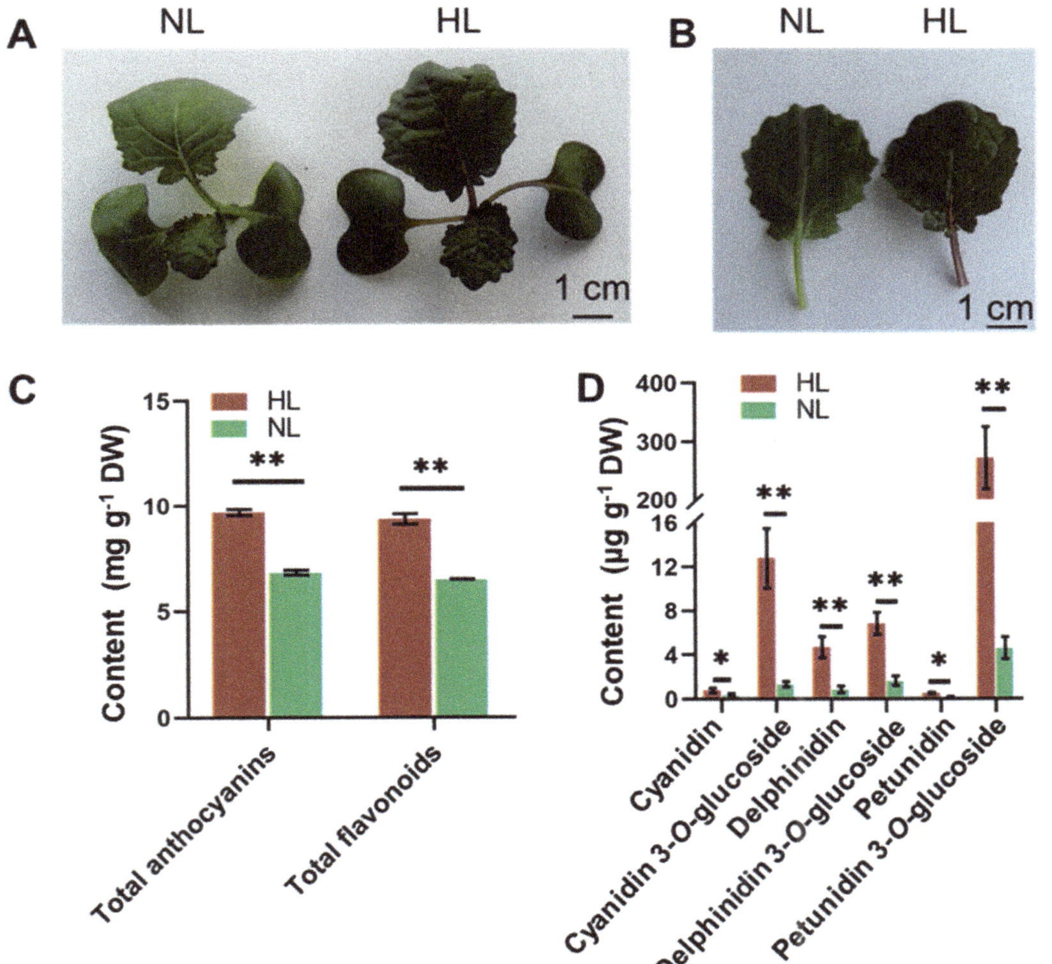


**Figure 1.** Phenotypic appearance and anthocyanin contents in rapeseed seedlings grown at different light-intensity conditions. (**A**,**B**) Colors of seedlings (**A**) and leaves (**B**) of rapeseed after 16 h exposure to normal-light (NL, 300 μmol $m^{-2}$ $s^{-1}$) and high-light (HL, 600 μmol $m^{-2}$ $s^{-1}$). (**C**) Total flavonoid and anthocyanin contents in NL and HL leaves after 16 h of exposure. (**D**) Total cyanidin, delphinidin and petunidin (three types of anthocyanins) contents in NL and HL leaves. Bars represent the means ± SDs ($n = 3$). * and ** indicate statistically significant differences between NL and HL leaves at $p < 0.05$ and $p < 0.01$, respectively, as determined by a Student's $t$-test.

### 2.2. Transcriptome Assembly and Analysis

Transcriptome analysis of rapeseed seedlings exposed to HL and NL for 16 h was conducted using RNA-sequencing (RNA-seq) to characterize the molecular response of rapeseed to HL intensity. A total of six cDNA libraries, representing three replicates of rapeseed seedling leaves from each light intensity group (HL group: HL1, HL2 and HL3; NL: NL1, NL2 and NL3), were constructed. High-throughput sequencing of the six libraries using the Illumina HiSeq 4000 platform generated a total of 42,937,020, 40,774,268, 42,601,798, 40,199,332, 42,171,552, and 62,116,104 raw reads for each of the six samples, respectively (Supplementary Table S2). After filtering (removal of low-quality reads and adapter sequences), approximately 6.64 GB of clean data per sample were obtained. The clean reads were aligned to the *B. napus* (PRJNA293435) reference genome using the HISAT2 software [58], resulting in an average alignment of 86% for each of the libraries, indicating that the high-throughput sequencing was of high quality. Ultimately, 198,694 genes were collectively identified from all of the samples. Furthermore, the correlation coefficient of gene expression levels among the samples (based on the fragments per kilobase per million reads) (FPKM) in each sample) was ≥0.86 (Supplementary Figure S1), indicating that the experimental samples and results were reliable and could be used for further analyses.

### 2.3. *Differentially Expressed Genes (DEGs) Identified in Leaves of Rapeseed Seedlings Exposed to HL and NL Conditions*

A total of 7390 differentially expressed genes (DEGs), including 4393 up- and 2997 down-regulated genes, were identified in the HL/NL comparison using a threshold of $|\log_2$ (fold change, FC)$| \geq 1$ and *FDR* (false discovery rate) < 0.05 (Figure 2A and Supplementary Table S3). Eleven DEGs shown in Supplementary Table S4 were randomly selected for the RT-qPCR analysis to validate the accuracy of the RNA-seq data. Results revealed that the relative expression levels determined by RT-qPCR were generally consistent with the transcript levels obtained by the RNA-seq data with a correlation coefficient of 0.7630 (Figure 2B), indicating that the RNA-seq data could be reliably used in further analyses.

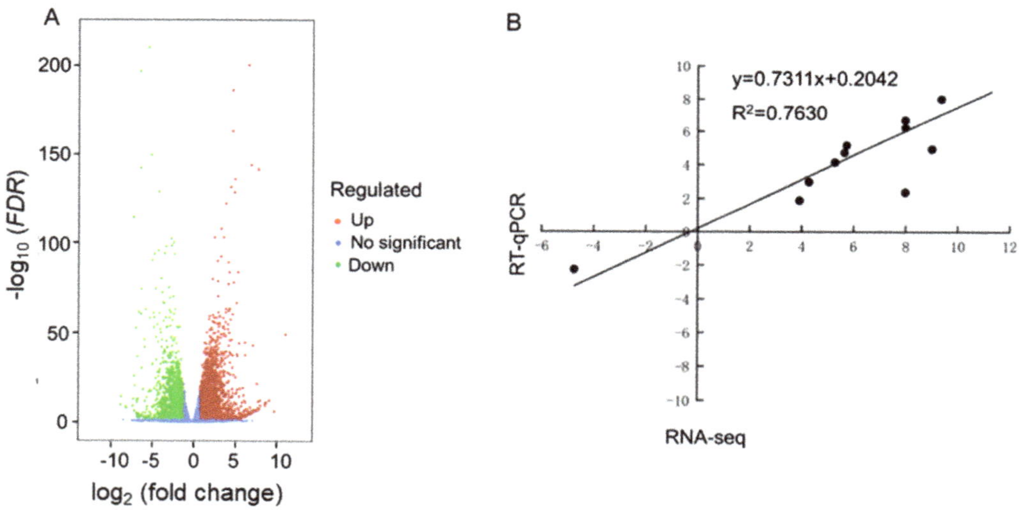


**Figure 2.** A volcano plot of differentially expressed genes (DEGs) identified in the RNA-sequencing (RNA-seq) data, and a correlation analysis of the RT-qPCR and RNA-seq data using 11 randomly-selected DEGs. (**A**) Number of DEGs in the HL/NL comparison. The *X*-axis and *Y*-axis indicate values of the $\log_2$ (fold change) and the $-\log_{10}$ (*FDR*) of all annotated genes in rapeseed seedlings grown under HL and NL conditions. (**B**) RT-qPCR validation of 11 DEGs. The *X*-axis and *Y*-axis represent the RT-qPCR and RNA-seq data, respectively, of the 11 randomly selected DEGs obtained from the HL/NL comparison. Data are expressed in $\log_2$ (fold change). HL and NL indicate high-light and normal-light samples, respectively.

### 2.4. *Functional Enrichment Analysis of DEGs*

Gene ontology (GO) enrichment analysis was conducted on the DEGs identified in the HL/NL comparison to obtain information on their potential role in the responses of rapeseed seedlings to HL. Results classified the DEGs into all three of the primary GO categories: 'Biological Process (BP)' (27 subcategories), 'Cellular Component (CC)' (16 subcategories), and 'Molecular Function (MF)' (12 subcategories) (Figure 3). Among the BP category, the 'cellular process', 'metabolic process' and 'response to stimulus' were the most significantly enriched. There were more functional terms for BP, and fewer terms of CC and MF categories assigned with transcripts. The subcategories 'cellular process', 'metabolic process', and 'response to stimulus' in BP, 'binding' and 'catalytic activity' in MF, and 'cell', 'cell part', and 'organelle' in CC categories were the most highly enriched (Figure 3 and Supplementary Table S5).

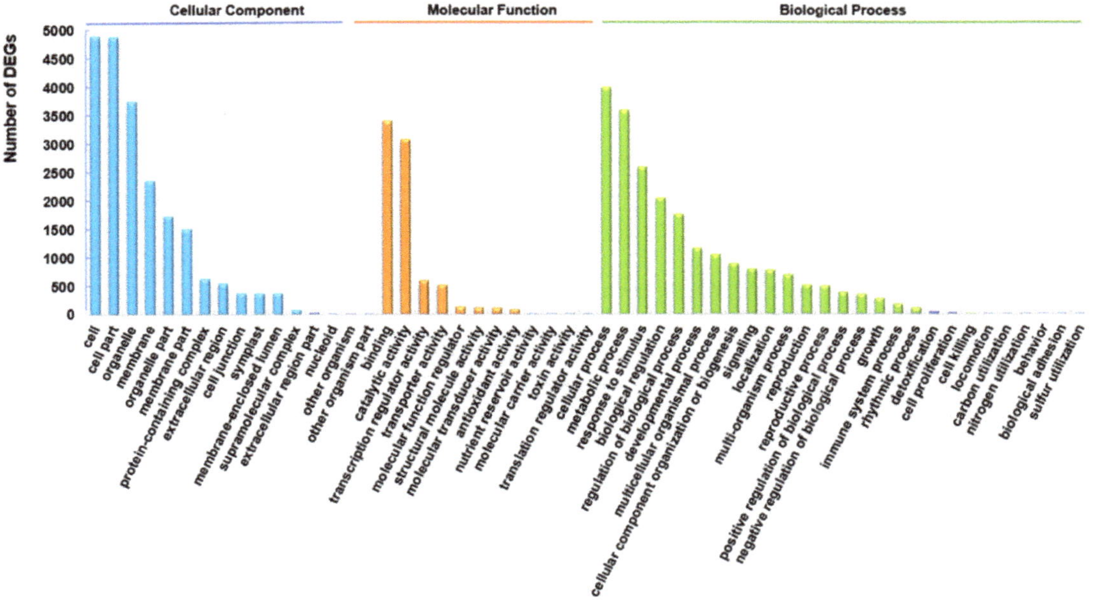


**Figure 3.** Gene ontology (GO) classification of differentially expressed genes (DEGs) obtained in the HL/NL comparison of rapeseed seedling leaves. The *X*-axis indicates the most highly-enriched 45 GO subcategories within the three primary categories. HL and NL indicate high-light and normal-light samples, respectively.

Many GO terms reflected a response to light, including 'response to light intensity' (GO:0009642), 'cellular response to light stimulus (GO:0071482), 'photosynthesis, light reaction' (GO:0019684), 'response to high light intensity' (GO:0009644) and 'response to red light' (GO:0010114) (Supplementary Table S5), which confirmed that the gene expression in rapeseed leaves was significantly influenced by HL intensity. Additionally, numerous genes were annotated to be involved in biological responses to light stimuli. For example, *BnaA03g36810D* ($log_2$ (FC) = 11.30), *BnaC09g00770D* ($log_2$ (FC) = 9.97), *BnaC06g20460D* ($log_2$ (FC) = 9.37), *BnaCnng37300D* ($log_2$ (FC) = 7.99) and *BnaA01g24440D* ($log_2$ (FC) = 7.98) were all significantly up-regulated in rapeseed leaves in response to the HL treatment. Among these genes, *BnaA03g36810D,* which encodes the early light-induced protein 1 (ELIP1), was annotated in the subcategories 'cellular response to high light intensity' (GO:0071486), 'response to blue light' (GO:0071483), 'chloroplast thylakoid membrane' (GO:0009535), 'photoprotection' (GO:0010117), 'photosynthesis' (GO:0015979), 'regulation of chlorophyll biosynthetic process' (GO:0010380), and 'photosystem I and II' (GO:0009522 and GO:0009523) of the BP category (Supplementary Table S5). This finding strongly suggests that *BnaA03g36810D* may be involved in rapeseed responses to photosynthesis and photoprotection and may represent a potential target for manipulation to improve the tolerance of rapeseed seedlings to HL stress.

Kyoto encyclopedia of genes and genomes (KEGG) enrichment analysis was performed to further understand the relationship between the identified DEGs and different aspects of metabolism. In total, five categories with 19 KEGG pathways were enriched with DEGs, namely 'Genetic Information Processing' (four terms, 772 DEGs), 'Metabolism' (11 terms, 5323 DEGs), 'Organism Systems' (one term, 134 DEGs), 'Cellular Processes' (one term, 160 DEGs) and 'Environmental Information Processing' (two terms, 516 DEGs), suggesting that the most enriched metabolic processes could be the main pathways involved in the responses of rapeseed seedlings to HL intensity (Figure 4 and Supplementary Table S6).

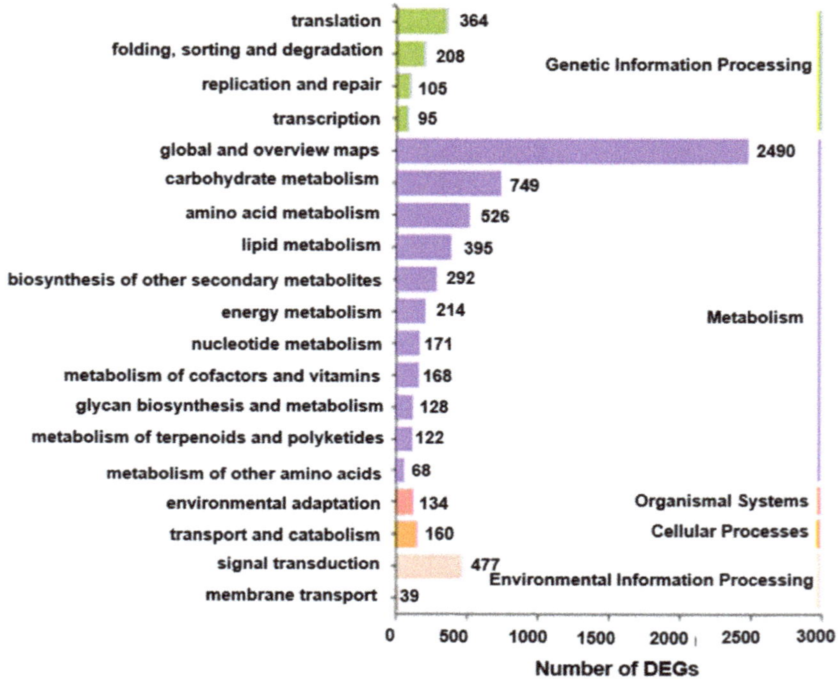


**Figure 4.** Kyoto encyclopedia of genes and genomes (KEGG) enrichment analysis of differentially expressed genes (DEGs) obtained in the HL/NL comparison of rapeseed seedling leaves. The *Y*-axis and *X*-axis represent the KEGG terms and the numbers of DEGs (HL/NL comparison), respectively. The numbers listed along each KEGG term indicate enriched DEGs found in each term. The KEGG pathways were classified into five categories indicated by different colors. HL and NL indicate high-light and normal-light samples, respectively.

### 2.5. *MapMan Analysis of Pathways Responding to HL Stress*

A MapMan analysis of the identified DEGs was conducted to visualize the pathways linked to the responses of rapeseed seedlings to light stress and gain more information on their biological functions. The mapping file (X4.2 *Brassica rapa*) of annotated DEGs was downloaded from the website https://mapman.gabipd.org/mapmanstore [59]. In total, 7390 DEGs were mapped to 977 pathways. The number of pathways was reduced by using the threshold of $p < 0.05$ (Figure 5; Supplementary Table S7). Among the mapped pathways, several were highly enriched (Figure 5), such as 's-misc', 'flavonoids', and 'light reaction', indicating that these pathways may function predominantly in enabling rapeseed seedlings to tolerate HL stress.

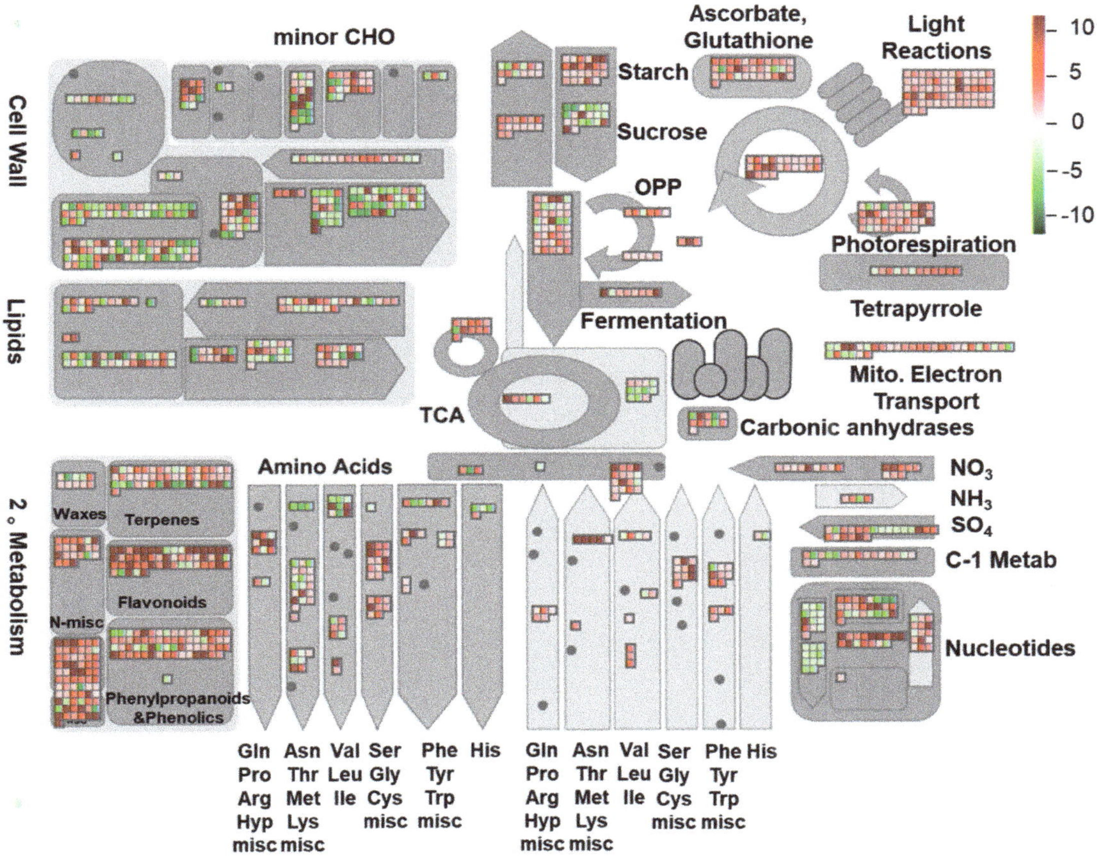


**Figure 5.** MapMan analysis of differentially expressed genes (DEGs) obtained in the HL/NL comparison of rapeseed seedling leaves, and their association with different metabolic pathways. Gray shapes indicate different metabolic pathways. The heatmaps in the gray boxes indicate up- (red-tinted boxes) or down-regulated (green-tinted boxes) genes in the HL/NL comparison. Expression changes were based on their $\log_2$ (fold change) values, and are presented by the colored scale. HL and NL indicate high-light and normal-light samples, respectively.

### *2.6. Anthocyanin Biosynthesis-Related Genes Are Up-Regulated in Leaves of Rapeseed Seedlings Exposed to HL Stress*

Since the content of total anthocyanins significantly increased in several individual plants, it is plausible that the genes participating in anthocyanin biosynthesis might exhibit a higher transcriptional abundance. Notably, the color change in plants exposed to HL was also closely correlated with anthocyanin accumulation (Figure 1). Therefore, elucidating the transcription of genes involved in anthocyanin biosynthesis will provide important information on how rapeseed responds to HL intensity. Results indicated that almost all of the anthocyanin biosynthesis-related genes were significantly up-regulated (Figure 6). In the initial stage of the general flavonoid pathway, transcripts encoding BnPAL, BnC4H, and Bn4CL were slightly up-regulated in rapeseed leaves in response to HL exposure (Figure 6). Since anthocyanin biosynthesis represents a branch of the flavonoid pathway, the biosynthesis and accumulation of flavonoids play a crucial role in anthocyanin production. All of the identified transcripts encoding BnCHS and BnCHI, which initialize flavonoid biosynthesis in rapeseed, were also significantly up-regulated. These included *BnaA03g04590D* ($log_2$ (FC) = 4.94), *BnaA10g19670D* ($log_2$ (FC) = 4.56) and *BnaC03g06120D* ($log_2$ (FC) = 4.08) encoding BnCHS, as well as *BnaA04g04230D* ($log_2$ (FC) = 5.24), *BnaC08g22640D* (*log2* (FC) = 2.51) and *BnaA09g31780D* ($log_2$ (FC) = 2.38) encoding BnCHIs (Figure 6; Supplementary Table S6). In the specific anthocyanin-biosynthetic pathway, BnDFR, BnANS, and BnUFGT represent key enzymes in the production of anthocyanin [60]. Four genes encoding BnANSs, namely *Bnac07g37670D* ($log_2$ (FC) = 3.87), *Bnac01g14310D* (log2 (FC) = 3.49), *Bnaa03g45610D* (log2 (FC) = 3.33), and *Bnaa01g12530D*, $log_2$ (FC) = 3.11), and four BnDFR-encoding genes, including *Bnac09g17150D* ($log_2$ (FC) = 6.18), *Bnaa09g15710D* (log2 (FC) = 3.62), *Bnac07g40800D* (log2 (FC) = 2.47), and *Bnaa03g48520D* (log2 (FC) = 1.37) were significantly up-regulated in response to the HL treatment (Figure 6; Supplementary Table S6). Notably, all of the BnUFGT-encoding genes, *BnaA09g55800D*, *BnaAnng15830D*, *BnaC01g42100D*, *BnaC03g42350D*, and *BnaC04g14920D*, involved in the final step of anthocyanin biosynthesis, were highly induced by HL stress, exhibiting $log_2$ (FC) values ranged from 1.13 to 1.72 (Figure 6; Supplementary Table S8).

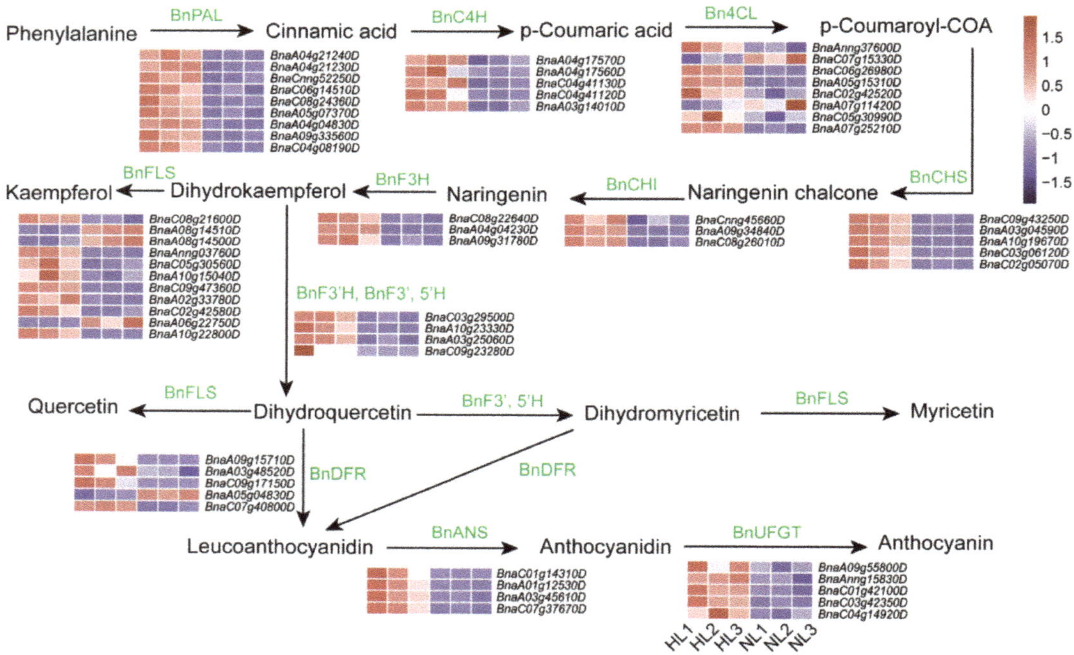


**Figure 6.** Change in the expression levels of anthocyanin biosynthesis-related genes in leaves of rapeseed seedlings exposed to high-light (HL) or normal-light (NL) intensity for 16 h. Green letters above/next to the arrows represent enzymes encoded by the indicated genes in each reaction step. The key biosynthetic enzymes are as follows: BnPAL, phenylalanine ammonium lyase; BnC4H, cinnamic acid 4-hydroxylase; Bn4CL, 4-coumarate-CoA ligase; BnCHS, chalcone synthase; BnCHI, chalcone isomerase; BnCHI, flavanone-3-hydroxylase; BnF3′H, flavonoid-3-hydroxylase; BnF3′, 5′H, flavonoid-3′,5′-hydroxylase; BnFLS, flavonol synthase; BnDFR, dihydroflavonol reductase; BnANS, anthocyanin synthase; BnANR, anthocyanin reductase; and BnUFGT, uridine diphosphate glucose-flavonoid-3-*O*-glycosyltranferase. Colored boxes and scale indicate expression levels in the HL and NL samples. Each column represents an independent biological replicate, labeled as 1, 2, and 3 following the HL or NL designation.

The expression levels of eight representative genes, namely *BnaC02g05070D*, *BnaA09g3 4840D*, *BnaC09g47360D*, *BnaA10g23330D*, *BnaC04g04230D*, *BnaC09g17150D*, *BnaC07g37670D*, and *BnaC01g42100D* encoding BnCHS, BnCHI, BnFLS, BnF3′H, BnF3H, BnDFR, BnANS and BnUFGT, respectively, were also studied by RT-qPCR in leaves of rapeseed seedlings exposed to HL stress (Figure 7). Notably, the relative expression level of *BnCHS*/*BnaC02g05070D* was up-regulated over 50-fold, while those of *BnANS*/*BnaC07g37670D* and *BnDFR*/*BnaC07g37670D* increased by 99.5- and 19.9-fold, respectively, after 16 h of HL exposure (Figure 7). These data strongly support the premise that anthocyanin biosynthesis and accumulation were enhanced in rapeseed seedlings in response to HL stress.

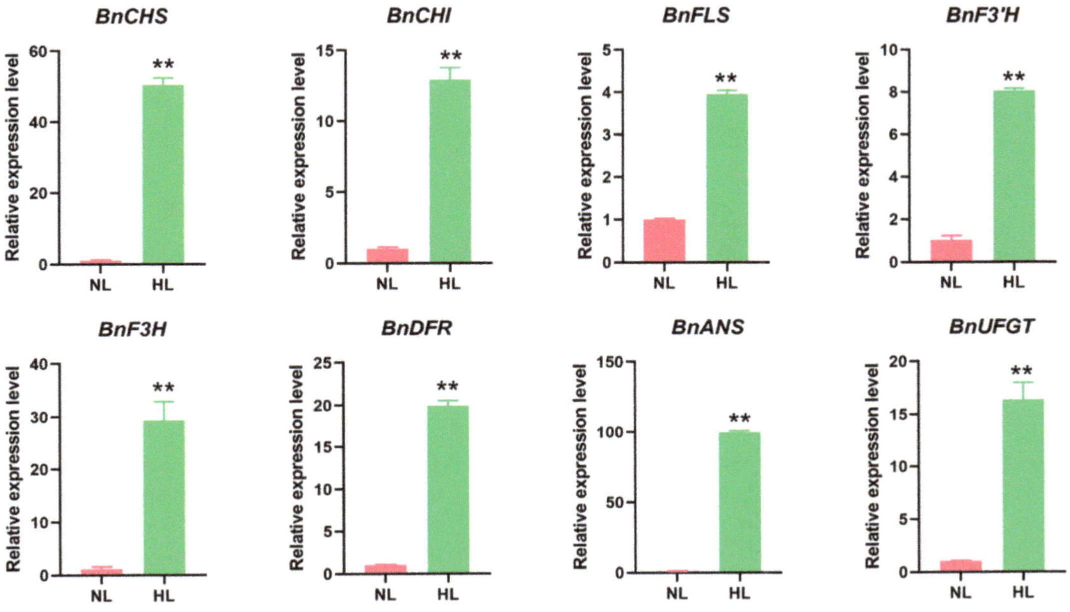


**Figure 7.** RT-qPCR analysis of 8 representative anthocyanin biosynthesis-related genes in leaves of rapeseed seedling exposed to normal-light (NL) and high-light (HL) conditions. Eight genes, namely *BnaC02g05070D*, *BnaA09g34840D*, *BnaC09g47360D*, *BnaA10g23330D*, *BnaC04g04230D*, *BnaC09g17150D*, *BnaC07g37670D* and *BnaC01g42100D* encoding BnCHS, BnCHI, BnFLS, BnF3′H, BnF3H, BnDFR, BnANS and BnUFGT, respectively, were selected. *BnACT* was used as an internal control for normalization. Data shown represent the means ± SEs ($n = 3$). ** indicates statistically significant differences between NL and HL leaves at $p < 0.01$ as determined by a Student's $t$-test.

### *2.7. Analysis of Phytohormone-Related Pathways Reveals Remarkable Up-Regulation of JA Biosynthesis-Related Genes in Rapeseed Seedlings under HL Stress*

Plant hormones have been demonstrated to play crucial roles in plant responses to environmental stresses [51]. In the present study, genes linked with eight different plant hormones, including auxin (IAA), abscisic acid (ABA), brassinolide (BR), ethylene (ET), cytokinin (CK), JA, salicylic acid (SA), and (gibberellin) GA, were found to respond to HL stress (Supplementary Table S9). Among these phytohormones, IAA metabolism and signal transduction pathways were associated with the greatest number of transcripts (114 transcripts), followed by the ethylene-related pathways (69 transcripts), the majority of which were down-regulated and up-regulated, respectively. Notably, all of 15 transcripts associated with JA biosynthesis identified by MapMan analysis were found to respond to HL stress and were up-regulated by HL (Figure 8A). Several studies have reported that JA mediates the biosynthesis and accumulation of anthocyanins in plants when they are exposed to adverse environmental conditions [40,42,44,45]. In our study, RT-qPCR analysis of five JA biosynthesis-associated genes also indicated that the expression of all of them was induced in rapeseed leaves by HL stress, supporting the results of the transcriptome analysis (Figure 8B). Therefore, it is plausible that JA may play a key role in plant response to HL stress and promote the accumulation of anthocyanins in rapeseed.

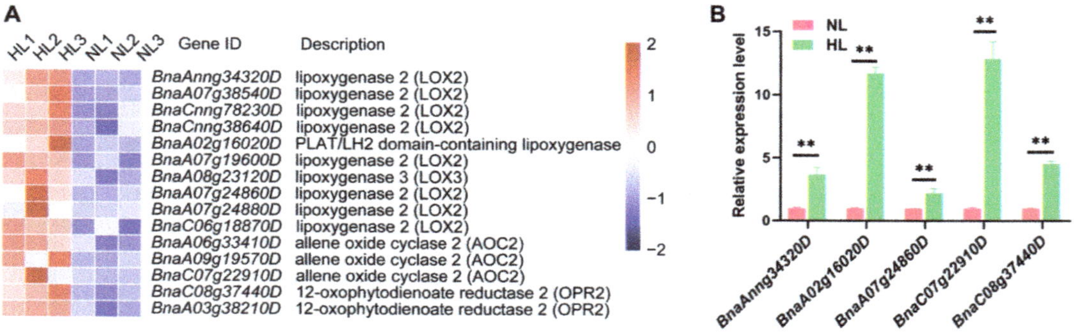


**Figure 8.** Expression and RT-qPCR analysis of JA biosynthesis-related genes in leaves of rapeseed seedlings exposed to high-light (HL) or normal-light (NL) intensity for 16 h. (**A**) Colored boxes and scale indicate normalized expression levels of all of the JA biosynthesis-related genes identified in the transcriptome data. Each column represents an independent biological replicate, labeled as 1, 2 and 3 following the HL or NL designation. (**B**) Relative expression levels of five genes related to JA-biosynthetic pathway were determined by RT-qPCR. *BnACT* was used as an internal control for normalization. Data shown represent the means ± SEs ($n$ = 3). ** indicate statistically significant differences between NL and HL leaves at $p < 0.01$ as determined by a Student's $t$-test.

### 2.8. Identification of TF-Encoding Genes Involved in Anthocyanin Biosynthesis in Rapeseed

In the present study, 707 DEGs encoding TFs assigned to 55 different families were identified in the HL/NL comparison of leaf transcriptomes of rapeseed seedlings. A total of 448 and 259 TF-encoding DEGs were up- and down-regulated, respectively (Supplementary Table S10). The top five abundant TF families were *apetala2/ethylene-responsive factor* (*AP2/ERF*, 74), *basic helix-loop-helix* (*bHLH*, 72), *NAC* (45), *MYB* (44), and *MYB-related* (38) (Figure 9A).

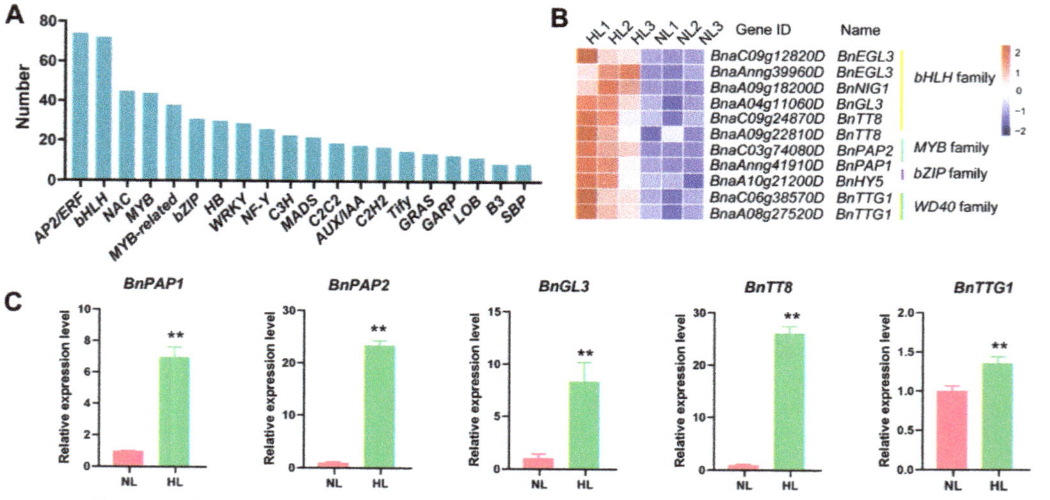


**Figure 9.** HL-responsive genes encoding transcription factors (TFs), and expression levels of HL-responsive TF-encoding genes involved in regulation of anthocyanin biosynthesis as identified in the HL/NL comparison of rapeseed seedling leaves.

(**A**) The top 20 abudant TF families. (**B**) Normalized expression levels of TF-encoding genes involved in regulation of anthocyanin biosynthesis as obtained from the transcriptome data. (**C**) Relative expression levels of five selected TF-encoding genes involved in the regulation of anthocyanin biosynthesis in rapeseed seedling leaves as determined by RT-qPCR. *BnACT* was used as an internal control for normalization. Data shown represent the means ± SEs ($n$ = 3). ** indicate statistically significant differences between NL and HL leaves at $p < 0.01$ as determined by a Student's $t$-test. HL and NL indicate high-light and normal-light samples, respectively.

Among the TF-encoding DEGs, nine genes belonging to *bHLH*, *MYB* and *bZIP* families, are known to participate in the regulation of anthocyanin biosynthesis. These anthocyanin biosynthesis-associated TF-encoding genes were significantly induced in rapeseed seedlings exposed to HL conditions, relative to the respective expression levels in seedlings exposed to NL conditions (Figure 9B). For example, members of the *bHLH* family, *BnaC09g12820D* ($\log_2$ (FC) = 1.67) and *BnaAnng39960D* ($\log_2$ (FC) = 1.84) encoding BnEGL3s, *BnaA04g11060D* ($\log_2$ (FC) = 2.02) encoding BnGL3, as well as *BnaC09g24870D* ($\log_2$ (FC) = 2.04) and *BnaA09g22810D* ($\log_2$ (FC) = 1.26) encoding BnTT8s were significantly up-regulated to a different extent. Two *MYB* genes, *BnaAnng41910D* and *BnaC03g74080D*, encoding BnPAP1 and BnPAP2, respectively, were significantly induced, exhibiting $\log_2$ (FC) values of 9.00 and 4.23, respectively, in seedlings exposed to HL conditions. Additionally, *BnaA10g21200D* ($\log_2$ (FC) = 1.86), belonging to the *bZIP* TF family, and encoding BnHY5, was also highly induced under HL conditions and annotated to be involved in the regulation of anthocyanin biosynthesis [61,62]. In addition, being one of the members of the MBW transcriptional complex involved in the anthocyanin-biosynthetic pathway [39], the *WD40* family of rapeseed has two members with high homology to *TRANSPARENT TESTA GLABRA 1* (*TTG1*) of *Arabidopsis*, which were identified to be differentially expressed under HL condition when compared with NL condition (Figure 9B). Specifically, the *WD40*-encoding genes *BnaC06g38570D* and *BnaA08g27520D* displayed the $\log_2$ (FC) values of 1.83 and 1.49, respectively, in the HL/NL comparison (Supplementary Table S3). The transcript levels of five selected TF-encoding genes, namely *BnPAP1*, *BnPAP2*, *BnGL3*, *BnTT8*, and *BnTTG1*, were also assessed by RT-qPCR, and these genes showed significant increases after 16 h of exposure to HL conditions (Figure 9C). Our analysis revealed that these 11 DEGs encoding TFs function as predominant regulators of anthocyanin biosynthesis and accumulation in rapeseed seedlings grown under HL conditions.

## 3. Discussion

Light is an essential requirement for the normal growth, development and reproduction of plants [63]. In their natural environment, plants have to continuously cope with daily and seasonal changes in light intensity. Alterations in plant color in response to light intensity represents one of the plants' adaptive responses to light signals [64]. Rapeseed is an oil crop that is widely planted in the Qinghai-Tibetan plateau, where the solar radiation is generally stronger than that in other low-altitude environments [65]. The molecular mechanisms underlying the response of rapeseed to light intensity, however, have not been fully elucidated, which has hindered the breeding of high-quality varieties that are tolerant to HL conditions. In the present study, the observation of color changes of rapeseed seedlings in response to HL stress encouraged us to further investigate underlying mechanisms using biochemical and transcriptomic approaches.

Transcriptome analysis of leaves of rapeseed seedlings exposed to HL stress for 16 h revealed 7390 DEGs (Figure 2), indicating a considerable change in gene expression under HL stress as observed in other plant species like *Panax ginseng*, *A. thaliana* and *Begonia semperflorens* [66–68]. Furthermore, GO analysis demonstrated that many GO terms related to photosynthesis were most enriched with DEGs (Figure 3). The majority of identified photosynthesis-associated genes were up-regulated (Figure 5), perhaps to prevent photooxidative stress through dissipating excess light energy by NPQ or electron flow [11]. Results of KEGG analysis revealed that the pathways associated with 'Metabolism' were significantly enriched with DEGs, especially the primary metabolic pathways related to carbohydrates, amino acids, and lipids (Figure 4). Such changes, as also supported by results of MapMan analysis (Figure 5), might help plants meet the demand of energy metabolism, and accumulate secondary metabolites like phenylpropanoids, flavonoids, and anthocyanins that were reported to be involved in preventing the HL-induced oxidative stress through balancing the levels of ROS [16,17,19,20,69].

Numerous studies have demonstrated that a wide range of environmental stresses can stimulate the production and accumulation of anthocyanins in plants [70–72], suggesting that these secondary metabolites may be closely correlated with the tolerance of plants to adverse environmental conditions as indicated by earlier studies [31,33]. Although light is required for plants to generate energy for growth and development, there are potential risks in capturing light energy by the photosynthetic apparatus of plants [73,74]. Excessive light absorption under HL conditions induces photoinhibition and increases the generation and accumulation of harmful ROS, which can result in oxidative stress. Excessive levels of ROS will damage proteins, nucleic acids and membrane lipids [9,75]. Therefore, photoprotection strategies have evolved in plants to detoxify excessive ROS and maintain normal photosynthesis under HL conditions. In this regard, the accumulation of anthocyanins is a primary strategy to shield against excessive light energy through optical masking chloroplasts, as well as avoid oxidative injury by directly scavenging excessive ROS owing to the antioxidant properties of anthocyanins [76–78]. In the present study, the accumulation of anthocyanins and other antioxidative flavonoids significantly increased in rapeseed seedlings exposed to HL for 16 h (Figure 1C), which might reflect their ability to provide photoprotection. Notably, the level of petunidin-3-*O*-glucodise was the highest (60-fold higher than in the normal-light, control group) among the differentially accumulated anthocyanins (Figure 1D), indicating its predominant role in the color change and protective responses observed in rapeseed seedlings exposed to HL stress.

Based on these observations, analysis of the expression patterns of candidate genes involved in anthocyanin biosynthesis was conducted using RNA-seq and RT-qPCR to provide deeper insights into the molecular mechanism underlying anthocyanin accumulation in response to HL stress in rapeseed seedlings. Results indicated that the majority of genes, including biosynthesis-related and those involved in the regulation of anthocyanin biosynthesis, exhibited a rapid response and high levels of transcription in rapeseed seedlings after 16 h of exposure to HL conditions (Figures 7 and 9C), resulting in enhanced anthocyanin accumulation. Numerous previous studies have reported on anthocyanin biosynthesis-associated genes and their role in the regulation of color phenotypes and tolerance to HL stress [79–81]. The functions of these genes have also been studied in detail using overexpression and/or silencing strategies in *Arabidopsis* or other plant species, such as apple [82], grape [42] and wheat [50]. Documenting the transcriptional expression of key anthocyanin biosynthesis-related genes may provide further insight into plant response to HL stress. In our study, the expression levels of eight representative genes *BnaC02g05070D*, *BnaA09g34840D*, *BnaC09g47360D*, *BnaA10g23330D*, *BnaC04g04230D*, *BnaC09g17150D*, *BnaC07g37670D*, and *BnaC01g42100D* encoding BnCHS, BnCHI, BnFLS, BnF3′H, BnF3H, BnDFR, BnANS, and BnUFGT were 50.4-, 12.9-, 3.95-, 8.1-, 29.3-, 19.9-, 99.5-, and 13.3-fold, respectively, higher in rapeseed seedlings under HL conditions (Figure 7). These results were consistent with previous studies, in which genes involved in the middle and late stages of anthocyanin biosynthesis were highly induced by HL stress [69,79,83]. Notably, the expression of *BnaC07g37670D* was the most induced, by 99.5-fold (Figure 7), and its encoding BnANS enzyme catalyzes the formation of colored anthocyanin precursors (anthocyanidins) [60], suggesting that this gene is of great significance in the green-to-purple color transition in rapeseed seedlings exposed to HL conditions. Additionally, the production of flavonoids, as a branch of the flavonoid-biosynthetic pathway, is crucial for anthocyanin biosynthesis [37]. The content of total flavonoids also increased dramatically in leaves of rapeseed seedlings under HL, relative to NL conditions (Figure 1C), which would contribute to the increased biosynthesis of anthocyanins (Figure 6).

*R2R3-MYB* TFs, a type of the ternary MBW complex, have been reported to play a dominant role in the regulation of anthocyanin biosynthesis [40]. Notably, *MYB75/PAP1* and *MYB90/PAP2* genes are induced by light in *Arabidopsis* [84], and their orthologs were found to be induced by 7- and 23.5-fold, respectively, in leaves of rapeseed seedlings exposed to HL conditions for 16 h in rapeseed under HL (Figure 9C) which is consistent with the observed anthocyanin accumulation under HL conditions in our study (Figure 1C,D). These data indicate that the up-regulation of both of the *PAP* genes was closely associated with the increase in anthocyanin biosynthesis and the development of purple pigmentation in rapeseed seedlings exposed to HL conditions. Similar to the color change observed in rapeseed seedlings in our study, *PAP1*- and *PAP2*-overexpressing *Arabidopsis* plants also displayed purple pigmentation in various organs, as well as anthocyanin accumulation [65]. In addition to the *R2R3-MYBs*, induction of other TF-encoding genes, such as *bHLH* and *bZIP* genes [85], in rapeseed seedlings under HL conditions, is required for anthocyanin accumulation. In our study, the transcripts of several *bHLH* TF-encoding genes, including *BnTT8*, *BnEGL3*, and *BnGL3*, that are involved in regulation of the anthocyanin biosynthesis, also significantly increased in rapeseed seedlings in response to HL stress (Figure 9C). A previous study reported that *GL3*-overexpressing *Arabidopsis* plants exhibited enhanced anthocyanin accumulation [85], indicating that *BnGL3* is involved in improving anthocyanin biosynthesis in plants in response to HL conditions. The *HY5* gene, another TF-encoding gene, whose ortholog was up-regulated in rapeseed by HL stress (Figure 9B), is well known to be involved in the light-dependent activation of anthocyanin biosynthesis-related downstream genes [86]. A recent study also reported that *HY5* plays a key role in controlling anthocyanin accumulation in plants during their response to light. For example, *HY5* is light-inducible and functions as a positive regulator of *FvbHLH9*-controlled anthocyanin accumulation in strawberry (*Fragaria* × *ananassa*) [61], and anthocyanin accumulation may also be positively regulated via activation of *PAP1* by HY5 in *Arabidopsis* [87]. Therefore, we propose that the *BnaA10g21200D/HY5*, with a $\log_2$ (FC) increase by 1.86 (Supplementary Table S3), is also involved in regulation of HL response and anthocyanin biosynthesis in rapeseed. In addition, the *TTG1/BnaC06g38570D* gene was also found to be induced in rapeseed leaves under HL stress. However, the induction level was lower in comparison with those of genes encoding bHLH or MYB TFs (Figure 9C). This result suggests that genes encoding *WD40* might not be as essential as those encoding *bHLH* and *MYB* TFs for the induction of anthocyanin biosynthesis under HL stress.

Numerous studies have shown that JA can positively impact anthocyanin accumulation in responses to light conditions. For example, Shan et al. reported that the molecular mechanism of JA-induced anthocyanin accumulation involves the anthocyanin biosynthesis-related regulators *PAP1*, *PAP2* and *GL3*, whose expression is induced through the activity of the JA-signaling pathway in *Arabidopsis* [87]. In our study, all of the key enzymes involved in the JA-biosynthetic pathway were up-regulated under HL stress (Figure 8A). Notably, *BnaA07g24880D* ($\log_2$ (FC) = 3.11), *BnaC07g22910D* ($\log_2$ (FC) = 3.11), and *BnaC08g37440D* ($\log_2$ (FC) = 1.50), which encode lipoxygenase 2 (LOX2), allene oxide cyclase 2 (AOC2), and 12-oxophytodienoate reductase 2 (OPR2), respectively, were all up-regulated in the leaves of rapeseed seedlings exposed to HL (Figure 8A), and the RT-qPCR data also support this finding (Figure 8B). Additionally, the gene *BnaC07g15660D* encoding a carboxyl methyltransferase that catalyzes the formation of methyl jasmonate (MeJA), another form of biologically active JA in addition to JA-Ile [88], was dramatically up-regulated with the $\log_2$ (FC) value of 4.77 (Supplementary Table S9). Previous studies have revealed that MeJA applications could enhance anthocyanin accumulation through up-regulating the transcription of related genes [89,90]. These finding together suggest that the up-regulation of JA biosynthesis by HL is required for the activation of anthocyanin biosynthesis and accumulation by modulating the genes encoding TFs (e.g., *bHLHs* and *MYBs*) that participate in the regulation of anthocyanin biosynthesis.

## 4. Materials and Methods

### 4.1. Plant Material

The inbred spring *Brassica napus* cultivar "GLH4" was used in this study. "GLH4" is an inbred line belonging to the spring-planted *Brassica napus* with characteristics of short stalk, compact plant architecture, short and upright branches, and lodging resistance. The growth period of "GLH4" is about 155 days. It begins to flower at about 60th day after germination, and the flowering period will extend to be 50 days. It is well-fitted for planting in the agricultural areas of the eastern Qinghai Province.

### 4.2. Experimental Design

"GLH4" seeds were sown in plastic pots, which were then placed in an environmental chamber (Percival LT-36VL, Perry, IA, USA) with a 16-h light photoperiod (light intensity of 300 μmol $m^{-2}$ $s^{-1}$) followed by 8-h of darkness. The temperatures were set at 17 °C and 4 °C during the light and dark periods, respectively. The relative humidity was maintained at 90%. When the two cotyledons were fully extended, the temperature during the dark period was changed to 10 °C, while the other conditions remained the same. When two true leaves were fully extended, seedlings of similar size were divided into two groups: NL, which served as the control, and HL which served as the treatment group. The HL group was subjected to a light intensity of 600 μmol $m^{-2}$ $s^{-1}$, while the light intensity of the NL group was left unchanged. After 16 h, the color change of seedlings from two groups were observed and photographed. In addition, leaves from the two groups were collected, immediately frozen in liquid nitrogen, and subsequently stored at −80 °C until they were processed for the extraction of total RNA used for transcriptome sequencing and RT-qPCR, or used in analyses of chemical constituents. Each biological replicate represented a pool of leaves collected from six plants, and three biological replicates were included for the NL and HL plants.

### 4.3. Analysis of Total Flavonoids, Total Anthocyanins and Other Anthocyanin Compounds

The extraction of total flavonoids and anthocyanins from fresh leaves of rapeseed seedlings under NL and HL conditions was carried out as previously described [64]. The contents of total flavonoids and total anthocyanins were assessed following the aluminum colorimetric procedure using the 510 and 530 nm wave lengths, respectively, with rutin being used as a standard [91].

The qualitative and quantitative analyses of 6 anthocyanin compounds, cyanidin, delphinidin, petunidin and their glucosides, were performed on a liquid chromatography-mass spectrometer (Agilent Technologies, Palo Alto, CA, USA). Purchased substances (Chengdu Herbpurity Co., Ltd., Chengdu, China) were dissolved in methanol at the 0.37–0.60 mg $mL^{-1}$ and final concentrations were used as standards. Hence, 10 μL of each sample and standard were injected in each analysis. LC-MS analysis of anthocyanins was carried out using a reverse phase Hypersil GOLD™ C18 C18 column (150 mm × 2.1 mm, 2.6 μm, ThermoFisher, Waltham, MA, USA) with 1 mL $min^{-1}$ flow rate over a 60-min gradient program. Further, 0.5% formic acid and acetonitrile containing 0.1% formic acid were used as solvent A and solvent B, respectively, in a linear step described previously [65].

### 4.4. RNA Isolation, Library Construction and Sequencing

Total RNA was extracted from the leaf samples using TRIzol reagent (Invitrogen, Carlsbad, CA, USA) following the manufacturer's instructions, and the extracted RNA was subsequently treated with DNase. RNA concentration and integrity were estimated using a NanoDrop 2000 (Thermo Scientific, Waltham, MA, USA) and an Agilent 2100 BioAnalyzer (Agilent Technologies, Santa Clara, CA, USA), respectively. Enrichment of mRNA from total RNA was conducted using poly-T oligo-attached magnetic beads. The mRNA fragments were then randomly broken into short fragments that were used as a template to synthesize cDNA for construction of libraries for sequencing. The constructed six cDNA libraries including three biological replicates in NL and HL groups, respectively, were sequenced on an Illumina HiSeq 2000 platform, and paired-end reads were generated. The sequenced raw reads were submitted to the NCBI Sequence Read Archive (SRA) database (https://www.ncbi.nlm.nih.gov/sra) [92] under the SRA accession number PRJNA667589.

### 4.5. Assembly, Data Analysis and Functional Annotation

Clean reads were secured by removing adapter sequences, poly-N sequences, and low-quality (Q-value $\leq 10$) reads from the raw reads. Q20, Q30, N50 and GC content values were then calculated for the clean reads. All further analyses were based on the clean data with high quality. The *B. rapa* reference genome (PRJNA293435) was used for alignment and the clean reads were mapped to the reference genome using the Hisat2 program to obtain gene annotation and position information, and the unique sequence characteristics of each of the sequenced libraries [58].

### 4.6. Identification and Functional Annotation of DEGs

FPKM for each gene were determined based on the length of the gene and read counts mapped to it. Statistical comparison of FPKM values between the two sample groups was conducted using DEseq2 software to identify DEGs using a threshold of $|\log_2$ (fold change)$| \geq 1$ and $FDR < 0.05$ [93]. TFs were predicted by searching against the Plant Transcriptional Factor Database (http://www.plntfdb.bio.uni-potsdam.de/v3.0/) [94].

GO enrichment analysis of identified DEGs was performed using the GOseq routine in R, in which gene length bias was corrected, and GO terms with a $p$-value $< 0.05$ were considered as significantly enriched. KOBAS software was used to test the statistical significance of DEG enrichments in the KEGG pathways. Annotation and classification of DEGs were conducted using MapMan (http://mapman.gabipd.org/; version 3.6.0RC1) [59]. Annotated files used for MapMan analysis were downloaded from https://mapman.gabipd.org/MapManStore [59].

### 4.7. RT-qPCR Analysis

First-strand cDNA synthesis was performed with 2 μg total RNA from each replicate using the PrimeScriptRT reagent kit with gDNA Eraser (Takara, Dalian, China). The selected genes and their specific primers are provided in Supplementary Table S4. RT-qPCR was conducted in three independent biological replicates using a LightCycler 480 System (Roche, Basel, Switzerland) and TB Green Premix Ex TaqTM II (Tli RNaseH Plus) (Takara, Dalian, China). The reaction system and procedures for RT-qPCR were generally consistent with the parameters described in a previous study [63]. Relative expression levels were determined using the $2^{-\Delta\Delta CT}$ method [95]. *BnACT* was used as a reference gene in data analysis.

## 5. Conclusions

Light is a fundamental requirement for plant growth and development. However, excessive light capture can result in irreversible damage to chloroplasts and cell metabolism. Rapeseed is a widely grown oil crop planted in the QTP regions of China, where long-day solar and UV radiation, drought, low temperature, and low-oxygen content characterize the extreme environmental conditions. Until now, however, few studies have been conducted on the underlying molecular mechanisms involved in the response of rapeseed to HL stress. In the present study, 7390 DEGs, including 4393 up- and 2997 down-regulated genes, were identified in a transcriptome analysis of leaves of rapeseed seedlings in a HL/NL comparison. Further analyses suggested that the HL-induced DEGs were especially enriched in anthocyanin-biosynthetic and JA-biosynthetic pathways. Specially, late anthocyanin biosynthesis-related genes, such as those encoding BnANS and BnDFR, and genes involved in the regulation of anthocyanin biosynthesis, such as those encoding *BnPAP2* and *BnGL3* TFs, were highly induced by HL stress. Notably, the accumulation of anthocyanins, which may act as photoprotectants, was significantly promoted in rapeseed seedlings under HL conditions. In addition, genes related to the JA-biosynthetic pathway were also activated as part of the HL response and proposed to participate in the up-regulation of anthocyanin biosynthesis in rapeseed seedlings through JA-mediated activation of the expression of genes encoding TFs like *BnPAP1*, *BnPAP2*, and *BnGL3*. Collectively, the results of the present study provide new information pertaining to the crucial role of JA-anthocyanin biosynthesis cascade in the adaptation of rapeseed seedlings to HL stress. The insights gained from this study may help in the effort to generate HL-tolerant varieties through traditional breeding and/or genetic engineering approaches. Due to its beneficial impacts on human health, the rapeseed could be a potential source for edible anthocyanins, and rapeseed quality with high contents of diverse anthocyanins could be promoted by engineering these candidate genes involved in the anthocyanin biosynthesis.

**Supplementary Materials:** The following are available online at https://www.mdpi.com/article/10.3390/ijms222313027/s1.

**Author Contributions:** Conceptualization, Y.L. and H.Y.; methodology, Y.L. and H.Y.; software, Y.L. and H.Y.; validation, S.T., X.T. and S.Z.; formal analysis, Y.L. and H.Y.; investigation, Y.L.; resources, Y.L.; data curation, Y.L. and H.Y.; writing—original draft preparation, H.Y.; writing—review and editing, H.Y. and L.-S.P.T.; visualization, H.Y. and L.-S.P.T.; supervision, Y.L. and H.Y.; project administration, Y.L.; funding acquisition, Y.L. and H.Y. All authors have read and agreed to the published version of the manuscript.

**Funding:** This research was funded by the Program from Science and Technology Department of Qinghai Province (2020-ZJ-745) and National Natural Science Foundation of China (31860060).

**Institutional Review Board Statement:** Not applicable.

**Informed Consent Statement:** Not applicable.

**Data Availability Statement:** Not applicable.

**Acknowledgments:** We would like to thank Haiqing Wang of Northwest Institute of Plateau Biology, CAS, for providing support in data analysis.

**Conflicts of Interest:** The authors declare no conflict of interest.

## References

1. Wu, D.; Liang, Z.; Yan, T.; Xu, Y.; Xuan, L.; Tang, J.; Zhou, G.; Lohwasser, U.; Hua, S.; Wang, H.; et al. Whole-Genome Resequencing of a Worldwide Collection of Rapeseed Accessions Reveals the Genetic Basis of Ecotype Divergence. *Mol. Plant* **2019**, *12*, 30–43. [CrossRef] [PubMed]
2. Wang, J.; Jiao, J.; Zhou, M.; Jin, Z.; Yu, Y.; Liang, M. Physiological and Transcriptional Responses of Industrial Rapeseed (*Brassica napus*) Seedlings to Drought and Salinity Stress. *Int. J. Mol. Sci.* **2019**, *20*, 5604. [CrossRef]
3. Qu, C.; Yin, N.; Chen, S.; Wang, S.; Chen, X.; Zhao, H.; Shen, S.; Fu, F.; Zhou, B.; Xu, X.; et al. Comparative Analysis of the Metabolic Profiles of Yellow- versus Black-Seeded Rapeseed Using UPLC-HESI-MS/MS and Transcriptome Analysis. *J. Agric. Food Chem.* **2020**, *68*, 3033–3049. [CrossRef]
4. Hatzig, S.; Zaharia, L.I.; Abrams, S.; Hohmann, M.; Legoahec, L.; Bouchereau, A.; Nesi, N.; Snowdon, R.J. Early osmotic adjustment responses in drought-resistant and drought-sensitive oilseed rape. *J. Integr. Plant Biol.* **2014**, *56*, 797–809. [CrossRef]
5. Wang, P.; Yang, C.; Chen, H.; Song, C.; Zhang, X.; Wang, D. Transcriptomic basis for drought-resistance in *Brassica napus* L. *Sci. Rep.* **2017**, *7*, 40532. [CrossRef] [PubMed]
6. Huang, R.; Liu, Z.; Xing, M.; Yang, Y.; Wu, X.; Liu, H.; Liang, W. Heat Stress Suppresses *Brassica napus* Seed Oil Accumulation by Inhibition of Photosynthesis and BnWRI1 Pathway. *Plant Cell Physiol.* **2019**, *60*, 1457–1470. [CrossRef]
7. Pu, Y.; Liu, L.; Wu, J.; Zhao, Y.; Bai, J.; Ma, L.; Yue, J.; Jin, J.; Niu, Z.; Fang, Y.; et al. Transcriptome Profile Analysis of Winter Rapeseed (*Brassica napus* L.) in Response to Freezing Stress, Reveal Potentially Connected Events to Freezing Stress. *Int. J. Mol. Sci.* **2019**, *20*, 2771. [CrossRef] [PubMed]
8. Zhang, Z.H.; Zhou, T.; Tang, T.J.; Song, H.X.; Guan, C.Y.; Huang, J.Y.; Hua, Y.P. A multiomics approach reveals the pivotal role of subcellular reallocation in determining rapeseed resistance to cadmium toxicity. *J. Exp. Bot.* **2019**, *70*, 5437–5455. [CrossRef] [PubMed]
9. Qiu, Z.; Zhu, L.; He, L.; Chen, D.; Zeng, D.; Chen, G.; Hu, J.; Zhang, G.; Ren, D.; Dong, G.; et al. DNA damage and reactive oxygen species cause cell death in the rice local lesions 1 mutant under high light and high temperature. *New Phytol.* **2019**, *222*, 349–365. [CrossRef]
10. Pinnola, A.; Bassi, R. Molecular mechanisms involved in plant photoprotection. *Biochem. Soc. Trans.* **2018**, *46*, 467–482. [CrossRef] [PubMed]

11. Ruban, A.V. Nonphotochemical Chlorophyll Fluorescence Quenching: Mechanism and Effectiveness in Protecting Plants from Photodamage. *Plant Physiol.* **2016**, *170*, 1903–1916. [CrossRef]
12. Kasahara, M.; Kagawa, T.; Oikawa, K.; Suetsugu, N.; Miyao, M.; Wada, M. Chloroplast avoidance movement reduces photodamage in plants. *Nature* **2002**, *420*, 829–832. [CrossRef] [PubMed]
13. Wada, M. Chloroplast movement. *Plant Sci.* **2013**, *210*, 177–182. [CrossRef]
14. Evans, J.R.; Poorter, H. Photosynthetic acclimation of plants to growth irradiance: The relative importance of specific leaf area and nitrogen partitioning in maximizing carbon gain. *Plant Cell Environ.* **2001**, *24*, 755–767. [CrossRef]
15. Huang, J.; Zhao, X.; Chory, J. The *Arabidopsis* Transcriptome Responds Specifically and Dynamically to High Light Stress. *Cell Rep.* **2019**, *29*, 4186–4199.e3. [CrossRef]
16. Foyer, C.H.; Bloom, A.J.; Queval, G.; Noctor, G. Photorespiratory metabolism: Genes, mutants, energetics, and redox signaling. *Annu. Rev. Plant Biol.* **2009**, *60*, 455–484. [CrossRef]
17. Nishizawa, A.; Yabuta, Y.; Shigeoka, S. Galactinol and raffinose constitute a novel function to protect plants from oxidative damage. *Plant Physiol.* **2008**, *147*, 1251–1263. [CrossRef]
18. Jung, H.S.; Crisp, P.A.; Estavillo, G.M.; Cole, B.; Hong, F.; Mockler, T.C.; Pogson, B.J.; Chory, J. Subset of heat-shock transcription factors required for the early response of *Arabidopsis* to excess light. *Proc. Natl. Acad. Sci. USA* **2013**, *110*, 14474–14479. [CrossRef]
19. Dyson, B.C.; Allwood, J.W.; Feil, R.; Xu, Y.; Miller, M.; Bowsher, C.G.; Goodacre, R.; Lunn, J.E.; Johnson, G.N. Acclimation of metabolism to light in *Arabidopsis thaliana*: The glucose 6-phosphate/phosphate translocator *GPT2* directs metabolic acclimation. *Plant Cell Environ.* **2015**, *38*, 1404–1417. [CrossRef]
20. Nägele, T.; Heyer, A.G. Approximating subcellular organisation of carbohydrate metabolism during cold acclimation in different natural accessions of *Arabidopsis thaliana*. *New Phytol.* **2013**, *198*, 777–787. [CrossRef]
21. Patzke, K.; Prananingrum, P.; Klemens, P.A.W.; Trentmann, O.; Rodrigues, C.M.; Keller, I.; Fernie, A.R.; Geigenberger, P.; Bölter, B.; Lehmann, M.; et al. The Plastidic Sugar Transporter pSuT Influences Flowering and Affects Cold Responses. *Plant Physiol.* **2019**, *179*, 569–587. [CrossRef]
22. Garcia-Molina, A.; Leister, D. Accelerated relaxation of photoprotection impairs biomass accumulation in *Arabidopsis*. *Nat Plants* **2020**, *6*, 9–12. [CrossRef] [PubMed]
23. Kromdijk, J.; Głowacka, K.; Leonelli, L.; Gabilly, S.T.; Iwai, M.; Niyogi, K.K.; Long, S.P. Improving photosynthesis and crop productivity by accelerating recovery from photoprotection. *Science* **2016**, *354*, 857–861. [CrossRef] [PubMed]
24. Alappat, B.; Alappat, J. Anthocyanin Pigments: Beyond Aesthetics. *Molecules* **2020**, *25*, 5500. [CrossRef]
25. Jaakola, L. New insights into the regulation of anthocyanin biosynthesis in fruits. *Trends Plant Sci.* **2013**, *18*, 477–483. [CrossRef]
26. Jezek, M.; Zörb, C.; Merkt, N.; Geilfus, C.M. Anthocyanin Management in Fruits by Fertilization. *J. Agric. Food Chem.* **2018**, *66*, 753–764. [CrossRef]
27. Butelli, E.; Titta, L.; Giorgio, M.; Mock, H.P.; Matros, A.; Peterek, S.; Schijlen, E.G.; Hall, R.D.; Bovy, A.G.; Luo, J.; et al. Enrichment of tomato fruit with health-promoting anthocyanins by expression of select transcription factors. *Nat. Biotechnol.* **2008**, *26*, 1301–1308. [CrossRef]
28. Lila, M.A.; Burton-Freeman, B.; Grace, M.; Kalt, W. Unraveling Anthocyanin Bioavailability for Human Health. *Annu. Rev. Food Sci. Technol.* **2016**, *7*, 375–393. [CrossRef]
29. Sun, C.; Deng, L.; Du, M.; Zhao, J.; Chen, Q.; Huang, T.; Jiang, H.; Li, C.B.; Li, C. A Transcriptional Network Promotes Anthocyanin Biosynthesis in Tomato Flesh. *Mol. Plant* **2020**, *13*, 42–58. [CrossRef]
30. Li, P.; Li, Y.J.; Zhang, F.J.; Zhang, G.Z.; Jiang, X.Y.; Yu, H.M.; Hou, B.K. The *Arabidopsis* UDP-glycosyltransferases UGT79B2 and UGT79B3, contribute to cold, salt and drought stress tolerance via modulating anthocyanin accumulation. *Plant J.* **2017**, *89*, 85–103. [CrossRef]
31. Li, W.; Nguyen, K.H.; Chu, H.D.; Watanabe, Y.; Osakabe, Y.; Sato, M.; Toyooka, K.; Seo, M.; Tian, L.; Tian, C.; et al. Comparative functional analyses of *DWARF14* and *KARRIKIN INSENSITIVE 2* in drought adaptation of *Arabidopsis thaliana*. *Plant J.* **2020**, *103*, 111–127. [CrossRef] [PubMed]
32. Naing, A.H.; Ai, T.N.; Lim, K.B.; Lee, I.J.; Kim, C.K. Overexpression of Rosea1 From Snapdragon Enhances Anthocyanin Accumulation and Abiotic Stress Tolerance in Transgenic Tobacco. *Front. Plant Sci.* **2018**, *9*, 1070. [CrossRef] [PubMed]
33. Nguyen, K.H.; Ha, C.V.; Nishiyama, R.; Watanabe, Y.; Leyva-González, M.A.; Fujita, Y.; Tran, U.T.; Li, W.; Tanaka, M.; Seki, M.; et al. *Arabidopsis* type B cytokinin response regulators *ARR1*, *ARR10*, and *ARR12* negatively regulate plant responses to drought. *Proc. Natl. Acad. Sci. USA* **2016**, *113*, 3090–3095. [CrossRef]
34. Chen, C.; Li, H.; Zhang, D.; Li, P.; Ma, F. The role of anthocyanin in photoprotection and its relationship with the xanthophyll cycle and the antioxidant system in apple peel depends on the light conditions. *Physiol. Plant* **2013**, *149*, 354–366. [CrossRef] [PubMed]
35. Zheng, X.T.; Chen, Y.L.; Zhang, X.H.; Cai, M.L.; Yu, Z.C.; Peng, C.L. ANS-deficient *Arabidopsis* is sensitive to high light due to impaired anthocyanin photoprotection. *Funct. Plant Biol.* **2019**, *46*, 756–765. [CrossRef]
36. Saigo, T.; Wang, T.; Watanabe, M.; Tohge, T. Diversity of anthocyanin and proanthocyanin biosynthesis in land plants. *Curr. Opin. Plant Biol.* **2020**, *55*, 93–99. [CrossRef]
37. Liu, Y.; Tikunov, Y.; Schouten, R.E.; Marcelis, L.F.M.; Visser, R.G.F.; Bovy, A. Anthocyanin Biosynthesis and Degradation Mechanisms in Solanaceous Vegetables: A Review. *Front. Chem.* **2018**, *6*, 52. [CrossRef] [PubMed]

38. Khoo, H.E.; Azlan, A.; Tang, S.T.; Lim, S.M. Anthocyanidins and anthocyanins: Colored pigments as food, pharmaceutical ingredients, and the potential health benefits. *Food Nutr. Res.* **2017**, *61*, 1361779. [CrossRef] [PubMed]
39. Naing, A.H.; Kim, C.K. Roles of *R2R3-MYB* transcription factors in transcriptional regulation of anthocyanin biosynthesis in horticultural plants. *Plant Mol. Biol.* **2018**, *98*, 1–18. [CrossRef]
40. Yan, H.; Pei, X.; Zhang, H.; Li, X.; Zhang, X.; Zhao, M.; Chiang, V.L.; Sederoff, R.R.; Zhao, X. MYB-Mediated Regulation of Anthocyanin Biosynthesis. *Int. J. Mol. Sci.* **2021**, *22*, 3103. [CrossRef]
41. Xu, W.; Dubos, C.; Lepiniec, L. Transcriptional control of flavonoid biosynthesis by MYB-bHLH-WDR complexes. *Trends Plant Sci.* **2015**, *20*, 176–185. [CrossRef] [PubMed]
42. Xie, S.; Lei, Y.; Chen, H.; Li, J.; Chen, H.; Zhang, Z. R2R3-MYB Transcription Factors Regulate Anthocyanin Biosynthesis in Grapevine Vegetative Tissues. *Front. Plant Sci.* **2020**, *11*, 527. [CrossRef]
43. Carletti, G.; Lucini, L.; Busconi, M.; Marocco, A.; Bernardi, J. Insight into the role of anthocyanin biosynthesis-related genes in Medicago truncatula mutants impaired in pigmentation in leaves. *Plant Physiol. Biochem.* **2013**, *70*, 123–132. [CrossRef]
44. Lin, Y.; Jiang, L.; Chen, Q.; Li, Y.; Zhang, Y.; Luo, Y.; Zhang, Y.; Sun, B.; Wang, X.; Tang, H. Comparative Transcriptome Profiling Analysis of Red- and White-Fleshed Strawberry (*Fragaria × ananassa*) Provides New Insight into the Regulation of the Anthocyanin Pathway. *Plant Cell Physiol.* **2018**, *59*, 1844–1859. [CrossRef] [PubMed]
45. Chen, W.; Zhang, M.; Zhang, G.; Li, P.; Ma, F. Differential Regulation of Anthocyanin Synthesis in Apple Peel under Different Sunlight Intensities. *Int. J. Mol. Sci.* **2019**, *20*, 6060. [CrossRef] [PubMed]
46. Cominelli, E.; Gusmaroli, G.; Allegra, D.; Galbiati, M.; Wade, H.K.; Jenkins, G.I.; Tonelli, C. Expression analysis of anthocyanin regulatory genes in response to different light qualities in *Arabidopsis thaliana*. *J. Plant Physiol.* **2008**, *165*, 886–894. [CrossRef]
47. Li, M.; Sun, L.; Gu, H.; Cheng, D.; Guo, X.; Chen, R.; Wu, Z.; Jiang, J.; Fan, X.; Chen, J. Genome-wide characterization and analysis of *bHLH* transcription factors related to anthocyanin biosynthesis in spine grapes (*Vitis davidii*). *Sci. Rep.* **2021**, *11*, 6863. [CrossRef]
48. Zhao, P.; Li, X.; Jia, J.; Yuan, G.; Chen, S.; Qi, D.; Cheng, L.; Liu, G. *bHLH92* from sheepgrass acts as a negative regulator of anthocyanin/proanthocyandin accumulation and influences seed dormancy. *J. Exp. Bot.* **2019**, *70*, 269–284. [CrossRef]
49. Peng, Y.; Lin-Wang, K.; Cooney, J.M.; Wang, T.; Espley, R.V.; Allan, A.C. Differential regulation of the anthocyanin profile in purple kiwifruit (*Actinidia species*). *Hortic. Res.* **2019**, *6*, 3. [CrossRef]
50. Deng, J.; Li, J.; Su, M.; Lin, Z.; Chen, L.; Yang, P. A *bHLH* gene *NnTT8* of Nelumbo nucifera regulates anthocyanin biosynthesis. *Plant Physiol. Biochem.* **2021**, *158*, 518–523. [CrossRef]
51. Verma, V.; Ravindran, P.; Kumar, P.P. Plant hormone-mediated regulation of stress responses. *BMC Plant Biol.* **2016**, *16*, 86. [CrossRef] [PubMed]
52. Gupta, A.; Sinha, R.; Fernandes, J.L.; Abdelrahman, M.; Burritt, D.J.; Tran, L.P. Phytohormones regulate convergent and divergent responses between individual and combined drought and pathogen infection. *Crit. Rev. Biotechnol.* **2020**, *40*, 320–340. [CrossRef] [PubMed]
53. Kazan, K. Diverse roles of jasmonates and ethylene in abiotic stress tolerance. *Trends Plant Sci.* **2015**, *20*, 219–229. [CrossRef] [PubMed]
54. Suzuki, N.; Devireddy, A.R.; Inupakutika, M.A.; Baxter, A.; Miller, G.; Song, L.; Shulaev, E.; Azad, R.K.; Shulaev, V.; Mittler, R. Ultra-fast alterations in mRNA levels uncover multiple players in light stress acclimation in plants. *Plant J.* **2015**, *84*, 760–772. [CrossRef] [PubMed]
55. Qi, T.; Song, S.; Ren, Q.; Wu, D.; Huang, H.; Chen, Y.; Fan, M.; Peng, W.; Ren, C.; Xie, D. The Jasmonate-ZIM-domain proteins interact with the WD-Repeat/bHLH/MYB complexes to regulate Jasmonate-mediated anthocyanin accumulation and trichome initiation in *Arabidopsis thaliana*. *Plant Cell* **2011**, *23*, 1795–1814. [CrossRef]
56. Wasternack, C.; Strnad, M. Jasmonates are signals in the biosynthesis of secondary metabolites—Pathways, transcription factors and applied aspects—A brief review. *New Biotechnol.* **2019**, *48*, 1–11. [CrossRef]
57. Li, W.; Nguyen, K.H.; Chu, H.D.; Ha, C.V.; Watanabe, Y.; Osakabe, Y.; Leyva-González, M.A.; Sato, M.; Toyooka, K.; Voges, L.; et al. The karrikin receptor *KAI2* promotes drought resistance in *Arabidopsis thaliana*. *PLoS Genet.* **2017**, *13*, e1007076. [CrossRef]
58. Kim, D.; Langmead, B.; Salzberg, S.L. HISAT: A fast spliced aligner with low memory requirements. *Nat. Methods* **2015**, *12*, 357–360. [CrossRef]
59. MapManStore. Available online: https://mapman.gabipd.org/mapmanstore (accessed on 5 July 2021).
60. Springob, K.; Nakajima, J.; Yamazaki, M.; Saito, K. Recent advances in the biosynthesis and accumulation of anthocyanins. *Nat. Prod. Rep.* **2003**, *20*, 288–303. [CrossRef]
61. Li, Y.; Xu, P.; Chen, G.; Wu, J.; Liu, Z.; Lian, H. FvbHLH9 Functions as a Positive Regulator of Anthocyanin Biosynthesis by Forming a *HY5-bHLH9* Transcription Complex in Strawberry Fruits. *Plant Cell Physiol.* **2020**, *61*, 826–837. [CrossRef]
62. Liu, C.C.; Chi, C.; Jin, L.J.; Zhu, J.; Yu, J.Q.; Zhou, Y.H. The bZip transcription factor *HY5* mediates *CRY1*a-induced anthocyanin biosynthesis in tomato. *Plant Cell Environ.* **2018**, *41*, 1762–1775. [CrossRef]
63. De Wit, M.; Galvão, V.C.; Fankhauser, C. Light-Mediated Hormonal Regulation of Plant Growth and Development. *Annu. Rev. Plant Biol.* **2016**, *67*, 513–537. [CrossRef] [PubMed]
64. Schneider, S.; Ziegler, C.; Melzer, A. Growth towards light as an adaptation to high light conditions in Chara branches. *New Phytol.* **2006**, *172*, 83–91. [CrossRef] [PubMed]

65. Yin, H.; Zhou, H.; Wang, W.; Tran, L.P.; Zhang, B. Transcriptome Analysis Reveals Potential Roles of Abscisic Acid and Polyphenols in Adaptation of *Onobrychis viciifolia* to Extreme Environmental Conditions in the Qinghai-Tibetan Plateau. *Biomolecules* **2020**, *10*, 967. [CrossRef] [PubMed]
66. Jung, J.H.; Kim, H.Y.; Kim, H.S.; Jung, S.H. Transcriptome analysis of *Panax ginseng* response to high light stress. *J. Ginseng Res.* **2020**, *44*, 312–320. [CrossRef] [PubMed]
67. Anna, B.B.; Grzegorz, B.; Marek, K.; Piotr, G.; Marcin, F. Exposure to High-Intensity Light Systemically Induces Micro-Transcriptomic Changes in *Arabidopsis thaliana* Roots. *Int. J. Mol. Sci.* **2019**, *20*, 5131. [CrossRef] [PubMed]
68. Wang, J.; Guo, M.; Li, Y.; Wu, R.; Zhang, K. High-throughput Transcriptome Sequencing Reveals the Role of Anthocyanin Metabolism in *Begonia semperflorens* Under High Light Stress. *Photochem. Photobiol.* **2018**, *94*, 105–114. [CrossRef]
69. Xu, Z.; Mahmood, K.; Rothstein, S.J. ROS Induces Anthocyanin Production via Late Biosynthetic Genes and Anthocyanin Deficiency Confers the Hypersensitivity to ROS-Generating Stresses in *Arabidopsis*. *Plant Cell Physiol.* **2017**, *58*, 1364–1377. [CrossRef]
70. Hinojosa-Gómez, J.; San Martín-Hernández, C.; Heredia, J.B.; León-Félix, J.; Osuna-Enciso, T.; Muy-Rangel, M.D. Anthocyanin Induction by Drought Stress in the Calyx of Roselle Cultivars. *Molecules* **2020**, *25*, 1555. [CrossRef]
71. Naing, A.H.; Kim, C.K. Abiotic stress-induced anthocyanins in plants: Their role in tolerance to abiotic stresses. *Physiol. Plant* **2021**, *172*, 1711–1723. [CrossRef]
72. Nakabayashi, R.; Yonekura-Sakakibara, K.; Urano, K.; Suzuki, M.; Yamada, Y.; Nishizawa, T.; Matsuda, F.; Kojima, M.; Sakakibara, H.; Shinozaki, K.; et al. Enhancement of oxidative and drought tolerance in *Arabidopsis* by overaccumulation of antioxidant flavonoids. *Plant J.* **2014**, *77*, 367–379. [CrossRef] [PubMed]
73. Gururani, M.A.; Venkatesh, J.; Tran, L.S. Regulation of Photosynthesis during Abiotic Stress-Induced Photoinhibition. *Mol. Plant* **2015**, *8*, 1304–1320. [CrossRef]
74. Kimura, M.; Yamamoto, Y.Y.; Seki, M.; Sakurai, T.; Sato, M.; Abe, T.; Yoshida, S.; Manabe, K.; Shinozaki, K.; Matsui, M. Identification of *Arabidopsis* genes regulated by high light-stress using cDNA microarray. *Photochem. Photobiol.* **2003**, *77*, 226–233. [CrossRef] [PubMed]
75. Choudhury, F.K.; Rivero, R.M.; Blumwald, E.; Mittler, R. Reactive oxygen species, abiotic stress and stress combination. *Plant J.* **2017**, *90*, 856–867. [CrossRef] [PubMed]
76. Li, W.; Nguyen, K.H.; Tran, C.D.; Watanabe, Y.; Tian, C.; Yin, X.; Li, K.; Yang, Y.; Guo, J.; Miao, Y.; et al. Negative Roles of Strigolactone-Related SMXL6, 7 and 8 Proteins in Drought Resistance in *Arabidopsis*. *Biomolecules* **2020**, *10*, 607. [CrossRef]
77. Yu, Z.C.; Zheng, X.T.; Lin, W.; He, W.; Shao, L.; Peng, C.L. Photoprotection of *Arabidopsis* leaves under short-term high light treatment: The antioxidant capacity is more important than the anthocyanin shielding effect. *Plant Physiol. Biochem.* **2021**, *166*, 258–269. [CrossRef]
78. Zheng, X.T.; Yu, Z.C.; Tang, J.W.; Cai, M.L.; Chen, Y.L.; Yang, C.W.; Chow, W.S.; Peng, C.L. The major photoprotective role of anthocyanins in leaves of *Arabidopsis thaliana* under long-term high light treatment: Antioxidant or light attenuator? *Photosynth. Res.* **2020**, *149*, 25–40. [CrossRef] [PubMed]
79. Albert, N.W.; Lewis, D.H.; Zhang, H.; Irving, L.J.; Jameson, P.E.; Davies, K.M. Light-induced vegetative anthocyanin pigmentation in *Petunia*. *J. Exp. Bot.* **2009**, *60*, 2191–2202. [CrossRef] [PubMed]
80. Yang, Y.N.; Yao, G.F.; Zheng, D.; Zhang, S.L.; Wang, C.; Zhang, M.Y.; Wu, J. Expression differences of anthocyanin biosynthesis genes reveal regulation patterns for red pear coloration. *Plant Cell Rep.* **2015**, *34*, 189–198. [CrossRef] [PubMed]
81. Yuan, Y.W.; Sagawa, J.M.; Frost, L.; Vela, J.P.; Bradshaw, H.D., Jr. Transcriptional control of floral anthocyanin pigmentation in monkeyflowers (Mimulus). *New Phytol.* **2014**, *204*, 1013–1027. [CrossRef]
82. An, X.H.; Tian, Y.; Chen, K.Q.; Liu, X.J.; Liu, D.D.; Xie, X.B.; Cheng, C.G.; Cong, P.H.; Hao, Y.J. *MdMYB9* and *MdMYB11* are involved in the regulation of the JA-induced biosynthesis of anthocyanin and proanthocyanidin in apples. *Plant Cell Physiol.* **2015**, *56*, 650–662. [CrossRef] [PubMed]
83. Zhang, T.J.; Zheng, J.; Yu, Z.C.; Huang, X.D.; Zhang, Q.L.; Tian, X.S.; Peng, C.L. Functional characteristics of phenolic compounds accumulated in young leaves of two subtropical forest tree species of different successional stages. *Tree Physiol.* **2018**, *38*, 1486–1501. [CrossRef] [PubMed]
84. Maier, A.; Schrader, A.; Kokkelink, L.; Falke, C.; Welter, B.; Iniesto, E.; Rubio, V.; Uhrig, J.F.; Hülskamp, M.; Hoecker, U. Light and the E3 ubiquitin ligase COP1/SPA control the protein stability of the *MYB* transcription factors *PAP1* and *PAP2* involved in anthocyanin accumulation in Arabidopsis. *Plant J.* **2013**, *74*, 638–651. [CrossRef]
85. Gonzalez, A.; Zhao, M.; Leavitt, J.M.; Lloyd, A.M. Regulation of the anthocyanin biosynthetic pathway by the TTG1/bHLH/Myb transcriptional complex in *Arabidopsis* seedlings. *Plant J.* **2008**, *53*, 814–827. [CrossRef]
86. Shin, D.H.; Choi, M.; Kim, K.; Bang, G.; Cho, M.; Choi, S.B.; Choi, G.; Park, Y.I. *HY5* regulates anthocyanin biosynthesis by inducing the transcriptional activation of the *MYB75/PAP1* transcription factor in *Arabidopsis*. *FEBS Lett.* **2013**, *587*, 1543–1547. [CrossRef]
87. Shan, X.; Zhang, Y.; Peng, W.; Wang, Z.; Xie, D. Molecular mechanism for jasmonate-induction of anthocyanin accumulation in *Arabidopsis*. *J. Exp. Bot.* **2009**, *60*, 3849–3860. [CrossRef]
88. Gupta, A.; Bhardwaj, M.; Tran, L.P. Jasmonic Acid at the Crossroads of Plant Immunity and Pseudomonas syringae Virulence. *Int. J. Mol. Sci.* **2020**, *21*, 7482. [CrossRef] [PubMed]

89. Wang, Y.; Liu, W.; Jiang, H.; Mao, Z.; Wang, N.; Jiang, S.; Xu, H.; Yang, G.; Zhang, Z.; Chen, X. The *R2R3-MYB* transcription factor *MdMYB24*-like is involved in methyl jasmonate-induced anthocyanin biosynthesis in apple. *Plant Physiol. Biochem.* **2019**, *139*, 273–282. [CrossRef] [PubMed]
90. Premathilake, A.T.; Ni, J.; Shen, J.; Bai, S.; Teng, Y. Transcriptome analysis provides new insights into the transcriptional regulation of methyl jasmonate-induced flavonoid biosynthesis in pear calli. *BMC Plant Biol.* **2020**, *20*, 388. [CrossRef] [PubMed]
91. Jia, Z.S.; Tang, M.C.; Wu, J.M. The determination of flavonoid contents in mulberry and their scavenging effects on superoxide radicals. *Food Chem.* **1999**, *64*, 555–559. [CrossRef]
92. NCBI SRA Database. Available online: https://www.ncbi.nlm.nih.gov/sra (accessed on 16 May 2021).
93. Love, M.I.; Huber, W.; Anders, S. Moderated estimation of fold change and dispersion for RNA-seq data with DESeq2. *Genome Biol.* **2014**, *15*, 550. [CrossRef] [PubMed]
94. PlnTFDB. Available online: http://www.plntfdb.bio.uni-potsdam.de/v3.0 (accessed on 20 June 2021).
95. Livak, K.J.; Schmittgen, T.D. Analysis of relative gene expression data using real-time quantitative PCR and the 2(-Delta Delta C(T)) Method. *Methods* **2001**, *25*, 402–408. [CrossRef] [PubMed]

International Journal of
*Molecular Sciences*

MDPI

*Article*

# Alleles of the GRF3-2A Gene in Wheat and Their Agronomic Value

**Mikhail S. Bazhenov [1,*], Anastasiya G. Chernook [1,2], Ludmila A. Bespalova [3], Tatiana I. Gritsay [3], Nadezhda A. Polevikova [3], Gennady I. Karlov [1], Lubov A. Nazarova [1] and Mikhail G. Divashuk [1,2]**

1 Laboratory of Applied Genomics and Crop Breeding, All-Russia Research Institute of Agricultural Biotechnology, 127550 Moscow, Russia; irbis-sibri@yandex.ru (A.G.C.); karlovg@gmail.com (G.I.K.); lpukhova@yandex.ru (L.A.N.); divashuk@gmail.com (M.G.D.)
2 Kurchatov Genomics Center-ARRIAB, All-Russia Research Institute of Agricultural Biotechnology, 127550 Moscow, Russia
3 National Center of Grain Named after P.P. Lukyanenko, Department of Breeding and Seed Production of Wheat and Triticale, Central Estate of KNIISH, 350012 Krasnodar, Russia; bespalova_l_a@rambler.ru (L.A.B.); t_i_gritsay@mail.ru (T.I.G.); nadyapolevikova@gmail.com (N.A.P.)
* Correspondence: mikhabazhenov@gmail.com

**Abstract:** The Growth-regulating factors (GRF) are a family of plant-specific transcription factors that have roles in plant growth, development and stress response. In this study the diversity of the *TaGRF3-2A* (TraesCS2A02G435100) gene was investigated in Russian bread wheat germplasm by means of next generation sequencing and molecular markers, and the results compared with those from multiple wheat genome and exome sequencing projects. The results showed that an allele possessing c.495G>T polymorphism found in Bezostaya 1 and designated as *TaGRF3-2Ab*, is connected with earlier heading and better grain filling under conditions of the Krasnodar Krai. *TaGRF3-2Ab* is more frequent among Russian winter wheat cultivars than in other germplasms found in the world, implying that it is adaptive for the Chernozem region. A new rare mutation of the *TaGRF3-2A* was found in the spring wheat cultivar Novosibirskaya 67. The molecular markers developed will facilitate utilization of *TaGRF3-2A* mutations in future agronomic studies and wheat improvement. Albeit *GRF3-2Ab* may be good at maintaining high milling quality of the grain, it should be used with caution in breeding of winter wheat cultivars in the perspective of climate change.

**Keywords:** GRF; transcription factor; diversity; NGS; earliness; kernel weight; test weight

**Citation:** Bazhenov, M.S.; Chernook, A.G.; Bespalova, L.A.; Gritsay, T.I.; Polevikova, N.A.; Karlov, G.I.; Nazarova, L.A.; Divashuk, M.G. Alleles of the GRF3-2A Gene in Wheat and Their Agronomic Value. *Int. J. Mol. Sci.* **2021**, *22*, 12376. https://doi.org/10.3390/ijms222212376

Academic Editors: Isabel Marques, Ana I. Ribeiro-Barros and José C. Ramalho

Received: 27 September 2021
Accepted: 15 November 2021
Published: 16 November 2021



## 1. Introduction

Growth-Regulating Factors (GRF) are a family of plant-specific transcription factors (TF) that have roles in plant growth, development and stress response [1]. The first *GRF* gene was characterized in rice as a factor involved in stem elongation under flooding or in response to gibberellic acid [2]. Later the *GRF* TF genes were identified in the model plant species *Arabidopsis thaliana*, as well as a number of agricultural crops [1,3,4]. These genes are mainly expressed in young growing tissues and organs [1,3,5].

The protein sequence of the GRF TF contains two conserved domains, QLQ (Gln, Leu, Gln) and WRC (Trp, Arg, Cys) in the N-terminus. The C-terminal region of GRF proteins is variable and may have transcriptional activation activity. The WRC domain consists of a nuclear localization signal and a DNA-binding motif and is involved in binding to cis-acting regions of gene promoters [2,3]. The QLQ domain is involved in interaction with GRF-interacting factor proteins (GIF) [6]. GIF proteins through their SNH (SYT N-terminal homology) domains interact with GRF proteins to form functional complexes that participate in regulation of expression of downstream genes [5]. DELLA proteins, accumulation of which is characteristic of green revolution varieties, can interfere with OsGRF4-GIF1 interaction [7].

The level of *GRF* gene expression is in part regulated by miRNA396 in the post-transcriptional stage. Thus, *GRF* genes participate in regulation of the same growth, development and stress-response processes as miRNA396 [8].

GRF TFs are involved not only in gibberellin, but also in other plant hormone signaling pathways. Auxin-response factors regulates expression of some *GRF* genes, like *AtGRF5* and *AtGRF6* in *Arabidopsis*, that lack miRNA396 regulation [9]. On the other hand, upregulation of *OsGRF6* in rice plants, having the miRNA396 blocked by target mimicry, activates auxin biosynthesis and signaling, stimulating development of auxiliary branches and spikelets in panicle [10]. Enhanced expression of *OsGRF4* in rice activates brassinosteroid-responsive genes that enhances the growth of seedlings and source leaves, and promotes grain development. Enhanced brassinosteroid responsiveness may in turn change the level of gibberellins in plants [11].

In *Arabidopsis, GRF* genes positively regulate leaf growth and are involved in regulation of the stress response to heat, drought, salinity and diseases [3,12]. In rice higher expression of *OsGRF4* was shown to be connected to higher 1000 kernel weight and grain yield [11,13], higher nitrogen uptake and assimilation in plants carrying the *Slr1* gibberellin-insensitive reduced height gene [7]. In rapeseed *GRF2* was found to enhance seed oil production by increasing leaf area and photosynthetic efficiency [4]. Some rare alleles of *TtGRF4-A* (a homolog of *TaGRF9-6A*) associated with increased grain weight were found in wild emmer wheat [14]. Thus, gain-of-function mutations of the *GRF* genes have great potential for increasing yields of agricultural crops through increasing the leaf area, the size of fruit organs and nutrient use efficiency [15].

In bread wheat (*Triticum aestivum* L.) 30 *TaGRF* genes were phylogenetically divided into four groups. They were designated using numbers from 1 to 12, each number being unique for a group of homeologs, and a chromosome name [1]. These genes are highly expressed in growing tissues including stem meristem and reproductive organs [1]. The level of expression of the *TaGRF* genes changes significantly under osmotic, drought or salt stress [1,5].

In this study we investigated the TraesCS2A02G435100 gene of bread wheat, which is *TaGRF3-2A* according to the nomenclature of Huang et al. (2021), and is known as *TaGRF3* according to Zan et al. (2020) [1,5]. This gene was chosen as one of the most similar to the rice *OsGRF4* gene with the highest percentage of query cover in BLAST output (however the most similar to *OsGRF4* are *TaGRF9-6A, -6B* and *-6D*). The *TaGRF3-2A* gene is primarily expressed in shoot apical meristem, stigma and ovary, seeds and young leaves. Also, it seems to be responsive to phosphorous starvation and drought stress [1].

## 2. Results

### 2.1. *TaGRF3-2A Alleles*

As a result of the next-generation sequencing of the polymerase-chain-rection-amplicons of the *TaGRF3-2A* gene (TraesCS2A01G435100) and its flanking regions in 19 winter bread wheat cultivars (Table S1), and comparing them with the genome sequences of 13 more varieties that were included in the Wheat 10+ Genomes Project [16] and the genome of the Chinese Spring bread wheat [17] (a total of 33 varieties of bread wheat), we found a total of 21 haplotypes. The considered region on wheat chromosome 2A covered 1,075 nucleotides before the start codon, and 1,389 nucleotides after the stop codon of the *TaGRF3-2A* gene, completely capturing the 5′ and 3′ untranslated regions (UTR) and part of the promoter. Also, in spelt wheat PI428198 (*Triticum spelta* L.), wild emmer wheat Zavitan (*Triticum dicoccoides* (Koern. ex Aschers. Et Graebn.) Schweinf.) and diploid wheat (*Triticum urartu* Thum. ex Gandil.), that were included in the comparison, three more haplotypes were found [16,18,19]. The *T. urartu* haplotype differed from the bread wheat haplotypes in multiple unique single-nucleotide variants, the haplotype of wild emmer had fewer differences, and the spelt haplotype did not have unique point mutations and differed only by a combination of the highly variable single nucleotide polymorphism (SNP) and the length of a microsatellite in the 5′UTR.

In the promoter of the gene, we found a frequently occurring polymorphism chr2A:g.687048627G>C (sequence variations are described according to HGVS recommendations, IWGSC RefSeq v1.0 wheat genome is used as reference) [20]. At this position, most varieties of bread wheat, as well as its wild relatives included in our study, have nucleotide variant 'C′, whereas wheat accessions Chinese Spring, Jagger, Lancer, CDC Stanley, Norin 61 and Novosibirskaya 67 have variant 'G′. Analysis of the promotor sequence using the PlantPAN 3.0 database showed that variant 'G′ gives an additional basic helix-loop-helix transcription factor binding site to the promoter. Several minor frequency SNPs were found in the promoter: an alternative variant of the polymorphism chr2A:g.687048382C>T is present in cultivars Grom and Altigo, variant 'T′ of the polymorphism chr2A:g.687048137G>T is present in varieties Velena and Paragon. Variant 'G′ of the polymorphism chr2A:g.687048328A>G is present only in *T. urartu* (PI428198) and *T. dicoccoides* (Zavitan). In the considered region of the promoter (chr2A:g.687047878_687048697) we found 15 unique single-nucleotide variants in the *T. urartu* haplotype.

In the 5′UTR of the *Grf3-2A* gene, there is a microsatellite repeat $(AG)_n$, where the number of repeats *n* varies from 11 to 37. Some variants of the microsatellite differ in loss of a single nucleotide A, resulting in a repeat length of 55 and 61 nucleotides (Figure 1). The wild emmer wheat Zavitan has a deletion of 12 nucleotides c.-102_-91del at a short distance before the microsatellite, and also a single nucleotide substitution c.-140A>G in the 5′UTR.

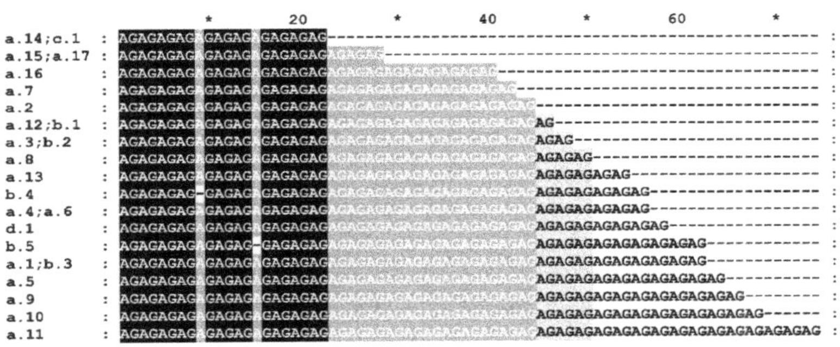


**Figure 1.** Microsatellite length variation in the 5′UTR of the *GRF3-2A* gene detected by sequencing. The *GRF3-2A* haplotypes are indicated in the left. An asterisk (*) in the upper line marks each 10-th nucleotide in the alignment between the numbers.

*T. urartu* PI428198 has two single-nucleotide substitutions c.*285A>G and c.*312A>T in the 3′UTR, and the bread wheat CDC Landmark has a substitution c.*8C>T.

In the 3′-end flanking region immediately after the transcription stop site, *T. urartu* PI428198 has a single nucleotide substitution chr2:g.687052487C>T, and the wild emmer Zavitan has a deletion of two nucleotides at the same point—chr2:g.687052487_687052488del.

Within the introns of the gene, differences from the reference Chinese Spring genome (IWGSC RefSeq v1.0) were found only in *T. urartu*. *T. urartu* accession PI428198 has an insertion of three nucleotides c.325+32_+33insTCC, as well as three single-nucleotide substitutions, c.325+169G>C, c.326-39A>G, and c.326-174A>G, in the second intron. Single-nucleotide substitutions c.689+136A>G and c.689+241T>C were found in the third intron.

In the protein-coding sequence of the *GRF3-2A* gene, a deletion of nine nucleotides c.126_134del was detected in the second exon in Novosibirskaya 67. This mutation led to the loss of three glutamine residues in the protein molecule in the polyglutamine region p.(Gln42_Gln44del). In the third exon, we found a c.495G>T polymorphism that is frequent among Russian winter wheat cultivars, leading to the replacement of amino acid residue glutamine with histidine p.(Gln165His). In the third exon of the *T. urartu* gene there are two adjacent single-nucleotide substitutions relative to the reference genome—c.528G>C and c.530C>G, which should result in the replacement of two consecutive amino acids in the protein p.(Gln176_Ala177delinsHisGly). In the 4th exon, *T. urartu* also has a c.729C>G missense variant, which results in a p.(Asp243Glu) amino-acid substitution.

### 2.2. Protein Isoforms

Following the above-described amino acid substitutions, 4 isoforms of the GRF3-2A protein were found among the examined wheat accessions. According to the frequencies of occurrence, we designated them as follows: A—the most frequent isoform, presents in Chinese Spring and most other cultivars, B—frequently found among Russian winter cultivars, C—the isoform characteristic of *T. urartu*, and D—a mutant form found only in Novosibirskaya 67 (Figure S1). Among accessions in which the *TaGRF3-2A* was completely sequenced, isoform B was present mainly among cultivars developed at Krasnodar. The correspondence of isoforms to amino acid substitutions, as well as the estimated significance of the amino acid changes for biological function of the protein, predicted by PROVEAN, are shown in Table 1. Isoform B, according to PROVEAN, has a functionally significant amino acid substitution. However, no mutation found in this study disturbed the WRC or QLQ conserved domains of the protein (Figure S1).

**Table 1.** GRF3-2A protein isoforms and functional significance of amino acid differences.

| Isoform. | Differences in Protein Sequence | PROVEAN Score | Representative Accessions * |
|---|---|---|---|
| A | - | - | Chinese Spring |
| B | Gln165His | −2.601 ** | Stan, Vassa, Vid |
| C | Gln176_Ala177delinsHisGly | −1.224 | PI428198 (*T. urartu*) |
| | Asp243Glu | 0.524 | |
| D | Gln42_Gln44del | 0.821 | Novosibirskaya 67 |

* If not specified, the species is *Triticum aestivum* L. ** Variants that have a score lower than −2.5 are assumed to be deleterious in protein biological function.

### 2.3. Phylogenetics of Protein Sequences

Phylogenetic analysis of the predicted amino-acid sequences of the GRF3-2A protein showed that isoform C characteristic of *T. urartu* is the most ancient, whereas isoform B is the newest (Figure 2). Isoform D, unique for Novosibirskaya 67, is a bit closer to the root of the tree than isoform A, apparently due to the proteins GRF3-2B and GRF3-2D, used as outgroup, having a smaller number of consecutive glutamine residues than GRF3-2A (5 vs. 7, beginning from 38 or 39 residue). Isoform D has 4 glutamine residues in that part of the molecule (Figure S1).

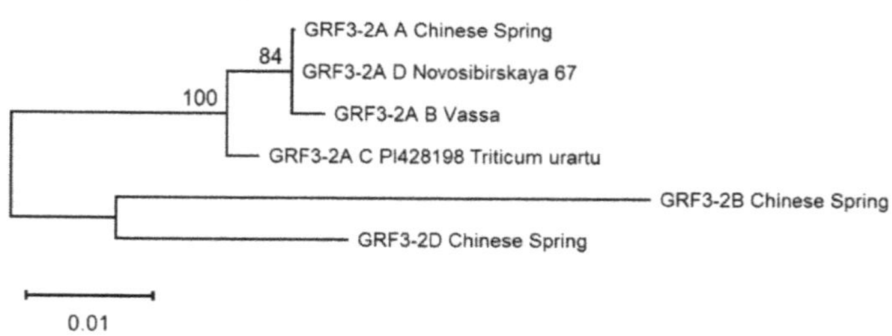


**Figure 2.** Molecular phylogenetic tree of GRF3-2A protein isoforms. GRF3-2B and GRF3-2D proteins of Chinese Spring were taken as outgroups.

### 2.4. Allele Designation

For further discussion, we designated the various haplotypes (alleles) of the *GRF3-2A* gene together with its flanking regions with small letters in accordance with the encoded protein isoforms (*a*, *b*, *c*, *d*) and the number after a dot assigned in accordance with the frequency of occurrence (1—for the most frequent, 2, 3, etc.—for rarer haplotypes) within each group of haplotypes encoding the same protein. The sequences of all haplotypes are represented in Supplementary Materials in FASTA format, the wheat accessions in which they were found are listed in Table S2, the frequency of occurrence among the sequenced accessions is listed in Table S3.

### *2.5. Allele Phylogenetic Analysis*

The phylogenetic analysis of the *Grf3-2A* and flanking region haplotypes was carried out using the SNPs only. Based on the alignment of the nucleotide sequences and constructed phylogenetic tree, it is clear that the *T. urartu* (PI428198) haplotype is evolutionarily distant from all others (Figure 3). The haplotype of wild emmer *T. dicoccoides* (Zavitan) was much closer to the haplotypes of bread wheat, although it has significant differences from them. Bread wheat haplotypes were divided into six groups. The older group includes *T. spelta* (haplotype *a.17*), as well as haplotypes *a.2*, *a.3*, *a.8*, *a.11* and *a.12* of bread wheat. There were also five younger groups: (1) a group that includes haplotypes of Chinese Spring (*a.1*) and Novosibirskaya 67 (*d.1*); (2) a group that includes all *b* haplotypes (*b.1* ... *b.5*); (3) a group that includes haplotypes *a.6* (Velena) and *a.13* (Paragon); (4) haplotype *a.10* (CDC Landmark); (5) haplotype *a.4* (Grom). The last two groups include only one haplotype. Judging by the alignment of nucleotide sequences, the differences between haplotypes within the same group are due either to the presence of indels (in case of *d.1*, *c.1* and *a.16*), or to a change in the length of the microsatellite in the 5′UTR.

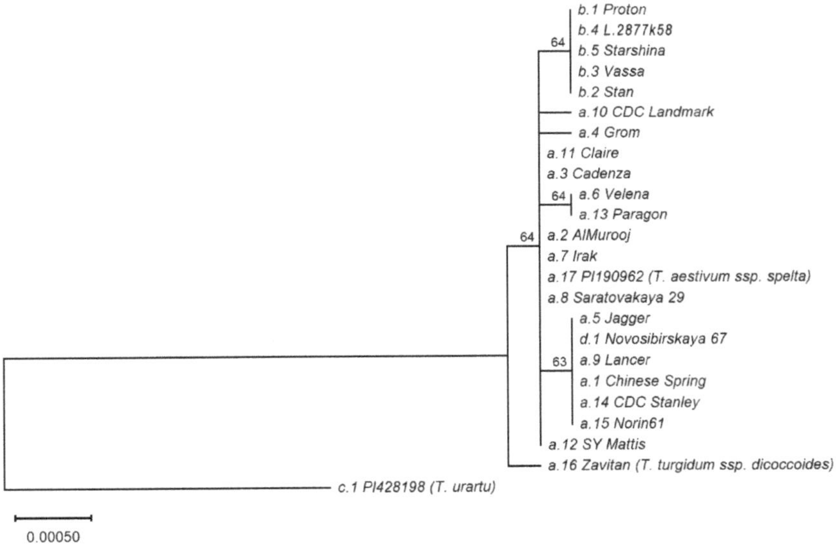


**Figure 3.** Molecular phylogenetic tree (subtree) of *GRF3-2A* haplotypes (gene + flanking sequences). To find a root, *GRF3-2B* gene of Chinese Spring was taken as an outgroup (not shown here). Bootstrap values are shown above nodes. At the tree leaves the *Grf3-2A* haplotypes and representative wheat accessions are indicated. If not specified, the accessions are *Triticum aestivum* L. ssp. *aestivum*.

### *2.6. Molecular Markers and Phenotype*

We designed subgenome-specific primers to detect the 9-nucleotide deletion in the second exon of the *GRF3-2A* gene that was found in Novosibirskaya 67 (see Section 4). The PCR with DNA of the Novosibirskaya 67 gave a fragment of 335 base pairs (bp) of expected size (Figure 4a). Screening of 199 winter bread wheat accessions from a collection of the National Center of Grain (Krasnodar, Table S4) showed that among them there was only the 344 bp variant of the marker. This suggests that the Novosibirskaya 67 cultivar carries a rare mutation of the *GRF3-2A* gene.

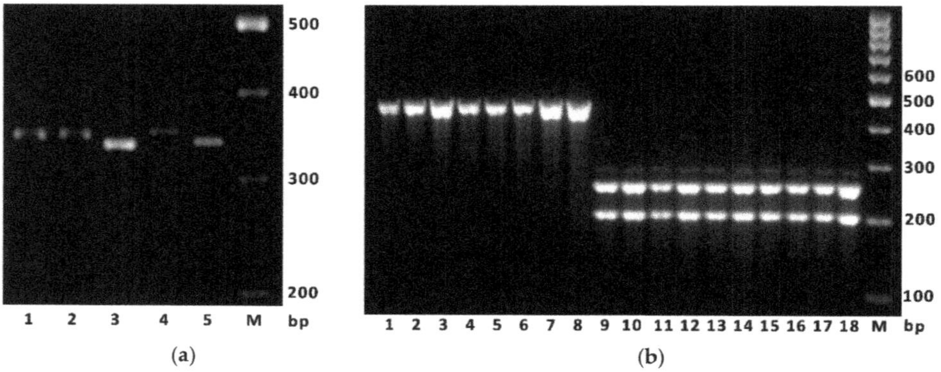


**Figure 4.** (**a**) An example of electrophoresis of the PCR marker designed for detecting the c.126_134del polymorphism. The PCR products were obtained using the primers GRF3A-Q42-F/R and DNA of the following wheat accessions: lane 1—Vid, 2—Stan, 4—Altigo, 3 and 5—Novosibirskaya 67, M—size standard M-100 (Syntol LLC, Moscow). (**b**) An example of electrophoresis of the marker detecting the c.495G>T polymorphism in *GRF3-2A*. The PCR products obtained using primers GRF3A-Q165-F/R and digested using *Sfa*N I endonuclease. Lanes 1–8—accessions having the *GRF3-2Aa* allele (nucleotide G), two lanes each: Grom, Altigo, Velena, Sila; lanes 9–18—accessions having the *GRF3-2Ab* allele (nucleotide T), two lanes each: Vassa, Proton, Alekseich, Vid, Stan. M—size standard M-100.

A marker designed for detection of the c.495G>T mutation resulting in amino-acid substitution, p.(Gln165His), was successfully validated on the DNA of wheat accession in which the *GRF3-2A* gene was sequenced (Figure 4b).

When a pool of 199 winter bread wheat accessions was analyzed using this marker, we found that the variant T of the polymorphism c.495G>T inherent to *GRF3-2Ab* allele was present in almost 39% of accessions (Table S5). *GRF3-2Ab*, found in Bezostaya 1 and Krasnodarskaya 6 (old Russian cultivars), was in 46% of modern Russian cultivars, and was also detected in some Bulgarian, Chinese, Polish, Romanian, Ukrainian, USA and Yugoslavian cultivars, most of which, but not all, have Bezostaya 1 in their pedigrees. Cultivars from Austria, the Czech Republic, France, Germany, Hungary, and the UK that were tested did not have *GRF3-2Ab*. Due to low numbers of accessions from countries other than Russia, we cannot assert statistically significant differences of allele frequencies between different geographical locations.

Statistical analysis of phenotypic data for 199 winter bread wheat accessions collected over three years (2018–2020) at Krasnodar showed that the *GRF3-2Ab* allele (T variant at c.495G>T polymorphism) in each of the three years was significantly associated with earlier heading ($p \leq 0.02$, Fisher's F-test) and higher test weight of the grain ($p \leq 0.01$) compared to the *GRF3-2Aa* allele (Figure 5). In one year, a positive association of the *GRF3-2Ab* allele with 1000 kernel weight (2019, $p < 0.01$) and a high grain protein content (2018, $p = 0.01$) was revealed. However, in 2018, the grain yield was significantly lower for winter wheat accessions with the *GRF3-2Ab* allele, which presumably was associated with lower either grain numbers per spike or tillers per plant. However, the protein yield per hectare was non-significantly decreased in accessions carrying *GRF3-2Ab* in 2018 (Figure S2). Mean values of agronomic traits among wheat accessions having different alleles of the c.495G>T polymorphism are represented in Table S7.

Comparison of the 5′UTR microsatellite among sequenced *GRF3-2A* haplotypes and haplotypes obtained from sequenced wheat genomes gave 18 different lengths of the tandem repeat (Figure 1). However, the GRF3-2AD-SSR marker tested on the 199 winter wheat accessions showed only nine of those variants. That could be explained by the different accessions that were sequenced and genotyped. Statistical analysis showed significant connection between the microsatellite marker and the agronomic traits (Figures S3 and S4). Mean values of agronomic traits among wheat accessions having different alleles of the microsatellite are represented in Table S8. However, the same microsatellite size was observed for different gene alleles, including those that code different protein isoforms (Figure 1).

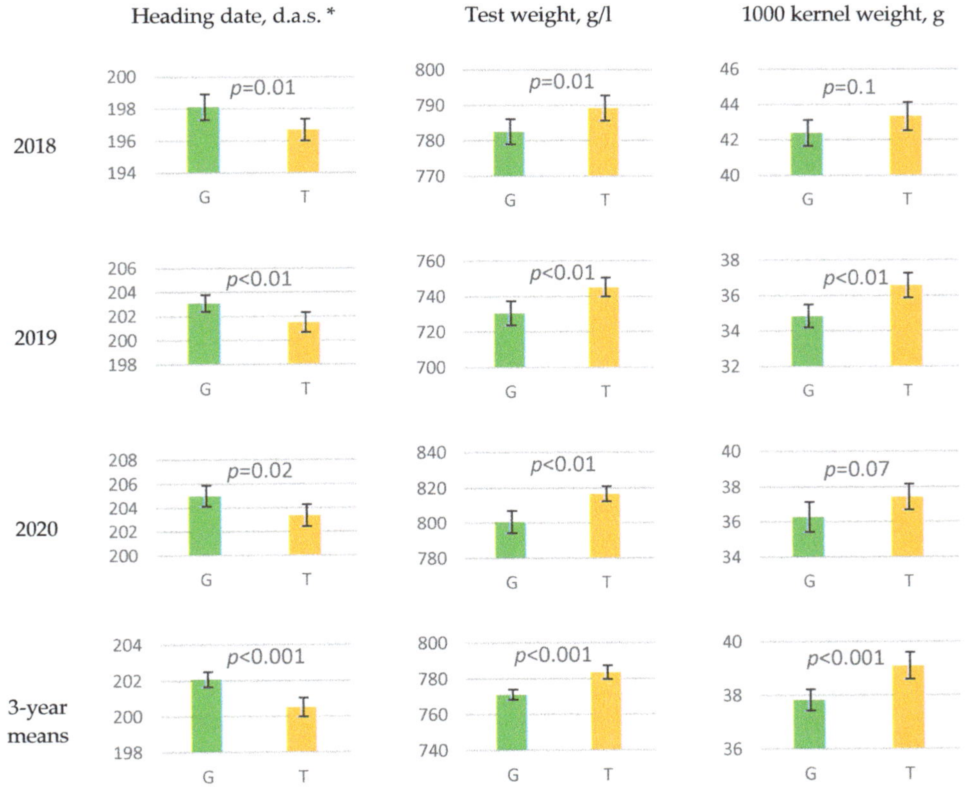


**Figure 5.** Heading date and grain traits for winter wheat accessions differing in missense mutation c.495G>T in *TaGRF3-2A* tested in 2018–2020 at Krasnodar. Bars indicate 95% confidence intervals. The *p*-values are calculated for Fisher's F-test. *—days after sowing.

## 3. Discussion

As expected for a functional gene prone to natural selection, most of the polymorphisms in *GRF3-2A* were detected in the non-coding regions and flanking sequences, while coding sequences were more conserved.

The *GRF3-2A* haplotypes of ancestral wild wheat species were evolutionary distant from the haplotypes of the bread wheat cultivars, indicating that wild species might serve as a source for the new alleles for investigation and crop improvement.

Although the observed polymorphisms in the promotor, 5′ and 3′UTR do not change the protein sequence, they could alter the level of the RNA transcripts and translated protein. Mutations in the promoter can alter the number and composition of the cis-acting elements, recognized by transcription factors [21], while mutations of the DNA sequence in the region of the polyadenylation signal can affect the length of the mRNA and the presence of targets for microRNA and other post-transcriptional regulation factors in it [22]. The length of the 5′UTR microsatellite could also influence the level of gene expression [23].

Spring bread wheat cultivar Novosibirskaya 67 had a rare mutation that was not reported for this gene before—a deletion of nine nucleotides in the second exon, leading to a deletion of three glutamine residues from the protein molecule. We designated this mutation as *TaGRF3-2Ad*, corresponding to protein isoform D. The PROVEAN prediction, based on the change in alignment score of the amino-acid sequence to itself and to other related sequences in the database, showed that the biological function of the GRF3-2A protein should not be significantly altered by this mutation. Novosibirskaya 67 was one of the most widely grown spring wheat cultivars in Siberia since 1974 until the end of the 20th century. It was bred from a population of mutants obtained from the cultivar Novosibirskaya 7 following radiation [24]. We did not find the *TaGRF3-2Ad* allele in any other wheat cultivar, thus we assume that this mutation resulted from artificial mutagenesis, and may confer some adaptive traits to Novosibirskaya 67. The agronomic value of the *TaGRF3-2Ad* allele would become the subject of further study.

The c.495G>T missense mutation found in our study was present in many Russian winter wheat cultivars. We designated the allele carrying this mutation as *TaGRF3-2Ab*. It is present in almost half of the modern cultivars developed at the National Center of Grain at Krasnodar and is present in the old cultivars Bezostaya 1 and Krasnodarskaya 6. Previously, this mutation was reported in the results of the 1000 wheat exomes project and mapped to the reference wheat genome (IWGSC RefSeq v1.0) at coordinate chr2A:687050412 [25]. Among 811 wheat accessions tested in that project, the c.495G>T missense mutation was found in only 3% of genotypes, including cultivars from the former USSR, Bulgaria, Argentina, Mexico, and some other countries.

Statistical analysis of the agronomic traits tested in 199 wheat accessions grown in Krasnodar during the three years showed that the *TaGRF3-2Ab* allele is associated with earlier heading and better grain filling as judged by the test weight and 1000 kernel weight. However, in some years *TaGRF3-2Ab* was associated with lower grain yield per hectare. In separate years, as in 2018, the mean yield of the cultivars and lines carrying *GRF3-2Ab* was even lower than in those carrying *GRF3-2Aa*. Most likely, *GRF3-2Ab* acts through the restriction of additional tillering at the late development stages of the plant, which releases the resources of assimilates for grain formation. This assumption has not been tested directly yet, but it can explain the observations. For winter wheat that may be exposed to severe winter conditions or even to drought in spring, which may become more often due to global warming, additional tillering may be favorable for the recovery of the crop. Although *GRF3-2Ab* may be good at maintaining the high milling quality of the grain, we think it should be used with caution in breeding of winter wheat cultivars from the perspective of climate change.

The *TaPPD1-2A* gene (TraesCS2A02G081900) in chromosome 2A that can affect earliness [26] is located at a considerable physical distance from *TaGRF3-2A*, probably on the other chromosome arm. Thus, the early heading associated with the *TaGRF3-2Ab* allele is unlikely to be explained by genetic linkage with the *PPD1* gene. Other genes that could affect earliness, such as *Vrn-1* (*Vernalization*) or *Eps-1* (*Earliness per se*) are located on other chromosomes [27,28]. However, the presence of a nearby gene linked to *TaGRF3-2A* and affecting earliness cannot be excluded. We screened numerous genome-wide association studies reporting SNP markers connected with agronomic traits, and found no one significant marker-trait association that hit exactly the locus of the *TaGRF3-2A*. However, we found two markers associated with heading time—BobWhite_c16923_64 and wsnp_Ex_c14953_23104041—which surround the *TaGRF3-2A* and are 128 and 42 mega base pairs (Mbp) distant from it correspondingly [29]. Also, a marker BS00009247 on chromosome 2B located 11 Mbp from *TaGRF3-2B*—a homeolog of the gene studied—was associated with heading time [30]. Further fine mapping and gene engineering experiments are needed to postulate the causal relationship between *TaGRF3-2A* and the phenotype.

Breeding for early maturity and plump grain, together with the initial germplasm used, can explain the high frequency of *TaGRF3-2Ab* in Russian cultivars. Bezostaya 1 is present in pedigrees of the most modern Russian winter wheat cultivars, and thus is the probable source of the *TaGRF3-2Ab* allele in them. Bezostaya 1 probably obtained this allele from the old cultivar Ukrainka [25], or from Krymka (a Crimean landrace) [31,32] both of which are in the pedigree of Bezostaya 1 and other cultivars carrying *TaGRF3-2Ab*, but not having Bezostaya 1 or Ukrainka in pedigree. The presence of the *TaGRF3-2Ab* in Krymka could explain its presence in some American (USA) cultivars [33]. Generally, we can assume that *TaGRF3-2Ab* was present in multiple local cultivars (landraces) grown in lands surrounding the Black Sea before the beginning of the scientific breeding.

## 4. Materials and Methods

### *4.1. Plant Material and Phenotyping*

Most of the wheat accessions used for sequencing *GRF3-2A*, genotyping and phenotyping were a part of the winter bread wheat collection maintained at the National Center of Grain named after P. P. Lukyanenko in Krasnodar (Table S4). Spring wheat accessions from Iraq were provided by Dr. Oleg G. Semenov (Department of Technosphere Safety, Agrarian-Technological Institute, RUDN University, Moscow). Other spring wheat accessions used for DNA extraction, PCR and sequencing were a part of a mini-collection maintained by the authors in Moscow (Table S1).

Yield testing and phenotyping for other parameters of the 199 winter wheat accessions was conducted according to the "Competitive variety testing methodologies" during the 2018–2020 harvest years at Krasnodar [34]. The experiments were laid in randomized complete block design on 25 $m^2$ field plots in 4 replications. 1000 kernel weight was determined using a SEED COUNTER S-25 device (DATA Detection Technologies Ltd., Kibbutz Tzora, Israel) and electronic scales. Test weight, grain moisture content, and grain protein content were measured on 700 g samples using an Infratec 1241 grain analyzer (FOSS, Hilleroed, Denmark). Grain yield was determined as grain mass at 14% moisture content divided by plot area excluding the area of damaged sites. The heading date was recorded when 75% of plants showed at least a 50% spike emergence from the flag leaf sheath. Mean values of the measurements in each year were used for statistics regarding molecular markers.

### *4.2. Weather Conditions*

The weather conditions during the vegetation periods in 2017–2020 for winter wheat at Krasnodar are represented in the Figures S5 and S6. The weather data were provided by the nearest meteorological station. The sowing of winter wheat was done on 21 October (both in 2017 and in 2018), and on 17 October in 2020. All three crop years were characterized by a dry period during seedbed preparation at the beginning of autumn and warm winters with no hard frosts and low precipitation, during which the plants maintained slow growth. The booting stage was observed in the last third of March, while heading and anthesis occurred at the beginning of May in each year. In 2018 March was wet, and from April to the end of vegetation the precipitations were lower than average. This resulted in partial loss of young tillers in 2018. The spring of the 2019 was close to the climatic average. March and April of the 2020 were characterized by severe rainfall deficiency and recurrent frosts that caused partial loss of the main stems of early-maturing lines, while in May rainfall was higher than average. Grain filling and maturation during all three years occurred under conditions of high temperature and low humidity.

### *4.3. DNA Extraction, PCR and Sequencing*

Genomic DNA was extracted from dried ground leaves of seedlings using the CTAB protocol [35]. The sequence of *TaGRF3-2A* (TraesCS2A02G435100) with about 1000 base pairs (bp) of flanking sequence, as well as the sequences of its homeologs on chromosome 2B and 2D (TraesCS2B02G458400 and TraesCS2D02G435200), were obtained from the annotated wheat genome assembly IWGSC RefSeq v1.0 using genome browser [17]. Specific primer pairs giving overlapping PCR products were designed using Primer-BLAST (NCBI) [36] to amplify the entire *TaGRF3-2A* sequence with a 1000 bp promotor region (Table 2). The specificity of the primers was rechecked using alignment of the three homoeologous genes.

**Table 2.** Primers used for PCR-amplification of the *TaGRF3-2A* fragments.

| Primer Sequence, 5′→3′ | Tm, °C | Expected Product Size, bp |
|---|---|---|
| *GRF-2A*-1F: AAATTGAAGGCTAGACAATCGGC<br>*GRF-2A*-1R: CCTTTTACTCCTACTTGCCTGGT | 60 | 1179 |
| *GRF-2A*-2F: CAAACGAACTTGACGGTACAGAT<br>*GRF-2A*-2R: CACATGAGGATGAGGCTTCTTGA | 60 | 1188 |
| *GRF-2A*-3F: AGATTTCAGGTGTACTCGACCTC<br>*GRF-2A*-3R: AGCATGCAGAAGATAAAAACGGC | 60 | 1185 |
| *GRF-2A*-4F: GCTCAGCTGCACATGGATAATG<br>*GRF-2A*-4R: CGAGTCAGATTTGCAGCATAGTG | 60 | 1119 |
| *GRF-2A*-5F: TGCAGCAACAATTGCTCGTATAG<br>*GRF-2A*-5R: CACCCCCACCCCTAAGATAGATA | 60 | 1250 |

PCR was performed in 25 µL reaction volumes, containing 70 mM Tris–HCl buffer (pH 9.3), 16.6 mM $(NH_4)_2SO_4$, 2.5 mM $MgCl_2$, 0.2 mM of each dNTP, 0.3 µM forward and reverse primers (Sintol Ltd., Moscow, Russia), 0.04 U/µL LR (long reading) Plus polymerase (Sileks Ltd., Moscow, Russia), 0.02 U/µL Taq polymerase (Sileks Ltd.), and 4 ng/µL DNA template. PCR conditions were as follows: (1) 95 °C for 10 min, (2) 45 cycles of 95 °C for 30 s, 60 °C for 30 s, 72 °C for 4 min; and (3) final extension step of 72 °C for 10 min. PCR products were separated in 1.5% agarose gels in TBE (90 mM Tris, pH 8.3, 90 mM boric acid, 0.1 mM EDTA) buffer using GeneRuler 100 bp Plus DNA Ladder (Thermo Fisher Scientific, Waltham, MA, USA) as a molecular weight marker, and stained with ethidium bromide for subsequent visualization in Gel Doc XR+ (Bio-Rad Laboratories, Inc., Hercules, CA, USA).

In cases of successful amplification, the PCR products obtained from DNA of the same wheat plant were mixed and submitted for NGS sequencing on Illumina MiSeq system. Sequencing was performed at "Genomed, Ltd." (Moscow, Russia). DNA libraries were prepared using Swift 2S™ Turbo DNA Library Kits. In the process of library preparation, the contents of each tube, corresponding to a single wheat plant, were labelled with individual DNA barcodes. The gene sequences for each wheat plant were reconstructed using the previously published algorithm [37,38]. The 19 wheat accessions in which the *TaGRF3-2A* gene was sequenced in this way are listed in Table S1.

### 4.4. Sequences Obtained from Genome Assemblies

*GRF3-2A* sequences were obtained from the assembled genomes of bread wheat [16], durum wheat [39], spelt [16], *T. urartu* [19] and wild emmer [18] wheat (Table S6). The rough coordinates of *GRF3-2A* sequences were found in the genomes using the BLAST+ command line tool [40]. Using these coordinates, the sequences with extended margins were extracted from the genome FASTA files using a program written in Python [41].

### 4.5. Sequence Analysis and Phylogenetics

The sequences of *TaGRF3-2A* gene obtained experimentally along with others found in assemblies of wheat genomes were aligned using the MUSCLE algorithm in MEGA X software [42,43]. The exons and protein-coding sequences were detected using alignment with such sequences annotated for TraesCS2A02G435100 in IWGSC RefSeq v1.0. The translation of coding DNA sequence to an amino acid sequence was performed in GeneDoc 2.7 software [44].

Protein domains were identified using a Conserved Domains Database search at the NCBI website [45]. The functional significance of amino acid substitutions was predicted using the PROVEAN online service [46].

Evolutionary analyses of the *TaGRF3-2A* DNA sequences including a 1000 bp promotor were conducted in MEGA X software using the maximum-likelihood method and Hasegawa–Kishino–Yano model [47]. All positions containing gaps and missing data were eliminated. The evolutionary tree of protein sequences was constructed using the maximum-likelihood method and Jones-Taylor-Thornton (JTT) model [48]. All sites, including gaps, were used. For both DNA and protein phylogenetic analyses, bootstrap support values were calculated using 500 replicates. The trees were drawn to scale, with branch lengths measured by numbers of substitutions per site. To establish a tree root, the homologs from wheat subgenome B and D were added as outgroups.

The analysis of the promoter sequence for the presence of transcription factor binding sites was done using the PlantPAN 3.0 database [21].

### 4.6. Pedigrees

The pedigrees of the wheat accessions studied were obtained from the Genetic Resources Information System for Wheat and Triticale website [49], or from the website of the State Commission for Selection Achievements Test and Protection (Russia) [50].

### 4.7. Molecular Markers

For detection of the 9-nucleotide deletion c.126_134del resulting in deletion of the three amino-acid residues p.(Gln42_Gln44del) of the GRF3 protein, we designed a pair of primers giving PCR products of 335 or 344 base pairs: GRF3A-Q42-F: 5′-CTTCTATCTGTAGCTCGAGGTGT-3′ and GRF3A-Q42-R: 5′-GTGGTAGGAGGAGGAGGAATCTA-3′.

PCR was performed in 25 μL reaction volumes, containing 70 mM Tris–HCl buffer (pH 8.6), 16.6 mM $(NH_4)_2SO_4$, 2.5 mM $MgCl_2$, 0.2 mM of each dNTP, 0.3 μM forward and reverse primers (Sintol Ltd., Moscow, Russia), 0.05 U/μL Taq polymerase (Sileks Ltd., Moscow, Russia), 4 ng/μL DNA template. The PCR conditions were as follows: (1) 95 °C for 10 min, (2) 36 cycles of 95 °C for 30 s, 60 °C for 30 s, 72 °C for 1 min; and (3) final extension step of 72 °C for 10 min. PCR products were separated in 2% agarose gels with TBE buffer for at least 1 h in an electric field intensity of 6 V/cm and visualized as described above.

To detect missense mutation c.495G>T leading to amino-acid change p.(Gln165His) we designed a pair of primers: GRF3A-Q165-F: 5′-GGGTTTTCTTAATTTGCTTGCAGT-3′, GRF3A-Q165-R: 5′-CAGAAGATAAAAACGGCAGGTGA-3′. The PCR conditions were as described above. The PCR product of 454 bp was subjected to endonuclease digestion using SfaN I enzyme (SibEnzyme Ltd., Novosibirsk, Russia) having recognition site GCATC(5/9)^. In a case of nucleotide T in the c.495G>T polymorphism (resulting in histidine in the protein chain), the PCR product were digested into 201 and 253 bp products, while in the case of nucleotide G, the PCR product remained intact. The products of digestion were separated into 1.5% agarose gels in a TBE buffer using GeneRuler 100 bp DNA Ladder (Thermo Fisher Scientific, Waltham, MA, USA) as a molecular weight marker, and stained with ethidium bromide for subsequent visualization in Gel Doc XR+ (Bio-Rad Laboratories, Inc., Hercules, CA, USA).

For detection of simple sequence repeat length polymorphism in the 5′UTR of the *TaGRF3-2A* gene we used primer pair GRF3-2AD-SSR-F: 5′-TCTCACCAGGCAGCAGATCG-3′ and GRF3-2AD-SSR-R: 5′-ACAGGGAGGCAAAGGGCATC-3′ which was also suited to the *TaGRF3-2D* gene. The reverse primer was 5′-labelled with 6-carboxyfluorescein. As predicted by gene sequencing, the length of the PCR products for the *TaGRF3-2A* gene was expected in range 230 to 282 bp, whereas for *TaGRF3-2D* the expected sizes were 211 to 223 bp [51]. Thus, the PCR-product sizes of *TaGRF3-2A* and *TaGRF3-2D* genes do not overlap. The PCR products were diluted 100 times and subjected to fragment analysis on a Nanofor-05 genetic analyzer (Sintol Ltd., Moscow) using a fluorescent fragment size standard SD-450 (Sintol Ltd.).

### 4.8. Statistical Analysis

Calculation of means, analysis of variance and confidence intervals were done in the Statistica 6.0 software package. Three-year means of agronomic traits for genotypes were calculated as least square means in two-way analysis of variance, where a year was one of the factors. Fisher's exact test for score traits was performed in the R programming language [52].

## 5. Conclusions

We studied the allelic diversity of the *GRF3-2A* gene in bread wheat and compared bread wheat alleles with those in some wild ancestral species. The allele designated as *TaGRF3-2Ab* was rare in a world wheat collection, but was quite common among Russian winter wheat cultivars. This allele was associated with earlier heading and better grain filling, while keeping almost the same yield per hectare. We can assume that this allele is adaptive for the steppes of the Black Sea region. We discovered a unique mutation of *TaGRF3-2A* in spring wheat cultivar Novosibirskaya 67, the agronomic value of which is yet to be established.

**Supplementary Materials:** The following are available online at https://www.mdpi.com/article/10.3390/ijms222212376/s1.

**Author Contributions:** Conceptualization, M.G.D.; methodology, M.G.D. and M.S.B.; software, M.S.B.; formal analysis, M.S.B.; investigation, A.G.C., N.A.P., T.I.G. and L.A.N.; resources, L.A.B.; data curation, L.A.B., N.A.P., T.I.G.; writing—original draft preparation, M.S.B.; writing—review and editing, M.G.D. and L.A.B.; visualization, A.G.C. and L.A.N.; supervision, G.I.K.; project administration, M.G.D.; funding acquisition, M.G.D. All authors have read and agreed to the published version of the manuscript.

**Funding:** This research was funded by the Ministry of Science and Higher Education of the Russian Federation, state task number 0431-2019-0004.

**Institutional Review Board Statement:** Not applicable.

**Informed Consent Statement:** Not applicable.

**Data Availability Statement:** The data presented in this study are available in Supplementary Materials to this article. The sequences of the *TaGRF3-2A* gene obtained in this study were deposited to NCBI GenBank under accession numbers OK094721-OK094732.

**Acknowledgments:** We are greatly thankful to Oleg G. Semenov (Department of Technosphere Safety, Agrarian-Technological Institute, RUDN University, Moscow), who provided spring wheat accessions from Iraq for our studies.

**Conflicts of Interest:** The authors declare no conflict of interest.

## References


1. Huang, W.; He, Y.; Yang, L.; Lu, C.; Zhu, Y.; Sun, C.; Ma, D.; Yin, J. Genome-Wide Analysis of Growth-Regulating Factors (GRFs) in *Triticum Aestivum*. *PeerJ* **2021**, *9*, e10701. [CrossRef]
2. van der Knaap, E.; Kim, J.H.; Kende, H. A Novel Gibberellin-Induced Gene from Rice and Its Potential Regulatory Role in Stem Growth. *Plant. Physiol.* **2000**, *122*, 695–704. [CrossRef]
3. Kim, J.H.; Choi, D.; Kende, H. The *AtGRF* Family of Putative Transcription Factors Is Involved in Leaf and Cotyledon Growth in *Arabidopsis*. *Plant. J.* **2003**, *36*, 94–104. [CrossRef]
4. Liu, J.; Hua, W.; Yang, H.-L.; Zhan, G.-M.; Li, R.-J.; Deng, L.-B.; Wang, X.-F.; Liu, G.-H.; Wang, H.-Z. The *BnGRF2* Gene (*GRF2*-like Gene from *Brassica napus*) Enhances Seed Oil Production through Regulating Cell Number and Plant Photosynthesis. *J. Exp. Bot.* **2012**, *63*, 3727–3740. [CrossRef] [PubMed]
5. Zan, T.; Zhang, L.; Xie, T.; Li, L. Genome-Wide Identification and Analysis of the Growth-Regulating Factor (*GRF*) Gene Family and GRF-Interacting Factor Family in *Triticum aestivum L. Biochem. Genet.* **2020**, *58*, 705–724. [CrossRef]
6. Kim, J.H.; Kende, H. A Transcriptional Coactivator, *AtGIF1*, Is Involved in Regulating Leaf Growth and Morphology in *Arabidopsis*. *Proc. Natl. Acad. Sci. USA* **2004**, *101*, 13374–13379. [CrossRef]
7. Li, S.; Tian, Y.; Wu, K.; Ye, Y.; Yu, J.; Zhang, J.; Liu, Q.; Hu, M.; Li, H.; Tong, Y.; et al. Modulating Plant Growth–Metabolism Coordination for Sustainable Agriculture. *Nature* **2018**, *560*, 595–600. [CrossRef]
8. Liebsch, D.; Palatnik, J.F. MicroRNA MiR396, GRF Transcription Factors and GIF Co-Regulators: A Conserved Plant Growth Regulatory Module with Potential for Breeding and Biotechnology. *Curr. Opin. Plant Biol.* **2020**, *53*, 31–42. [CrossRef] [PubMed]
9. Beltramino, M.; Debernardi, J.M.; Ferela, A.; Palatnik, J.F. ARF2 Represses Expression of Plant GRF Transcription Factors in a Complementary Mechanism to MicroRNA MiR396. *Plant Physiol.* **2021**, *185*, 1798–1812. [CrossRef] [PubMed]

10. Gao, F.; Wang, K.; Liu, Y.; Chen, Y.; Chen, P.; Shi, Z.; Luo, J.; Jiang, D.; Fan, F.; Zhu, Y.; et al. Blocking MiR396 Increases Rice Yield by Shaping Inflorescence Architecture. *Nat. Plants* **2016**, *2*, 15196. [CrossRef]
11. Che, R.; Tong, H.; Shi, B.; Liu, Y.; Fang, S.; Liu, D.; Xiao, Y.; Hu, B.; Liu, L.; Wang, H.; et al. Control of Grain Size and Rice Yield by GL2-Mediated Brassinosteroid Responses. *Nat. Plants* **2016**, *2*, 15195. [CrossRef] [PubMed]
12. Kim, J.-S.; Mizoi, J.; Kidokoro, S.; Maruyama, K.; Nakajima, J.; Nakashima, K.; Mitsuda, N.; Takiguchi, Y.; Ohme-Takagi, M.; Kondou, Y.; et al. *Arabidopsis* GROWTH-REGULATING FACTOR7 Functions as a Transcriptional Repressor of Abscisic Acid– and Osmotic Stress–Responsive Genes, Including *DREB2A*. *Plant. Cell* **2012**, *24*, 3393–3405. [CrossRef]
13. Duan, P.; Ni, S.; Wang, J.; Zhang, B.; Xu, R.; Wang, Y.; Chen, H.; Zhu, X.; Li, Y. Regulation of *OsGRF4* by OsmiR396 Controls Grain Size and Yield in Rice. *Nat. Plants* **2015**, *2*, 1–5. [CrossRef]
14. Avni, R.; Oren, L.; Shabtay, G.; Assili, S.; Pozniak, C.; Hale, I.; Ben-David, R.; Peleg, Z.; Distelfeld, A. Genome Based Meta-QTL Analysis of Grain Weight in Tetraploid Wheat Identifies Rare Alleles of *GRF4* Associated with Larger Grains. *Genes* **2018**, *9*, 636. [CrossRef]
15. Tsukaya, H. Yield Increase: GRFs Provide the Key. *Nat. Plants* **2015**, *2*, 1–2. [CrossRef] [PubMed]
16. Walkowiak, S.; Gao, L.; Monat, C.; Haberer, G.; Kassa, M.T.; Brinton, J.; Ramirez-Gonzalez, R.H.; Kolodziej, M.C.; Delorean, E.; Thambugala, D.; et al. Multiple Wheat Genomes Reveal Global Variation in Modern Breeding. *Nature* **2020**, *588*, 277–283. [CrossRef]
17. Alaux, M.; Rogers, J.; Letellier, T.; Flores, R.; Alfama, F.; Pommier, C.; Mohellibi, N.; Durand, S.; Kimmel, E.; Michotey, C.; et al. Linking the International Wheat Genome Sequencing Consortium Bread Wheat Reference Genome Sequence to Wheat Genetic and Phenomic Data. *Genome Biol.* **2018**, *19*, 111. [CrossRef] [PubMed]
18. Avni, R.; Nave, M.; Barad, O.; Baruch, K.; Twardziok, S.O.; Gundlach, H.; Hale, I.; Mascher, M.; Spannagl, M.; Wiebe, K.; et al. Wild Emmer Genome Architecture and Diversity Elucidate Wheat Evolution and Domestication. *Science* **2017**, *357*, 93–97. [CrossRef]
19. Ling, H.-Q.; Ma, B.; Shi, X.; Liu, H.; Dong, L.; Sun, H.; Cao, Y.; Gao, Q.; Zheng, S.; Li, Y.; et al. Genome Sequence of the Progenitor of Wheat A Subgenome *Triticum Urartu*. *Nature* **2018**, *557*, 424–428. [CrossRef]
20. den Dunnen, J.T.; Dalgleish, R.; Maglott, D.R.; Hart, R.K.; Greenblatt, M.S.; McGowan-Jordan, J.; Roux, A.-F.; Smith, T.; Antonarakis, S.E.; Taschner, P.E.M. HGVS Recommendations for the Description of Sequence Variants: 2016 Update. *Hum. Mutat.* **2016**, *37*, 564–569. [CrossRef]
21. Chow, C.-N.; Lee, T.-Y.; Hung, Y.-C.; Li, G.-Z.; Tseng, K.-C.; Liu, Y.-H.; Kuo, P.-L.; Zheng, H.-Q.; Chang, W.-C. PlantPAN3.0: A New and Updated Resource for Reconstructing Transcriptional Regulatory Networks from ChIP-Seq Experiments in Plants. *Nucleic Acids Res.* **2019**, *47*, D1155–D1163. [CrossRef]
22. Hong, L.; Ye, C.; Lin, J.; Fu, H.; Wu, X.; Li, Q.Q. Alternative Polyadenylation Is Involved in Auxin-Based Plant Growth and Development. *Plant J.* **2018**, *93*, 246–258. [CrossRef] [PubMed]
23. Kumar, S.; Bhatia, S. A Polymorphic (GA/CT)n- SSR Influences Promoter Activity of Tryptophan Decarboxylase Gene in *Catharanthus roseus* L. Don. *Sci. Rep.* **2016**, *6*, 33280. [CrossRef]
24. Artemova, G.V.; Likhenko, I.E. Historical aspects and main results of scientific research in SIBNIIRS—Branch of the Institute of Cytology and Genetics SO RAN. *Lett. Vavilov J. Genet. Breed.* **2016**, *2*, 3. Available online: http://pismavavilov.ru/2016-2-1/ (accessed on 16 November 2021).
25. He, F.; Pasam, R.; Shi, F.; Kant, S.; Keeble-Gagnere, G.; Kay, P.; Forrest, K.; Fritz, A.; Hucl, P.; Wiebe, K.; et al. Exome Sequencing Highlights the Role of Wild-Relative Introgression in Shaping the Adaptive Landscape of the Wheat Genome. *Nat. Genet.* **2019**, *51*, 896–904. [CrossRef]
26. Guo, W.; Xin, M.; Wang, Z.; Yao, Y.; Hu, Z.; Song, W.; Yu, K.; Chen, Y.; Wang, X.; Guan, P.; et al. Origin and Adaptation to High Altitude of Tibetan Semi-Wild Wheat. *Nat. Commun.* **2020**, *11*, 5085. [CrossRef] [PubMed]
27. Alvarez, M.A.; Tranquilli, G.; Lewis, S.; Kippes, N.; Dubcovsky, J. Genetic and Physical Mapping of the Earliness per Se Locus *Eps-Am1* in *Triticum monococcum* Identifies *EARLY FLOWERING 3* (*ELF3*) as a Candidate Gene. *Funct. Integr. Genom.* **2016**, *16*, 365–382. [CrossRef] [PubMed]
28. Santra, D.K.; Santra, M.; Allan, R.E.; Campbell, K.G.; Kidwell, K.K. Genetic and Molecular Characterization of Vernalization Genes *Vrn-A1*, *Vrn-B1*, and *Vrn-D1* in Spring Wheat Germplasm from the Pacific Northwest Region of the U.S.A. *Plant. Breed.* **2009**, *128*, 576–584. [CrossRef]
29. Zanke, C.; Ling, J.; Plieske, J.; Kollers, S.; Ebmeyer, E.; Korzun, V.; Argillier, O.; Stiewe, G.; Hinze, M.; Beier, S.; et al. Genetic Architecture of Main Effect QTL for Heading Date in European Winter Wheat. *Front. Plant. Sci.* **2014**, *5*, 217. [CrossRef] [PubMed]
30. Gao, L.; Zhao, G.; Huang, D.; Jia, J. Candidate Loci Involved in Domestication and Improvement Detected by a Published 90K Wheat SNP Array. *Sci. Rep.* **2017**, *7*, 44530. [CrossRef]
31. KRYMKA. Available online: http://wheatpedigree.net/sort/show/32936 (accessed on 29 July 2021).
32. CRIMEAN. Available online: http://wheatpedigree.net/sort/show/14818 (accessed on 29 July 2021).
33. Moon, D. Introduction of Russian wheat in the Great Plains of the United States of America. In *American Yearbook 2011*; Ves' Mir: Moscow, Russia, 2011; pp. 161–176.
34. Competitive Variety Testing Methodologies. Available online: https://gossortrf.ru/en/metodiki_ksi/ (accessed on 6 September 2021).
35. Doyle, P.J. DNA Protocols for Plants. In *Molecular Techniques in Taxonomy*; Hewitt, G.M., Johnston, A.W.B., Young, J.P.W., Eds.; NATO ASI Series; Springer: Berlin/Heidelberg, Germany, 1991; ISBN 978-3-642-83964-1.

36. Ye, J.; Coulouris, G.; Zaretskaya, I.; Cutcutache, I.; Rozen, S.; Madden, T.L. Primer-BLAST: A Tool to Design Target-Specific Primers for Polymerase Chain Reaction. *BMC Bioinform.* **2012**, *13*, 134. [CrossRef] [PubMed]
37. Bazhenov, M.S.; Chernook, A.G.; Goncharov, N.P.; Chikida, N.N.; Belousova, M.K.; Karlov, G.I.; Divashuk, M.G. The Allelic Diversity of the Gibberellin Signaling Pathway Genes in *Aegilops tauschii* Coss. *Plants* **2020**, *9*, 1696. [CrossRef] [PubMed]
38. MikhailBazhenov Consensus: A Program for Making Consensus Sequence from CAP3 Output. Available online: https://github.com/MikhailBazhenov/Consensus (accessed on 27 September 2021).
39. Fruzangohar, M.; Kalashyan, E.; Kalambettu, P.; Ens, J.; Wiebe, K.; Pozniak, C.J.; Tricker, P.J.; Baumann, U. Novel Informatic Tools to Support Functional Annotation of the Durum Wheat Genome. *Front. Plant. Sci.* **2019**, *10*, 1244. [CrossRef] [PubMed]
40. Camacho, C.; Coulouris, G.; Avagyan, V.; Ma, N.; Papadopoulos, J.; Bealer, K.; Madden, T.L. BLAST+: Architecture and Applications. *BMC Bioinform.* **2009**, *10*, 421. [CrossRef] [PubMed]
41. MikhailBazhenov Fastfinder: A Program for Extracting Some Part of a Sequence from a Genome-Large FASTA File. Available online: https://github.com/MikhailBazhenov/fastfinder (accessed on 27 September 2021).
42. Edgar, R.C. MUSCLE: Multiple Sequence Alignment with High Accuracy and High Throughput. *Nucleic Acids Res.* **2004**, *32*, 1792–1797. [CrossRef] [PubMed]
43. Kumar, S.; Stecher, G.; Li, M.; Knyaz, C.; Tamura, K. MEGA X: Molecular Evolutionary Genetics Analysis across Computing Platforms. *Mol. Biol. Evol.* **2018**, *35*, 1547–1549. [CrossRef]
44. NICHOLAS, K.B. GeneDoc: Analysis and Visualization of Genetic Variation. *EMBNEW News* **1997**, *4*, 14.
45. Marchler-Bauer, A.; Derbyshire, M.K.; Gonzales, N.R.; Lu, S.; Chitsaz, F.; Geer, L.Y.; Geer, R.C.; He, J.; Gwadz, M.; Hurwitz, D.I.; et al. CDD: NCBI's Conserved Domain Database. *Nucleic Acids Res.* **2015**, *43*, D222226. [CrossRef]
46. Choi, Y.; Chan, A.P. PROVEAN Web Server: A Tool to Predict the Functional Effect of Amino Acid Substitutions and Indels. *Bioinformatics* **2015**, *31*, 2745–2747. [CrossRef]
47. Hasegawa, M.; Kishino, H.; Yano, T. Dating of the Human-Ape Splitting by a Molecular Clock of Mitochondrial DNA. *J. Mol. Evol.* **1985**, *22*, 160–174. [CrossRef]
48. Jones, D.T.; Taylor, W.R.; Thornton, J.M. The Rapid Generation of Mutation Data Matrices from Protein Sequences. *Comput. Appl. Biosci. Cabios* **1992**, *8*, 275–282. [CrossRef] [PubMed]
49. Genetic Resources Information System for Wheat and Triticale. Available online: http://wheatpedigree.net/ (accessed on 16 August 2021).
50. State Commission for Selection Achievements Test and Protection. Available online: https://gossortrf.ru/en/ (accessed on 16 August 2021).
51. Kroupin, P.Y.; Chernook, A.G.; Bazhenov, M.S.; Karlov, G.I.; Goncharov, N.P.; Chikida, N.N.; Divashuk, M.G. Allele Mining of *TaGRF-2D* Gene 5′-UTR in *Triticum aestivum* and *Aegilops tauschii* Genotypes. *PLoS ONE* **2020**, *15*, e0231704. [CrossRef] [PubMed]
52. R Core Team. *R: A Language and Environment for Statistical Computing*; R Foundation for Statistical Computing: Vienna, Austria, 2020.

MDPI
St. Alban-Anlage 66
4052 Basel
Switzerland
www.mdpi.com

*International Journal of Molecular Sciences* Editorial Office
E-mail: ijms@mdpi.com
www.mdpi.com/journal/ijms

www.ingramcontent.com/pod-product-compliance
Lightning Source LLC
Chambersburg PA
CBHW061515210926
51850CB00008B/1930
*9783036599380*